H. Wackernagel

Multivariate Geostatistics

Springer
Berlin
Heidelberg
New York
Hong Kong
London
Milan
Paris
Tokyo

Hans Wackernagel

Multivariate Geostatistics

An Introduction with Applications

Third, completely revised edition

with 117 Figures and 7 Tables

DR. HANS WACKERNAGEL
Centre de Géostatistique
Ecole des Mines de Paris
35 rue Saint Honoré
77305 Fontainebleau
France

ISBN 3-540-44142-5 Springer-Verlag Berlin Heidelberg New York
ISBN 3-540-64721-X 2nd edition Springer-Verlag Berlin Heidelberg New York

Library of Congress Cataloging-in-Publication Data

Wackernagel, Hans. Multivariate geostatistics : an introduction with applications / Hans Wackernagel. –3rd, completely rev.ed.
p.cm. Includes bibliographical references
ISBN 3540441425 (alk.paper)
1. Geology – Staistical ,methods. 2. Multivariate analysis. I. Title.

Springer-Verlag Berlin Heidelberg New York
a member of Springer Science+Business Media

http://www.springer.de

Camera ready by author
Cover design: E. Kirchner, Heidelberg
Printed on acid-free paper SPIN 11398189 32/3111/as 5 4 3 2 1

L'analyse des données est "un outil pour dégager de la gangue des données le pur diamant de la véridique nature".

JP BENZÉCRI (according to [333])

Multivariate analysis is "a tool to extract from the gangue of the data the pure diamond of truthful nature".

Preface to the 3rd edition

Geostatistics has become an important methodology in environmental, climatological and ecological studies. So a new chapter demonstrating in a simple manner the application of cokriging to conductivity, salinity and chlorophyll measurements as well as numerical model output has been incorporated. Otherwise some recent material on collocated cokriging has been added, leading to a separate chapter on this topic. Time was too short to include further results at the interface between geostatistics and data assimilation, which is a promising area of future developments (see reference [27]).

The main addition, however, is a detailed treatment of geostatistics for selection problems, which features five new chapters on non linear methods.

Fontainebleau, September 2002 *HW*

Preface to the 2nd edition

"Are you a statistician?" I have been asked. "Sort of..." was my answer. A geostatistician is probably as much a statistician as a geophysicist is a physicist. The statistician grows up, intellectually speaking, in the culture of the iid (independent identically distributed random variables) model, while the geostatistician is confronted with spatially/temporally correlated data right from the start. This changes radically the basic attitude when approaching a data set.

The present new edition has benefited from a dozen reviews in statistical, applied mathematics, earth and life science journals. The following principal changes have been made. The statistical introductory chapter has been split into three separate chapters for improved clarity. The ordinary kriging and cokriging chapters have been reshaped. The part on non-stationary geostatistics was entirely rewritten and rearranged after fruitful discussions with Dietrich Stoyan. I have also received interesting comments from Vera Pawlowsky, Tim Haas, Gérard Biau and Laurent Bertino. Last but not least I wish to thank Wolfgang Engel from Springer-Verlag for his editorial advice.

Fontainebleau, May 1998 *HW*

Preface to the first edition

Introducing geostatistics from a multivariate perspective is the main aim of this book. The idea took root while teaching geostatistics at the Centre de Géostatistique (Ecole des Mines de Paris) over the past ten years in the two postgraduate programs DEA and CFSG. A first script of lecture notes in French originated from this activity.

A specialized course on *Multivariate and Exploratory Geostatistics* held in September 1993 in Paris (organized in collaboration with the Department of Statistics of Trinity College Dublin) was the occasion to test some of the material on a pluridisciplinary audience. Another important opportunity arose last year when giving a lecture on *Spatial Statistics* during the summer term at the Department of Statistics of the University of Washington at Seattle, where part of this manuscript was distributed in an early version. Short accounts were also given during COMETT and TEMPUS courses on geostatistics for environmental studies in Fontainebleau, Freiberg, Rome and Prague, which were sponsored by the European Community.

I wish to thank the participants of these various courses for their stimulating questions and comments. Among the organizers of these courses, I particularly want to acknowledge the support received from Georges Matheron, Pierre Chauvet, Margaret Armstrong, John Haslett and Paul Sampson. Michel Grzebyk has made valuable comments on Chapters 29 and 30, which partly summarize some of his contributions to the field.

Fontainebleau, May 1995 *HW*

Contents

1 Introduction

Geostatistics is a rapidly evolving branch of applied mathematics which originated in the mining industry in the early fifties to help improve ore reserve calculation. The first steps were taken in South Africa, with the work of the mining engineer DG KRIGE and the statistician HS SICHEL (see reference number [164] in the bibliography).

The techniques attracted the attention of French mining engineers and in particular of Georges MATHERON (1930–2000), who put together KRIGE's innovative concepts and set them in a single powerful framework with his *Theory of Regionalized Variables* [195, 196, 197, 198, 200, 63, 2].

Originally developed for solving ore reserve estimation problems the techniques spread in the seventies into other areas of the earth sciences with the advent of high-speed computers. They are nowadays popular in many fields of science and industry where there is a need for evaluating spatially or temporally correlated data. A first international meeting on the subject was organized in Rome, Italy in 1975 [132]. Further congresses were held at Lake Tahoe, U.S.A. in 1983 [331], at Avignon, France in 1988 [11], at Troia, Portugal in 1992 [305], at Montréal, Canada in 1994 [92], at Wollongong, Australia in 1996 [19], at Cape Town, South Africa in 2000 [161]. Three well attended European Conferences on Geostatistics for Environmental Applications took place at Lisbon in 1996 [306], at Valencia in 1998 [116] and at Avignon in 2000 [225].

As geostatistics is now incorporating an increasing number of methods, theories and techniques, it sounds like an impossible task to give a full account of all developments in a single volume which is not intended to be encyclopedic.[1] So a selection of topics had to be made for the sake of convenience and we start by presenting the contents of the book from the perspective of a few general categories.

The analysis of spatial and temporal phenomena will be discussed keeping three issues in mind

Data description. The data need to be explored for spatial, temporal and multivariate structure and checked for outlying values which mask structure. Modern computer technology with its high-power graphic screens displaying multiple, linkable windows allows for dynamic simultaneous views on the data. A map of the position of samples in space or representations along time can be linked with histograms, correlation diagrams, variogram clouds and experimental variograms. First ideas about the spatial, time and multivariate structure emerge

[1] CHILÈS & DELFINER [51] have nevertheless managed this *tour de force* and produced a major reference text that gives a complete picture of the whole range of geostatistical techniques in one volume.

from a variety of such simple displays.

Interpretation. The graphical displays gained from the numerical information are evaluated by taking into account past experience on similar data and scientific facts related to the variables under study. The interpretation of the spatial or time structure, the associations and the causal relations between variables are built into a model which is fitted to the data. This model not only describes the phenomenon at sample locations, but it is usually also valid for the spatial or time continuum in the sampled region and it thus represents a step beyond the information contained in the numerical data.

Estimation. Armed with a model of the variation in the spatial or temporal continuum, the next objective can be to estimate values of the phenomenon under study at various scales and at locations different from the sample points. The methods to perform this estimation are based on least squares and need to be adapted to a wide variety of model formulations in different situations and for different problems encountered in practice.

We have decided to deal only with these three issues, leaving aside questions of simulation and control which would have easily doubled the length of the book and changed its scope. To get an idea of what portion of geostatistics is actually covered it is convenient to introduce the following common subdivision into

1. Linear stationary geostatistics,
2. Non-linear geostatistics,
3. Non-stationary geostatistics,
4. Geostatistical simulation.

We shall mainly cover the first topic, examining single- and multi-variate methods based on linear combinations of the sample values and we shall assume that the data stem from the realization of a set of random functions which are stationary or, at least, whose spatial or time increments are stationary.

The second topic deals with the estimation of the proportion of values above a fixed value for distributions that are generally not Gaussian. This requires non linear techniques based on multivariate models for bivariate distributions and an introductory treatment has therefore been included. While starting with lognormal kriging it focuses on isofactorial models which provide a coherent model of the coregionalization of indicators and explicit change of support models. The more primitive methods of indicator cokriging, which give only unsatisfactory solutions to these problems are exposed in [121, 51] and will not be treated in this book. An important omission is also the cokriging of orthogonal indicator residuals [267, 270].

A short review of the third topic is given in the last three chapters of the book with the aim of providing a better understanding of the status of drift functions which are not translation invariant. We had no intention of giving an extensive treatment of non-stationary geostatistics which would justify a monograph on its own.

The fourth topic, i.e. stochastic simulation of regionalized variables, will not at all be treated in this volume. A monograph entirely devoted to geostatistical simulation has just been published by LANTUÉJOUL [178].

Multivariate Geostatistics consists of thirty-nine short chapters which each on average represent the contents of a two hour lecture. The material is subdivided into six parts.

Part A reviews the basic concepts of mean, variance, covariance, variance-covariance matrix, mathematical expectation, linear regression, multiple linear regression. The transposition of multiple linear regression into a spatial context is explained, where regression receives the name of *kriging*. The problem of estimating the mean of spatially or temporally correlated data is then solved by kriging.

Part B offers a detailed introduction to linear geostatistics for a single variable. After presenting the random function model and the concept of stationarity, the display of spatial variation with a variogram cloud is discussed. The necessity of replacing the experimental variogram, obtained from the variogram cloud, by a theoretical variogram is explained. The theoretical variogram and the covariance function are introduced together with the assumptions of stationarity they imply. As variogram models are frequently derived from covariance functions, a few basic isotropic covariance models are presented. Stationarity means translation-invariance of the moments of the random function, while isotropy is a corresponding rotation-invariance. In the case of geometric anisotropy a linear transformation of space is defined to adapt the basically isotropic variogram models to this situation.

An important feature of spatial or temporal data is that a measurement refers to a given volume of space or an interval of time, which is called the *support* of the measurement. Extension and dispersion variances take account of the support of the regionalized variable and furthermore incorporate the description of spatial correlation provided by the variogram model.

Spatial regression techniques known as *kriging* draw on the variogram or the covariance function for estimating either the mean in a region or values at particular locations of the region. The weights computed by kriging to estimate these quantities are distributed around the estimation location in a way that can be understood by looking at simple sample configurations.

The linear model of regionalization characterizes distinct spatial or time scales of a phenomenon. Kriging techniques are available to extract the variation pertaining to a specific scale and to map a corresponding component. As a byproduct the theory around the analysis and filtering of characteristic scales gives a better understanding of how and why ordinary kriging provides a smoothed image of a regionalized variable which has been sampled with irregularly spaced data.

Part C presents three well-known methods of multivariate analysis. Principal component analysis is the simplest and most widely used method to define factors explaining the multivariate correlation structure. Canonical analysis generalizes the method to the case of two groups of variables. Correspondence analysis is an application of canonical analysis to two qualitative variables coded into disjunctive tables. The transposition of the latter, by coding a quantitative variable into disjunctive tables, has yielded models used in disjunctive kriging, a technique of non-linear geostatistics.

Part D extends linear geostatistics to the multivariate case. The properties of the cross variogram and the cross covariance function are discussed and compared. The characterization of matrices of covariance functions is a central problem of multivariate geostatistics. Two models, the intrinsic correlation model and the nested multivariate model, are examined in the light of two multivariate random function models, the linear and the bilinear coregionalization models. Cokriging of real data is discussed in detail in three separate chapters, depending on whether or not all variables involved have been sampled at the same locations and whether the auxiliary variables are densely sampled in space. A detailed cokriging case study using data from the Ebro estuary in Spain is proposed in a separate chapter. The cokriging of a complex variable is based on a bivariate coregionalization model between the real and the imaginary part and its comparison with complex kriging provides a rich reservoir of problems for teasing students. The modeling of the complex covariance function in complex kriging opens the gate to the bilinear coregionalization model which allows for non-even cross covariance functions between real random functions.

Part E introduces to non-linear methods of geostatistics. The selection problem posed by a threshold that serves as basis of a decision is discussed and several statistics based on thresholds, the selectivity curves, are presented. Lognormal kriging is a simple entry point to non linear geostatistical methods and permits with little effort to discuss and exercise the central questions of selective geostatistics. The severe limitations of the lognormal model are then resolved in the following chapters by Gaussian and gamma anamorphosis, isofactorial change of support and kriging methods based on discrete point-block models.

Part F discusses phenomena involving a non-stationary component called the *drift*. When the drift functions are translation-invariant, generalized covariance functions can be defined in the framework of the rich theory of intrinsic random functions of order k. In multivariate problems auxiliary variables can be incorporated into universal kriging as external drift functions which however are not translation-invariant.

The *Appendix* contains two additional chapters on matrix algebra and linear regression theory in a notation consistent with the rest of the material. It also contains a list of common covariance functions and variograms, additional exercises and solutions to the exercises. References classified according to topics of theory and fields of applications are found at the end of the book, together with a list of sources of geostatistical computer software, the bibliography and a subject index.

Geostatistics has become a branch of its own of applied *stochastics* (Greek: στοχαστικός, the art of aiming, skillful guessing, estimation), which encompasses probability theory and mathematical statistics. Like in other areas of applied mathematics, three levels can be distinguished: pure theory, sound application of theory and data cooking. This book is dedicated to the second aspect, keeping at the one end the mathematics as elementary as possible, but seeking at the other end to avoid simplified recipes, which anyway may change with the rapidly evolving practice of geostatistics.

Part A

From Statistics to Geostatistics

2 Mean, Variance, Covariance

In this chapter the elementary concepts of mean, variance and covariance are presented. The expectation operator is introduced, which serves to compute these quantities in the framework of probabilistic models.

The mean: center of mass

To introduce the notion of mean value let us take an example from physics.

Seven weights are hanging on a bar whose own weight is negligible. The locations z on the bar at which the weights are suspended are denoted by

$$z \quad = \quad 5,\ 5.5,\ 6,\ 6.5,\ 7,\ 7.5,\ 8,$$

as shown on Figure 2.1. The mass $w(z)$ of the weights is

$$w(z) \quad = \quad 3,\ 4,\ 6,\ 3,\ 4,\ 4,\ 2.$$

The location $\overline{z}$ where the bar, when suspended, stays in equilibrium is evidently calculated using a weighted average

$$\overline{z} \quad = \quad \frac{1}{\left(\sum\limits_k w(z_k)\right)} \sum_{k=1}^{7} z_k\, w(z_k) = \sum_{k=1}^{7} z_k\, p(z_k), \tag{2.1}$$

where

$$p(z_k) \quad = \quad \frac{w(z_k)}{\left(\sum\limits_k w(z_k)\right)} \tag{2.2}$$

are normed weights with

$$\sum_k p(z_k) \quad = \quad 1. \tag{2.3}$$

In this example the weights $w(z_k)$ can be disassembled into $n=7$ elementary weights $v(z_\alpha)$ of unit mass.

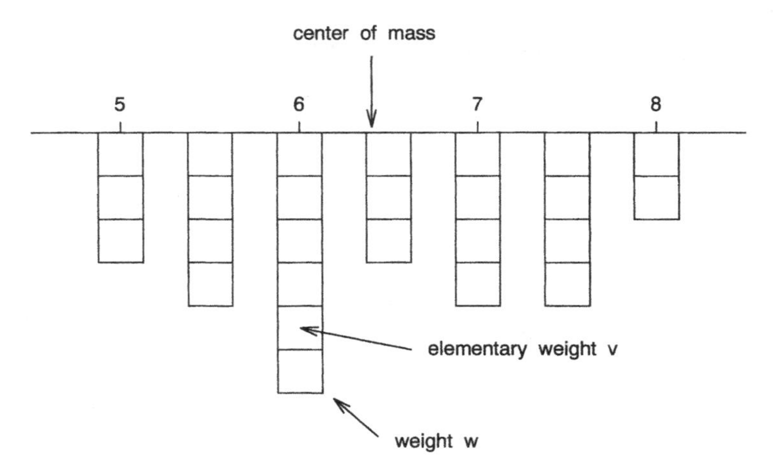

Figure 2.1: Bar with suspended weights w, subdivided into elementary weights v.

The normed weights $p(z_\alpha)$ corresponding to the elementary weights are equal to $1/n$ and the location of equilibrium of the bar, its center of mass, is computed as

$$\overline{z} \;=\; \sum_{\alpha=1}^{n} z_\alpha \, p(z_\alpha) = \frac{1}{n}\sum_{\alpha=1}^{n} z_\alpha = 6.4. \tag{2.4}$$

Transposing the physical problem at hand into a probabilistic context, we realize that $\overline{z}$ is the mean value $m^\star$ of data z_α,

$$m^\star \;=\; \frac{1}{n}\sum_{\alpha=1}^{n} z_\alpha, \tag{2.5}$$

and that the normed weights $p(z_k)$ or $p(z_\alpha)$ can be interpreted as probabilities, i.e. the frequency of appearance of the values z_k or z_α. The z_k represent a grouping of the z_α and have the meaning of classes of values z_α. The weights p can be called probabilities as they fulfill the requirements: $0 \leq p \leq 1$ and $\sum p = 1$.

Another characteristic value which can be calculated is the average squared distance to the center of mass

$$\begin{aligned} \mathrm{dist}^2(m^\star) \;&=\; \sum_{\alpha=1}^{n} (z_\alpha - m^\star)^2 \, p(z_\alpha) \\ &=\; .83. \end{aligned} \tag{2.6}$$

This average squared distance will be called the *experimental variance,*

$$s^2 \;=\; \frac{1}{n}\sum_{\alpha=1}^{n} (z_\alpha - m^\star)^2, \tag{2.7}$$

which gives an indication about the dispersion of the data around the center of mass $m^\star$ of the data.

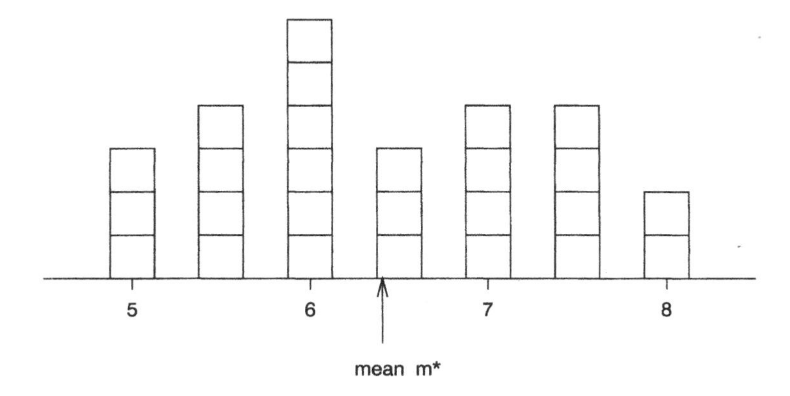

Figure 2.2: Histogram.

COMMENT 2.1 *In the framework of classical statistics based on a hypothesis of independence (i.e. the samples are considered as realizations of independent identically distributed random variables) the sample variance would usually be defined as*

$$s^2 = \frac{1}{n-1} \sum_{\alpha=1}^{n} (z_\alpha - m^\star)^2, \tag{2.8}$$

dividing by $n-1$ *instead of* n. *Dealing with spatially or temporally correlated data, we favour in this text (like e.g.* PRIESTLEY *[251], p48) the more straightforward definition (2.7).*

In fact, what has been introduced here under the cover of a weightless bar with weights of different size attached to it, can be seen as an upside down *histogram*, as represented on Figure 2.2.

An alternate way to represent the frequencies of the values z is by cumulating the frequencies from left to right as on Figure 2.3, where a *cumulative histogram* is shown.

Distribution function

Suppose we draw randomly values z from a set of values Z. We call each value z a realization of a random variable Z. The mathematical idealization of the cumulative histogram, for a random variable Z that takes values in $\mathbb{R}$, is the *probability distribution function* $F(z)$ defined as

$$F(z) = P(Z < z) \qquad \text{with} \qquad -\infty < z < \infty. \tag{2.9}$$

The distribution function indicates the probability P that a value of the random variable Z is below z. The probability P actually tells the proportion of values of Z that are below a given value z.

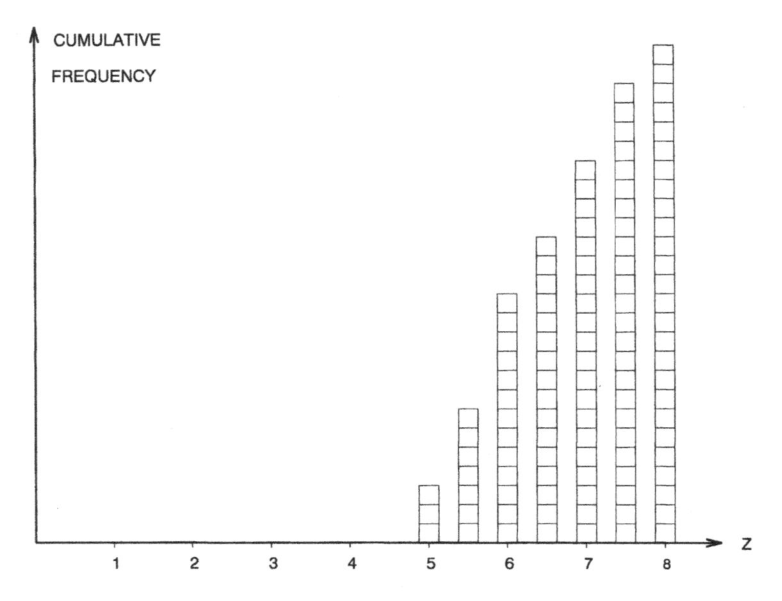

Figure 2.3: Cumulative histogram.

If we partition z into intervals of infinitesimal length dz, the probability that a realization of Z belongs to such an interval is $F(dz)$. We shall only consider differentiable distribution functions. The derivative of the frequency distribution is the density function $p(z)$

$$F(dz) \quad = \quad p(z)\, dz. \tag{2.10}$$

Expectation

The idealization of the concept of mean value is the ***mathematical expectation*** or ***expected value.*** The expected value of Z, also called the ***first moment*** of the random variable, is defined as the integral over the realizations z of Z weighted by the density function

$$\mathrm{E}[\,Z\,] \quad = \quad \int\limits_{z \in \mathbb{R}} z\, p(z)\, dz \; = m. \tag{2.11}$$

The expectation is a linear operator, so that computations are easy. Let a and b be constants; as they are deterministic (i.e. non random) we denote them with small letters, reserving capitals for random variables. From the definition (2.11) we easily deduce that

$$\mathrm{E}[\,a\,] \quad = \quad a, \tag{2.12}$$

$$\mathrm{E}[\,b\,Z\,] \quad = \quad b\,\mathrm{E}[\,Z\,], \tag{2.13}$$

and that

$$\boxed{\mathrm{E}[\,a + b\,Z\,] \;=\; a + b\,\mathrm{E}[\,Z\,]}\,. \tag{2.14}$$

The second moment of the random variable is the expectation of its squared value

$$\mathrm{E}[\,Z^2\,] \;=\; \int\limits_{z\,\in\,\mathbb{R}} z^2\,p(z)\,dz \tag{2.15}$$

and the k-th moment is defined as the expected value of the k-th power of Z

$$\mathrm{E}[\,Z^k\,] \;=\; \int\limits_{z\,\in\,\mathbb{R}} z^k\,p(z)\,dz. \tag{2.16}$$

When Z is a discrete random variable with states z_i, the integral in the definition (2.11) of the mathematical expectation is replaced by a sum

$$\mathrm{E}[\,Z\,] \;=\; \sum_i z_i\,p_i \;= m, \tag{2.17}$$

where $p_i = P(Z = z_i)$ is the probability that Z takes the value z_i. We are back to the weighted average of expression (2.4).

Variance

The variance σ^2 of the random variable Z, called the *theoretical variance*, is defined as

$$\mathrm{var}(Z) \;=\; \mathrm{E}\Big[\,(Z - \mathrm{E}[\,Z\,])^2\,\Big] \;=\; \mathrm{E}[\,(Z-m)^2\,] \;=\; \sigma^2. \tag{2.18}$$

Multiplying out we get

$$\mathrm{var}(Z) \;=\; \mathrm{E}[\,Z^2 + m^2 - 2mZ\,]$$

and, as the expectation is a linear operator,

$$\boxed{\mathrm{var}(Z) \;=\; \mathrm{E}[\,Z^2\,] - (\mathrm{E}[\,Z\,])^2}\,. \tag{2.19}$$

The variance can thus be expressed as the difference between the second moment and the squared first moment.

EXERCISE 2.2 *Let a and b be constants. Show that*

$$\mathrm{var}(a\,Z) \;=\; a^2\,\mathrm{var}(Z) \tag{2.20}$$

and that

$$\mathrm{var}(Z + b) \;=\; \mathrm{var}(Z). \tag{2.21}$$

Covariance

The theoretical covariance σ_{ij} between two random variables Z_i and Z_j is defined as

$$\begin{aligned} \operatorname{cov}(Z_i, Z_j) &= \mathrm{E}[\,(Z_i - \mathrm{E}[\,Z_i\,]) \cdot (Z_j - \mathrm{E}[\,Z_j\,])\,] \\ &= \mathrm{E}[\,(Z_i - m_i) \cdot (Z_j - m_j)\,] \;=\; \sigma_{ij}, \end{aligned} \tag{2.22}$$

where m_i and m_j are the means of the two random variables.

Note that the covariance of Z_i with itself is the variance of Z_i,

$$\sigma_{ii} \;=\; \mathrm{E}[\,(Z_i - m_i)^2\,] \;=\; \sigma_i^2. \tag{2.23}$$

The covariance divided by the square root of the variances is called the theoretical *correlation coefficient*

$$\rho_{ij} \;=\; \frac{\sigma_{ij}}{\sqrt{\sigma_i^2\,\sigma_j^2}}. \tag{2.24}$$

EXERCISE 2.3 *Two random variables whose covariance is zero are said to be uncorrelated. Show that the variance of the sum of two uncorrelated variables is equal to the sum of the variances, that is*

$$\operatorname{var}(Z_i + Z_j) \;=\; \operatorname{var}(Z_i) + \operatorname{var}(Z_j), \tag{2.25}$$

when $\operatorname{cov}(Z_i, Z_j) = 0$.

3 Linear Regression and Simple Kriging

Linear regression is presented in the case of two variables and then extended to the multivariate case. Simple kriging is a transposition of multiple regression in a spatial context. The algebraic problems generated by missing values in the multivariate case serve as a motivation for introducing a covariance function in the spatial case.

Experimental covariance

In the case of two variables, z_1 and z_2 say, the data values can be represented on a scatter diagram like on Figure 3.1, which shows the cloud of data points in the plane spanned by two perpendicular axes, one for each variable. The center of mass of the data cloud is the point defined by the two means $(m_1^\star, m_2^\star)$. A way to measure the dispersion of the data cloud around its center of mass is to multiply the difference between a value of one variable and its mean, called a *residual*, with the corresponding residual of the other variable. The average of the products of residuals is the *experimental covariance*

$$s_{12} \;=\; \frac{1}{n}\sum_{\alpha=1}^{n}(z_1^\alpha - m_1^\star)(z_2^\alpha - m_2^\star). \tag{3.1}$$

When the residual of z_1 tends to have the same sign as the residual of z_2 on average, the covariance is positive, while when the two residuals are of opposite sign on average, the covariance is negative. When a large value of one residual is on average associated with a large value of the residual of the other variable, the covariance has a large positive or negative value. Thus the covariance measures on one hand the liking or disliking of two variables through its sign and on the other hand the strength of this relationship by its absolute value.

We see that when z_1 is identical with z_2, the covariance is equal to the variance.

It is often desirable to compare the covariances of pairs of variables. When the units of the variables are not comparable, especially when they are of a different type, e.g. cm, kg, %, ..., it is preferable to standardize each variable z, centering first its values around the center of mass by subtracting the mean, and subsequently norming the distances of the values to the center of mass by dividing them with the *standard deviation*, which is the square root of the experimental variance s^2. The standardized variable $\widetilde{z}$

$$\widetilde{z} \;=\; \frac{z - m^\star}{s} \tag{3.2}$$

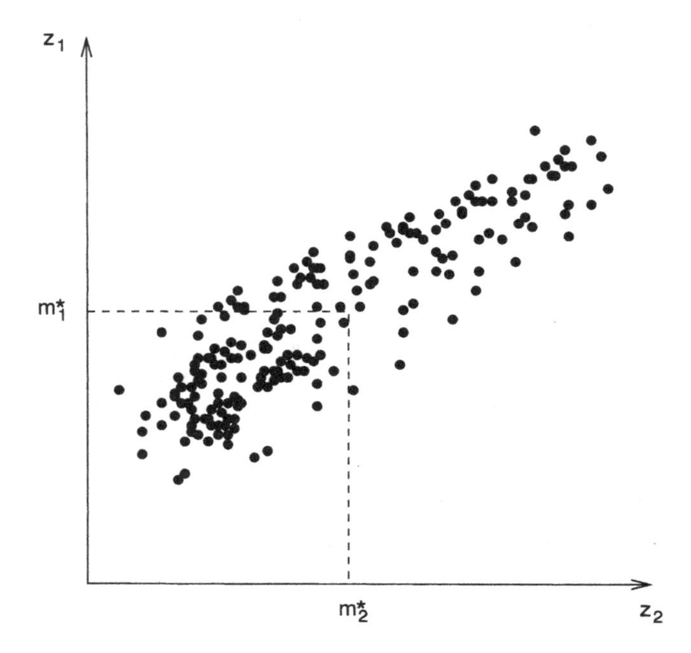

Figure 3.1: Scatter diagram showing the cloud of sample values and the center of mass $(m_2^\star, m_1^\star)$.

has a variance equal to 1. The covariance of two standardized variables $\widetilde{z}_1$ and $\widetilde{z}_2$ is a normed quantity r_{ij}, called the experimental *correlation coefficient,* with bounds

$$-1 \leq r_{ij} \leq 1. \tag{3.3}$$

The correlation coefficient r_{ij} can also be calculated directly from z_i and z_j dividing their covariance by the product of their standard deviations

$$r_{ij} = \frac{s_{ij}}{s_i\, s_j}. \tag{3.4}$$

Please note that the experimental variance of z_i is also the covariance of z_i with itself

$$s_i^2 = s_{ii}, \tag{3.5}$$

so that s_i stands for the standard deviation.

Linear regression

Two variables that have a correlation coefficient different from zero are said to be correlated. It is often reasonable to suppose that some of the information conveyed by the measured values is common to two correlated variables. Consequently it seems interesting to look for a function which, knowing a value of one variable, yields the best approximation to the unknown value of a second variable.

We shall call "best" or "optimal" a function $z^\star$ of a given type which minimizes the mean squared distance $\mathrm{dist}^2(z^\star)$ to the samples

$$\mathrm{dist}^2(z^\star) \;=\; \frac{1}{n}\sum_{\alpha=1}^{n}(z_\alpha - z_\alpha^\star)^2 . \tag{3.6}$$

This is intuitively appealing as using this criterion the best function $z^\star$ is the one which passes closest to the data values.

Let us take two variables z_1, z_2 and denote by $z_1^\star$ the function which approximates best unknown values of z_1.

The simplest type of approximation of z_1 is by a constant c, so let

$$z_1^\star \;=\; c, \tag{3.7}$$

and this does not involve z_2. The average distance between the data and the constant is

$$\mathrm{dist}^2(c) \;=\; \frac{1}{n}\sum_{\alpha=1}^{n}(z_1^\alpha - c)^2. \tag{3.8}$$

The minimum is achieved for a value of c for which the first derivative of the distance function $\mathrm{dist}^2(c)$ is zero

$$\frac{\partial\, \mathrm{dist}^2(c)}{\partial c} \;=\; 0 \tag{3.9}$$

$$\iff \quad \left(\frac{1}{n}\sum_{\alpha=1}^{n}(z_1^\alpha - c)^2\right)' \;=\; 0$$

$$\iff \quad \frac{1}{n}\sum_{\alpha=1}^{n}\left(\, (z_1^\alpha)^2 \,+\, c^2 \,-\, 2\,c\,z_1^\alpha \,\right)' \;=\; 0$$

$$\iff \quad \frac{1}{n}\sum_{\alpha=1}^{n}(2\,c \,-\, 2\,z_1^\alpha) \;=\; 0$$

$$\iff \quad \frac{1}{n}\sum_{\alpha=1}^{n} z_1^\alpha \;=\; c. \tag{3.10}$$

The constant c which minimizes the average square distance to the data of z_1 is the mean

$$\boxed{c \;=\; m_1^\star}\,. \tag{3.11}$$

The mean is the point nearest to the data in the least squares sense. Replacing c by $m_1^\star$ in expression (3.8), we see that the minimal distance for the optimal estimator $z_1^\star = m_1^\star$ is the experimental variance

$$\boxed{\text{dist}^2_{min}(c) \;=\; s_1^2}\,. \tag{3.12}$$

A more sophisticated function that can be chosen to approximate z_1 is a linear function of z_2

$$z_1^\star \;=\; a\,z_2 + b, \tag{3.13}$$

which defines a straight line through the data cloud with a slope a and an intercept b.

The distance between the data of z_1 and the straight line depends on the two parameters a and b:

$$\begin{aligned}\text{dist}^2(a,b) &= \frac{1}{n}\sum_{\alpha=1}^{n}(z_1^\alpha - a\,z_2^\alpha - b)^2 \qquad (3.14)\\ &= \frac{1}{n}\sum_{\alpha=1}^{n}\left((z_1^\alpha)^2 + a^2(z_2^\alpha)^2 + b^2 - 2\,a\,z_1^\alpha\,z_2^\alpha - 2\,b\,z_1^\alpha + 2\,a\,b\,z_2^\alpha\right)\\ &= -2\,b\,m_1^\star + 2\,a\,b\,m_2^\star + \frac{1}{n}\sum_{\alpha=1}^{n}\left((z_1^\alpha)^2 + a^2(z_2^\alpha)^2 + b^2 - 2\,a\,z_1^\alpha\,z_2^\alpha\right).\end{aligned}$$

If we shift the data cloud and the straight line to the center of mass, this translation does not change the slope of the straight line. We can thus consider, without loss of generality, the special case $m_1^\star = m_2^\star = 0$ to determine the slope of the straight line, introducing a new intercept b'. In the new translated coordinate system the distance is

$$\text{dist}^2(a,b') \;=\; s_1^2 + a^2\,s_2^2 + (b')^2 - 2\,a\,s_{12}, \tag{3.15}$$

where b' is the intercept of the shifted straight line.

At the minimum, the partial first derivative of the distance function with respect to a is zero

$$\begin{aligned}\frac{\partial\,\text{dist}^2(a,b')}{\partial a} &= 0 \qquad (3.16)\\ \Longleftrightarrow \qquad 2\,a\,s_2^2 - 2\,s_{12} &= 0,\end{aligned}$$

so that the slope is

$$\boxed{a \;=\; \frac{s_{12}}{s_2^2}}\,. \tag{3.17}$$

As the minimum with respect to b' is reached for

$$\frac{\partial\,\text{dist}^2(a,b')}{\partial b'} \;=\; 2\,b' \;=\; 0 \quad\Longleftrightarrow\quad b' = 0, \tag{3.18}$$

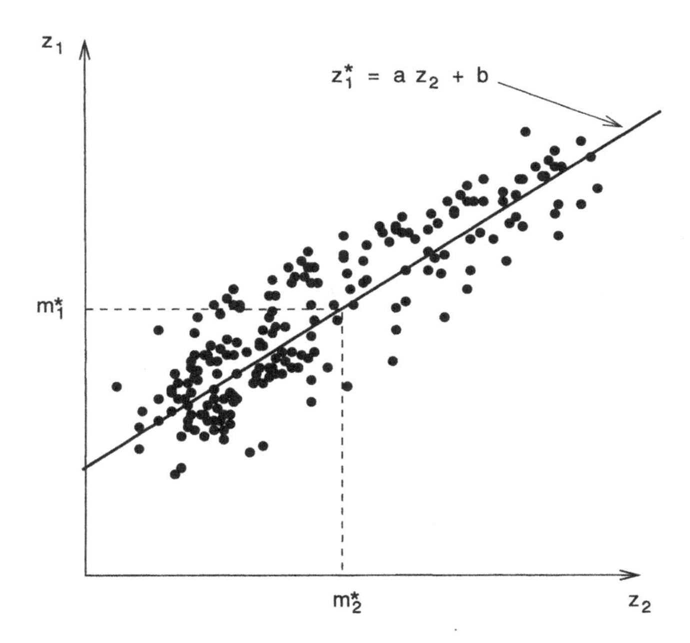

Figure 3.2: Data cloud with regression line $z_1^\star$.

we conclude that the optimal straight line passes through the center of mass.

Coming back to the more general case $m_1^\star \neq 0$, $m_2^\star \neq 0$, the optimal value of b (knowing a) is

$$\frac{\partial \operatorname{dist}^2(a,b)}{\partial b} = 0 \tag{3.19}$$

$$\iff \quad \frac{1}{n}\sum_{\alpha=1}^{n}(2\,b - 2\,z_1^\alpha + 2\,a\,z_2^\alpha) = 0,$$

so that the intercept is

$$\boxed{b = m_1^\star - a\,m_2^\star}\,. \tag{3.20}$$

The best straight line approximating the variable z_1 knowing z_2, called the *linear regression* of z_2 on z_1, is represented on Figure 3.2.

Rewriting the linear regression, using the values of the coefficients in (3.17) and (3.20) we have

$$\begin{aligned} z_1^\star &= \frac{s_{12}}{s_2^2}\,(z_2 - m_2^\star) + m_1^\star \qquad (3.21)\\ &= m_1^\star + \frac{s_1}{s_2}\,r_{12}\,(z_2 - m_2^\star) \end{aligned}$$

so that

$$z_1^\star \quad = \quad m_1^\star + s_1\, r_{12}\, \widetilde{z}_2, \tag{3.22}$$

where $\widetilde{z}_2$ is a standardized variable as defined in the expression (3.2).

The linear regression is an improvement over the approximation by a constant. The proportion r_{12} of the normed residual $\widetilde{z}_2$, rescaled with the standard deviation s_1, is added to the mean $m_1^\star$. The linear regression is constant and equal to mean in the case of a zero correlation coefficient.

The improvement can be evaluated by computing the minimal squared distance

$$\begin{aligned} \text{dist}^2_{min}(a,b) &= \frac{1}{n}\sum_{\alpha=1}^{n}(z_1^\alpha - m_1^\star - s_1\, r_{12}\, \widetilde{z}_2^\alpha)^2 \qquad (3.23) \\ &= \frac{1}{n}\sum_{\alpha=1}^{n}(s_1\, \widetilde{z}_1^\alpha - s_1\, r_{12}\, \widetilde{z}_2^\alpha)^2 \end{aligned}$$

and finally we get

$$\boxed{\text{dist}^2_{min}(a,b) \;=\; s_1^2\, \big(1-(r_{12})^2\big)}\,. \tag{3.24}$$

The stronger the correlation between two variables, the smaller the distance to the optimal straight line, the higher the explanative power of this straight line and the lower the expected average error. The improvement of the linear regression (3.22) with respect to the approximation by the constant $m_1^\star$ alone depends on the strength of the correlation r_{12} between the two variables z_1 and z_2. The higher the squared correlation, the closer (in the least squares sense) the regression line fits the bivariate data cloud.

Variance-covariance matrix

Take N different types of measurements z_i. Suppose they have been sampled n times, and arrange them into a rectangular table, the data matrix $\mathbf{Z}$:

$$\begin{array}{cc} & \textit{Variables} \\ \textit{Samples} & \begin{pmatrix} z_{11} & \ldots & z_{1i} & \ldots & z_{1N} \\ \vdots & & \vdots & & \vdots \\ z_{\alpha 1} & \ldots & z_{\alpha i} & \ldots & z_{\alpha N} \\ \vdots & & \vdots & & \vdots \\ z_{n1} & \ldots & z_{ni} & \ldots & z_{nN} \end{pmatrix} \end{array} \tag{3.25}$$

Define a matrix $\mathbf{M}$ with the same dimension $n \times N$ as $\mathbf{Z}$, replicating n times in its columns the mean value of each variable

$$\mathbf{M} \quad = \quad \begin{pmatrix} m_1^\star & \ldots & m_i^\star & \ldots & m_N^\star \\ \vdots & & \vdots & & \vdots \\ m_1^\star & \ldots & m_i^\star & \ldots & m_N^\star \\ \vdots & & \vdots & & \vdots \\ m_1^\star & \ldots & m_i^\star & \ldots & m_N^\star \end{pmatrix}. \tag{3.26}$$

A matrix $\mathbf{Z}_c$ of centered variables is obtained by subtracting $\mathbf{M}$ from the raw data matrix

$$\mathbf{Z}_c = \mathbf{Z} - \mathbf{M}. \tag{3.27}$$

The matrix $\mathbf{V}$ of experimental variances and covariances is calculated by premultiplying $\mathbf{Z}_c$ with its matrix transpose $\mathbf{Z}_c^\top$ (the $\top$ indicates matrix transposition, as defined in the appendix on Matrix Algebra) and dividing by the number of samples

$$\begin{aligned}\mathbf{V} &= \frac{1}{n}\mathbf{Z}_c^\top \mathbf{Z}_c \\ &= \begin{pmatrix} \mathrm{var}(\mathbf{z}_1) & \dots & \mathrm{cov}(\mathbf{z}_1,\mathbf{z}_j) & \dots & \mathrm{cov}(\mathbf{z}_1,\mathbf{z}_N) \\ \vdots & \ddots & & & \vdots \\ \mathrm{cov}(\mathbf{z}_i,\mathbf{z}_1) & \dots & \mathrm{var}(\mathbf{z}_i) & \dots & \mathrm{cov}(\mathbf{z}_i,\mathbf{z}_N) \\ \vdots & & & \ddots & \vdots \\ \mathrm{cov}(\mathbf{z}_N,\mathbf{z}_1) & \dots & \mathrm{cov}(\mathbf{z}_N,\mathbf{z}_j) & \dots & \mathrm{var}(\mathbf{z}_N) \end{pmatrix} \\ &= \begin{pmatrix} s_{11} & \dots & s_{1j} & \dots & s_{1N} \\ \vdots & \ddots & & & \vdots \\ s_{i1} & \dots & s_{ii} & \dots & s_{iN} \\ \vdots & & & \ddots & \vdots \\ s_{N1} & \dots & s_{Nj} & \dots & s_{NN} \end{pmatrix}. \end{aligned} \tag{3.28}$$

The variance-covariance matrix $\mathbf{V}$ is symmetric. As it is the result of a product $\mathbf{A}^\top\mathbf{A}$, where $\mathbf{A} = \frac{1}{\sqrt{n}}\mathbf{Z}_c$, it is nonnegative definite. The notion of nonnegative definiteness of the variance-covariance matrix plays an important role in theory and also has practical implications. The definition of nonnegative definiteness together with three criteria is given in the appendix on Matrix Algebra on p324.

Multiple linear regression

Having a set of N auxiliary variables z_i, $i = 1, \dots, N$, it is often desirable to estimate values of a main variable z_0 using a linear function of the N other variables.

A multiple regression plane $z_0^\star$ in $N+1$ dimensional space is given by a linear combination of the residuals of the N variables with a set of coefficients a_i

$$z_0^\star = m_0^\star + \sum_{i=1}^{N} a_i(z_i - m_i^\star), \tag{3.29}$$

where $m_i^\star$ are the respective means of the different variables.

For n samples we have the matrix equation

$$\mathbf{z}_0^\star = \mathbf{m}_0 + (\mathbf{Z} - \mathbf{M})\,\mathbf{a}. \tag{3.30}$$

The distance between $\mathbf{z}_0$ and the hyperplane is

$$\begin{aligned} \text{dist}^2(\mathbf{a}) &= \frac{1}{n}(\mathbf{z}_0 - \mathbf{z}_0^\star)^\top(\mathbf{z}_0 - \mathbf{z}_0^\star) \\ &= \text{var}(\mathbf{z}_0) + \mathbf{a}^\top \mathbf{V}\mathbf{a} - 2\,\mathbf{a}^\top \mathbf{v}_0, \end{aligned} \tag{3.31}$$

where $\mathbf{v}_0$ is the vector of covariances between $\mathbf{z}_0$ and $\mathbf{z}_i$, $i = 1, \ldots, N$.

The minimum is found for

$$\frac{\partial\, \text{dist}^2(\mathbf{a})}{\partial \mathbf{a}} = 0 \tag{3.32}$$

$$\iff \quad 2\,\mathbf{V}\mathbf{a} - 2\,\mathbf{v}_0 = 0$$

$$\iff \quad \mathbf{V}\mathbf{a} = \mathbf{v}_0. \tag{3.33}$$

This system of linear equations

$$\begin{pmatrix} \text{var}(\mathbf{z}_1) & \ldots & \text{cov}(\mathbf{z}_1, \mathbf{z}_N) \\ \vdots & \ddots & \vdots \\ \text{cov}(\mathbf{z}_N, \mathbf{z}_1) & \ldots & \text{var}(\mathbf{z}_N) \end{pmatrix} \begin{pmatrix} a_1 \\ \vdots \\ a_N \end{pmatrix} = \begin{pmatrix} \text{cov}(\mathbf{z}_0, \mathbf{z}_1) \\ \vdots \\ \text{cov}(\mathbf{z}_0, \mathbf{z}_N) \end{pmatrix} \tag{3.34}$$

has exactly one solution if the determinant of the matrix $\mathbf{V}$ is different from zero.

The minimal distance resulting from the optimal coefficients is

$$\text{dist}^2_{min}(\mathbf{a}) = \text{var}(\mathbf{z}_0) - \mathbf{a}^\top \mathbf{v}_0. \tag{3.35}$$

To make sure that this optimal distance is not negative, it has to be checked beforehand that the augmented matrix $\mathbf{V}_0$ of variances and covariances for all $N+1$ variables

$$\mathbf{V}_0 = \begin{pmatrix} \text{var}(\mathbf{z}_0) & \ldots & \text{cov}(\mathbf{z}_0, \mathbf{z}_N) \\ \vdots & \ddots & \vdots \\ \text{cov}(\mathbf{z}_N, \mathbf{z}_0) & \ldots & \text{var}(\mathbf{z}_N) \end{pmatrix} \tag{3.36}$$

is nonnegative definite.

EXAMPLE 3.1 (MISSING VALUES) *In the following situation a major problem can arise. Suppose that for some variables data values are missing for some samples. If the variances and covariances are naively calculated on the entire set, then the variances and covariances will have been obtained from subsets of samples with a varying size. The matrix* $\mathbf{V}_0$ *assembled with these variance-covariance values stemming from different sample subsets can be indefinite, i.e. not a variance-covariance matrix.*

This is illustrated with a simple numerical example involving three samples and three variables. A data value of z_0 *is missing for the first sample. The eight data values are*

	z_0	z_1	z_2
Sample 1	–	3.	5.
Sample 2	3.	2.	2.
Sample 3	1.	4.	5.

(3.37)

The means are $m_0^\star = 2,\ m_1^\star = 3,\ m_2^\star = 4$. *We center the three variables and obtain the following matrix* $\mathbf{V}_0$ *of variances and covariances*

$$\mathbf{V}_0 = \begin{pmatrix} 1 & -1 & -\frac{3}{2} \\ -1 & \frac{2}{3} & 1 \\ -\frac{3}{2} & 1 & 2 \end{pmatrix}, \tag{3.38}$$

which is not nonnegative definite because one of the minors of order two is negative (following a criterion given on p325)

$$\det\begin{pmatrix} 1 & -1 \\ -1 & \frac{2}{3} \end{pmatrix} = -\frac{1}{3}, \quad \det\begin{pmatrix} \frac{2}{3} & 1 \\ 1 & 2 \end{pmatrix} = \frac{1}{3}, \quad \det\begin{pmatrix} 1 & -\frac{3}{2} \\ -\frac{3}{2} & 2 \end{pmatrix} = -\frac{1}{4}.$$

The linear regression equations for $z_0^\star$

$$\mathbf{V}\mathbf{a} = \begin{pmatrix} \frac{2}{3} & 1 \\ 1 & 2 \end{pmatrix} \begin{pmatrix} a_1 \\ a_2 \end{pmatrix} = \begin{pmatrix} -1 \\ -\frac{3}{2} \end{pmatrix} = \mathbf{v}_0 \tag{3.39}$$

can be solved as the matrix $\mathbf{V}$ *is not singular. The solution is*

$$z_0^\star = m_0^\star - \frac{3}{2}\,(z_1 - m_1^\star) + 0\,(z_2 - m_2^\star). \tag{3.40}$$

The minimal distance achieved by this "optimal" regression line is however negative

$$\text{dist}^2_{min} = -\frac{1}{2} \tag{3.41}$$

and the whole operation turns out to be meaningless!

The lesson from this is that the experimental variances and covariances for a set of variables have to be calculated on the same subset of samples. This is the subset of samples with no values missing for any variable.

Similarly, in a spatial context we wish to estimate values at locations where no measurement has been made using the data from neighboring locations. We see immediately that there is a need for a *covariance function* defined in such a way that the corresponding matrix $\mathbf{V}_0$ is nonnegative definite. The spatial multiple regression based on a covariance function model is called *simple kriging*. We briefly sketch this in the next section.

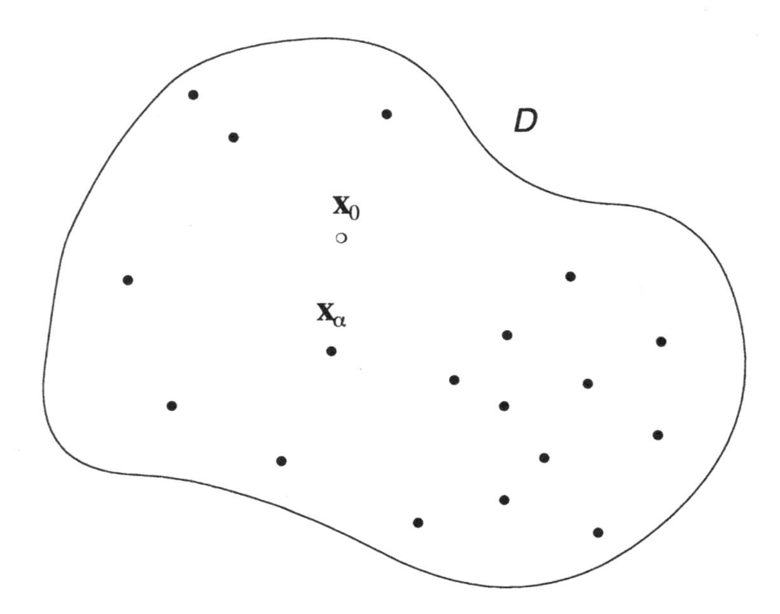

Figure 3.3: Data points $\mathbf{x}_\alpha$ and the estimation point $\mathbf{x}_0$ in a spatial domain $\mathcal{D}$.

Simple kriging

Suppose we have data for one type of measurement z at various locations $\mathbf{x}_\alpha$ of a spatial domain $\mathcal{D}$. Take the data locations $\mathbf{x}_\alpha$ and construct at each of the n locations a random variable $Z(\mathbf{x}_\alpha)$. Take an additional location $\mathbf{x}_0$ and let $Z(\mathbf{x}_0)$ be a random variable at $\mathbf{x}_0$. Further assume that these random variables are a subset of an infinite collection of random variables called a *random function* $Z(\mathbf{x})$ defined at any location $\mathbf{x}$ of the domain $\mathcal{D}$. We assume the random function is *second-order stationary*. This means that the expectation and the covariance are both translation invariant over the domain, i.e. for a vector $\mathbf{h}$ linking any two points $\mathbf{x}$ and $\mathbf{x}+\mathbf{h}$ in the domain

$$\mathrm{E}[\,Z(\mathbf{x}+\mathbf{h})\,] \;=\; \mathrm{E}[\,Z(\mathbf{x})\,], \tag{3.42}$$

$$\mathrm{cov}\Big[Z(\mathbf{x}+\mathbf{h}), Z(\mathbf{x})\Big] \;=\; C(\mathbf{h}). \tag{3.43}$$

The expected value $\mathrm{E}[\,Z(\mathbf{x})\,] = m$ is the same at any point $\mathbf{x}$ of the domain. The covariance between any pair of locations depends only on the vector $\mathbf{h}$ which separates them.

The problem of interest is to build a weighted average to make an estimation of a value at a point $\mathbf{x}_0$ using information at points $\mathbf{x}_\alpha$, $\alpha = 1, \ldots, n$, see Figure 3.3. This estimation procedure should be based on the knowledge of the covariances between the random variables at the points involved. The answer closely resembles multiple regression transposed into a spatial context where the $Z(\mathbf{x}_\alpha)$ play the role of regressors on a regressand $Z(\mathbf{x}_0)$. This spatial regression bears the name of *simple kriging* (after the mining engineer DG KRIGE). With a mean m supposed constant over the whole domain and calculated as the average of the data, simple kriging is used to estimate residuals from this reference value m given a priori. Therefore simple kriging is sometimes called *kriging with known mean.*

In analogy with multiple regression the weighted average for this estimation is defined as

$$Z^{\star}(\mathbf{x}_0) \;=\; m + \sum_{\alpha=1}^{n} w_\alpha (Z(\mathbf{x}_\alpha) - m), \tag{3.44}$$

where w_α are weights attached to the residuals $Z(\mathbf{x}_\alpha) - m$. Note that unlike in multiple regression the mean m is the same for all locations due to the stationarity assumption.

The estimation error is the difference between the estimated and the true value at $\mathbf{x}_0$

$$Z^{\star}(\mathbf{x}_0) - Z(\mathbf{x}_0). \tag{3.45}$$

The estimator is said to be unbiased if the estimation error is nil on average

$$\begin{aligned} \mathrm{E}[\,Z^{\star}(\mathbf{x}_0) - Z(\mathbf{x}_0)\,] \;&=\; m + \sum_{\alpha=1}^{n} w_\alpha \Big(\mathrm{E}[\,Z(\mathbf{x}_\alpha)\,] - m\Big) - \mathrm{E}[\,Z(\mathbf{x}_0)\,] \\ &=\; m + \sum_{\alpha=1}^{n} w_\alpha (m - m) - m \\ &=\; 0. \end{aligned} \tag{3.46}$$

The variance of the estimation error, the *estimation variance* σ^2_{E}, then is simply expressed as

$$\sigma^2_{\mathrm{E}} \;=\; \mathrm{var}(Z^{\star}(\mathbf{x}_0) - Z(\mathbf{x}_0)) = \mathrm{E}[\,(Z^{\star}(\mathbf{x}_0) - Z(\mathbf{x}_0))^2\,]. \tag{3.47}$$

Expanding this expression

$$\begin{aligned} \sigma^2_{\mathrm{E}} \;&=\; \mathrm{E}[\,(Z^{\star}(\mathbf{x}_0))^2 + (Z(\mathbf{x}_0))^2 - 2\,Z^{\star}(\mathbf{x}_0)\,Z(\mathbf{x}_0)\,] \\ &=\; \sum_{\alpha=1}^{n}\sum_{\beta=1}^{n} w_\alpha\, w_\beta\, C(\mathbf{x}_\alpha - \mathbf{x}_\beta) + C(\mathbf{x}_0 - \mathbf{x}_0) - 2\sum_{\alpha=1}^{n} w_\alpha\, C(\mathbf{x}_\alpha - \mathbf{x}_0). \end{aligned} \tag{3.48}$$

In computing the last equation we have written:

$$\mathrm{cov}[Z(\mathbf{x}_\alpha), Z(\mathbf{x}_\beta)] \;=\; C(\mathbf{x}_\alpha - \mathbf{x}_\beta), \tag{3.49}$$

because the spatial covariances depend only on the difference vector between points, following our stationarity assumptions (3.42) and (3.43).

The estimation variance is minimal where its first derivative is zero

$$\frac{\partial \sigma^2_{\mathrm{E}}}{\partial w_\alpha} \;=\; 0 \qquad \text{for } \alpha = 1, \ldots, n. \tag{3.50}$$

Explicitly we have, for each $\alpha = 1, \ldots, n$,

$$2\sum_{\beta=1}^{n} w_\beta \, C(\mathbf{x}_\alpha - \mathbf{x}_\beta) - 2\, C(\mathbf{x}_\alpha - \mathbf{x}_0) \;=\; 0, \tag{3.51}$$

and the equation system for simple kriging is written

$$\sum_{\beta=1}^{n} w_\beta \, C(\mathbf{x}_\alpha - \mathbf{x}_\beta) \;=\; C(\mathbf{x}_\alpha - \mathbf{x}_0) \qquad \text{for } \alpha = 1, \ldots, n. \tag{3.52}$$

The left hand side of the equations describes the covariances between the locations. The right hand side describes the covariance between each data location and the location where an estimate is sought. The resolution of the system yields the optimal kriging weights w_α.

The operation of simple kriging can be repeated at regular intervals displacing each time the location $\mathbf{x}_0$. A regular grid of kriging estimates is obtained which can be contoured for representation as a map.

A second quantity of interest is the optimal variance for each location $\mathbf{x}_0$. It is obtained by substituting the left hand of the kriging system by its right hand side in the first term of the expression of the estimation variance σ_{E}^2. This is the variance of simple kriging

$$\begin{aligned} \sigma_{\mathrm{SK}}^2 &= \sum_{\alpha=1}^{n} w_\alpha \boxed{C(\mathbf{x}_\alpha - \mathbf{x}_0)} + C(\mathbf{x}_0 - \mathbf{x}_0) - 2\sum_{\alpha=1}^{n} w_\alpha \, C(\mathbf{x}_\alpha - \mathbf{x}_0) \\ &= C(0) - \sum_{\alpha=1}^{n} w_\alpha \, C(\mathbf{x}_\alpha - \mathbf{x}_0). \end{aligned} \tag{3.53}$$

When sample locations are scattered irregularly in space it is worthwhile to produce a map of the kriging variance as a complement to the map of kriged estimates. It gives an appreciation of the varying precision of the kriged estimates due to the irregular disposition of informative points.

4 Kriging the Mean

The mean value of samples from a geographical space can be estimated, either using the arithmetic mean or by constructing a weighted average integrating the knowledge of the spatial correlation of the samples. The two approaches are developped and it turns out that the solution of kriging the mean reduces to the arithmetic mean when there is no spatial correlation between data locations.

Arithmetic mean and its estimation variance

We denote by z the realization of a random variable Z. Suppose we have n measurements z_α, where α is an index numbering the samples from 1 to n. This is our data. We introduce a probabilistic model by considering that each z_α is a realization of a random variable Z_α and we further assume that the random variables are *independent and identically distributed* (iid). This means that each sample is a realization of a separate random variable and that all these random variables have the same distribution. The random variables are assumed independent which implies that they are uncorrelated (the converse being true only for Gaussian random variables).

Let us assume we know the variance σ^2 of the random variables Z_α (they have the same variance as they all have the same distribution). We do not know the mean m and wish to estimate it from the data z_α using the arithmetic mean as an estimator

$$m_{\mathrm{A}}^{\star} \;=\; \frac{1}{n}\sum_{\alpha=1}^{n} z_\alpha. \tag{4.1}$$

Imagine we would take many times n samples z_α under unchanged conditions. As we assume some randomness we clearly would not get identical values each time and the arithmetic mean would each time different. Thus we can consider $m_{\mathrm{A}}^{\star}$ itself as a realization of a random variable and write

$$M_{\mathrm{A}}^{\star} \;=\; \frac{1}{n}\sum_{\alpha=1}^{n} Z_\alpha. \tag{4.2}$$

What is the average value of $M_{\mathrm{A}}^{\star}$? Our probabilistic model allows to compute it easily as

$$\mathrm{E}[\,M_{\mathrm{A}}^{\star}\,] \;=\; \mathrm{E}\Big[\frac{1}{n}\sum_{\alpha=1}^{n} Z_\alpha\Big] \;=\; \frac{1}{n}\sum_{\alpha=1}^{n}\mathrm{E}[\,Z_\alpha\,] \;=\; m. \tag{4.3}$$

So the estimated mean fluctuates around the true mean and is on average equal to it. If we call $M_A^\star - m$ the estimation error associated with the arithmetic mean, we see that the error is zero on average

$$\mathrm{E}[\, M_A^\star - m \,] \;=\; 0. \tag{4.4}$$

The use of the arithmetic mean as an estimator of the true mean does not lead to a systematic error and the estimator is said to be *unbiased*.

What is the average fluctuation of the arithmetic mean around the true mean? This can be characterized by the variance σ_A^2 of the estimator

$$\sigma_A^2 \;=\; \mathrm{var}(M_A^\star) = \mathrm{var}(\frac{1}{n}\sum_{\alpha=1}^{n} Z_\alpha) \;=\; \frac{1}{n^2}\,\mathrm{var}(\sum_{\alpha=1}^{n} Z_\alpha) \tag{4.5}$$

using relation (2.20). Applying relation (2.25) to the uncorrelated identically distributed random variables we get

$$\sigma_A^2 \;=\; \frac{1}{n^2}\sum_{\alpha=1}^{n} \mathrm{var}(Z_\alpha) \;=\; \frac{\sigma^2}{n}. \tag{4.6}$$

So the distribution of $M_A^\star$ has a variance n times smaller than that of the random variables Z. We can view the variance of the estimator also as an estimation variance σ_E^2, i.e. the variance of the estimation error $M_A^\star - m$, because by relation (2.21) we have

$$\sigma_E^2 \;=\; \mathrm{var}(M_A^\star - m) \;=\; \mathrm{var}(M_A^\star) \;=\; \frac{\sigma^2}{n}. \tag{4.7}$$

Estimating the mean with spatial correlation

When samples have been taken at irregular spacing in a domain $\mathcal{D}$ like on Figure 4.1, a quantity of interest is the value of the mean m. We assume again that each sample $z(\mathbf{x}_\alpha)$ is a realization of a random variable $Z(\mathbf{x}_\alpha)$ and that the random variables are identically distributed.

A first approach for estimating m is to use the arithmetic mean

$$M_A^\star = \frac{1}{n}\sum_{\alpha=1}^{n} Z(\mathbf{x}_\alpha). \tag{4.8}$$

However the samples from a spatial domain cannot in general be assumed independent. This implies that the random variables at two different locations are usually correlated, especially when they are near to each other in space.

A second approach to estimate m is to use a weighted average

$$M^\star = \sum_{\alpha=1}^{n} w_\alpha\, Z(\mathbf{x}_\alpha) \tag{4.9}$$

with unknown weights w_α.

How best choose the weights w_α? We have to specify the problem further.

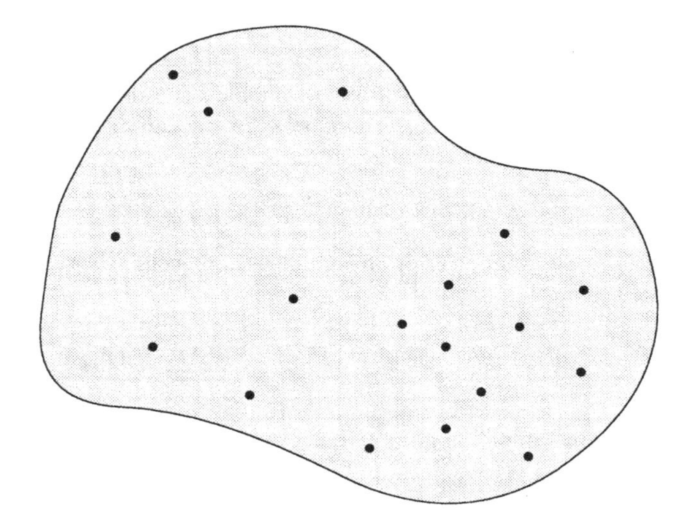

Figure 4.1: A domain with irregularly spaced sample points.

No systematic bias

We need to assume that the mean exists at all points of the region

$$\mathrm{E}[\, Z(\mathbf{x})\,] \;=\; m \qquad \text{for all } \mathbf{x} \in \mathcal{D}. \tag{4.10}$$

We want the estimation error

$$\underbrace{M^{\star}}_{\text{estimated value}} \;-\; \underbrace{m}_{\text{true value}} \tag{4.11}$$

to be zero on average

$$\mathrm{E}[\; M^{\star} - m \;] \;=\; 0, \tag{4.12}$$

i.e. we do not wish to have a systematic bias in our estimation procedure.

This can be achieved by constraining the weights w_α to sum up to one

$$\sum_{\alpha=1}^{n} w_\alpha \;=\; 1. \tag{4.13}$$

Indeed, if we replace $M^{\star}$ by the weighted average, the unbiasedness condition is fulfilled

$$\begin{aligned} \mathrm{E}[\, M^{\star} - m\,] \;&=\; \mathrm{E}\Big[\, \sum_{\alpha=1}^{n} w_\alpha\, Z(\mathbf{x}_\alpha) - m \,\Big] \;=\; \sum_{\alpha=1}^{n} w_\alpha \underbrace{\mathrm{E}[\, Z(\mathbf{x}_\alpha)\,]}_{m} - m \\ &=\; m \underbrace{\sum_{\alpha=1}^{n} w_\alpha}_{1} - m \;=\; 0. \end{aligned} \tag{4.14}$$

Variance of the estimation error

We need to assume that $Z(\mathbf{x})$ is second-order stationary and thus has a covariance function $C(\mathbf{h})$ which describes the correlation between any pair of points $\mathbf{x}$ and $\mathbf{x}+\mathbf{h}$ in the spatial domain

$$C(\mathbf{h}) \;=\; \operatorname{cov}(Z(\mathbf{x}), Z(\mathbf{x}+\mathbf{h})) \;=\; \mathrm{E}[\,Z(\mathbf{x}) \cdot Z(\mathbf{x}+\mathbf{h})\,] - m^2. \tag{4.15}$$

The variance of the estimation error in our unbiased procedure is simply the average of the squared error

$$\operatorname{var}(M^\star - m) \;=\; \mathrm{E}[\,(M^\star - m)^2\,] - \Big(\underbrace{\mathrm{E}[\,M^\star - m\,]}_{0}\Big)^2. \tag{4.16}$$

The variance of the estimation error can be expressed in terms of the covariance function

$$\begin{aligned}
\operatorname{var}(M^\star - m) &= \mathrm{E}[\;M^{\star 2} - 2\,mM^\star + m^2\;] \\
&= \sum_{\alpha=1}^{n}\sum_{\beta=1}^{n} w_\alpha\, w_\beta\, \mathrm{E}[\,Z(\mathbf{x}_\alpha)\, Z(\mathbf{x}_\beta)\,] - 2\,m \sum_{\alpha=1}^{n} w_\alpha\; \underbrace{\mathrm{E}[\,Z(\mathbf{x}_\alpha)\,]}_{m} + m^2 \\
&= \sum_{\alpha=1}^{n}\sum_{\beta=1}^{n} w_\alpha\, w_\beta\, C(\mathbf{x}_\alpha - \mathbf{x}_\beta).
\end{aligned} \tag{4.17}$$

The estimation variance is the sum of the cross-products of the weights assigned to the sample points. Each cross-product is weighted by the covariance between the corresponding sample points.

Minimal estimation variance

The criterion to define the "best" weights will be that we want the procedure to reduce as much as possible the variance of the estimation error. Additionally we want to respect the unbiasedness condition. We are thus looking for the

$$\boxed{\text{minimum of} \quad \operatorname{var}(M^\star - m) \quad \text{subject to} \quad \sum_{\alpha=1}^{n} w_\alpha = 1}\,. \tag{4.18}$$

The minimum of a positive quadratic function is found by setting the first order partial derivatives to zero. The condition on the weights is introduced into the problem by the method of Lagrange (see, for example, STRANG [319], p96).

An objective function φ is defined, consisting of the quadratic function plus a term containing a Lagrange multiplier μ,

$$\varphi(w_\alpha, \mu) \;=\; \operatorname{var}(M^\star - m) - 2\,\mu\,\Big(\sum_{\alpha=1}^{n} w_\alpha - 1\Big). \tag{4.19}$$

The new unknown μ is built into the objective function in such a way that the constraint on the weights is recovered when setting the partial derivative with respect to μ to zero. The Lagrange multiplier is multiplied by a factor -2 only for ease of subsequent computation.

Kriging equations

To solve the optimization problem the partial derivatives of the objective function $\varphi(w_\alpha, \mu)$ are set to zero

$$\frac{\partial\varphi(w_\alpha,\mu)}{\partial w_\alpha} = 0 \qquad \text{for} \quad \alpha = 1,\dots,n \tag{4.20}$$

$$\frac{\partial\varphi(w_\alpha,\mu)}{\partial \mu} = 0. \tag{4.21}$$

This yields a linear system of $n+1$ equations. The solution of this system provides the optimal weights w_α^{KM} for the estimation of the mean by a weighted average

$$\begin{cases} \displaystyle\sum_{\beta=1}^{n} w_\beta^{\mathrm{KM}}\, C(\mathbf{x}_\alpha - \mathbf{x}_\beta) - \mu_{\mathrm{KM}} = 0 \qquad \text{for} \quad \alpha = 1,\dots,n \\ \displaystyle\sum_{\beta=1}^{n} w_\beta^{\mathrm{KM}} = 1. \end{cases} \tag{4.22}$$

This is the system for the *kriging of the mean* (KM).

The minimal estimation variance σ^2_{KM} is computed by using the equations for the optimal weights

$$\sum_{\beta=1}^{n} w_\beta^{\mathrm{KM}}\, C(\mathbf{x}_\alpha - \mathbf{x}_\beta) = \mu_{\mathrm{KM}} \qquad \text{for } \alpha = 1,\dots,n. \tag{4.23}$$

in the expression of the estimation variance

$$\begin{aligned} \sigma^2_{\mathrm{KM}} &= \operatorname{var}(M^\star - m) = \sum_{\alpha=1}^{n}\sum_{\beta=1}^{n} w_\alpha^{\mathrm{KM}}\, w_\beta^{\mathrm{KM}}\, C(\mathbf{x}_\alpha - \mathbf{x}_\beta) \\ &= \sum_{\alpha=1}^{n} w_\alpha^{\mathrm{KM}}\, \mu_{\mathrm{KM}} = \mu_{\mathrm{KM}} \underbrace{\sum_{\alpha=1}^{n} w_\alpha^{\mathrm{KM}}}_{1} = \mu_{\mathrm{KM}}. \end{aligned} \tag{4.24}$$

The variance of the kriging of the mean is thus given by the Lagrange multiplier μ_{KM}.

Case of no correlation

It may happen that there is no correlation between different points in the spatial domain. This is described by the following covariance model

$$\mathrm{cov}(Z(\mathbf{x}_\alpha), Z(\mathbf{x}_\beta)) = \begin{cases} \sigma^2 & \text{if } \mathbf{x}_\alpha = \mathbf{x}_\beta, \\ 0 & \text{if } \mathbf{x}_\alpha \neq \mathbf{x}_\beta. \end{cases} \tag{4.25}$$

With this model the system for the kriging of the mean simplifies to

$$\begin{cases} w_\alpha^{\mathrm{KM}} \sigma^2 = \mu_{\mathrm{KM}} & \text{for } \alpha = 1, \ldots, n \\ \displaystyle\sum_{\beta=1}^{n} w_\beta^{\mathrm{KM}} = 1. \end{cases} \tag{4.26}$$

As all weights are equal and sum up to one, we have

$$w_\alpha^{\mathrm{KM}} = \frac{1}{n} \tag{4.27}$$

and the estimator of the kriging of the mean is equivalent to the arithmetic mean

$$M_{\mathrm{KM}}^\star = \sum_{\alpha=1}^{n} w_\alpha^{\mathrm{KM}} Z(\mathbf{x}_\alpha) = \frac{1}{n} \sum_{\alpha=1}^{n} Z(\mathbf{x}_\alpha) = M_{\mathrm{A}}^\star. \tag{4.28}$$

The estimation variance associated to $M_{\mathrm{KM}}^\star$ is

$$\sigma_{\mathrm{KM}}^2 = \mu_{\mathrm{KM}} = \frac{1}{n}\sigma^2. \tag{4.29}$$

When there is no correlation between points in space, the arithmetic mean is the best linear unbiased estimator and we recover also the corresponding estimation variance. The kriging of the mean is thus a generalization of the arithmetic mean estimator to the case of spatially correlated samples.

EXAMPLE 4.1 *Assuming a known population variance and* independent *identically normally distributed data, the 95% confidence interval for the true mean m is*

$$P(M_{\mathrm{A}}^\star - 1.96\,\frac{\sigma}{\sqrt{n}} < m < M_{\mathrm{A}}^\star + 1.96\,\frac{\sigma}{\sqrt{n}}). \tag{4.30}$$

With autocorrelated *data and assuming a multinormal distribution, the 95% confidence interval using the kriged mean estimator becomes*

$$P(M_{\mathrm{KM}}^\star - 1.96\,\sigma_{\mathrm{KM}} < m < M_{\mathrm{KM}}^\star + 1.96\,\sigma_{\mathrm{KM}}). \tag{4.31}$$

Let us consider a time series of autocorrelated trichloroethylene data, with samples averaged over 7mn measured during one afternoon near a degreasing machine (for more details see [342, 332]). The recommended maximum exposure at the time *of writing is 405 mg/m*^{3}. *Neglecting autocorrelation, i.e. assuming independence, the upper bound of the confidence interval is estimated at 145 mg/m*^{3}. *Taking account of the autocorrelation in a kriging of the mean we however get a value of 219 mg/m*^{3}. *We see that ignoring the autocorrelation could easily lead, in other situations, to underestimate severely the risk of exceeding legal limit values, as well as the associated health risks for the workers.*

Part B

Geostatistics

La chance a voulu que, d'entrée de jeu, deux circonstances imposent à la géostatistique d'utiliser des *variogrammes* (plutôt que des covariances), c'est-à-dire des *F.A. intrinsèques* (plutôt que des F.A. stationnaires).
La première circonstance, ce sont les résultats expérimentaux de l'Ecole d'Afrique du Sud, en particulier cette courbe d'allure logarithmique où DG KRIGE (1952) présentait la variance expérimentale d'échantillons de taille fixée dans des zones de plus en plus grandes du gisement (panneaux, quartiers, concessions, le Rand tout entier, etc...).
La seconde circonstance, c'était le modèle à homothétie interne élaboré par DE WIJS (1951), et qui permettait de retrouver cette loi logarithmique de la variance.

G MATHERON [208]

Chance has determined that, from the beginning, two circumstances were responsible for geostatistics to use variograms *(instead of covariances), that is to say* intrinsic Random Functions *(rather than stationary Random Functions).*
The first circumstance was given by the experimental results of the South African School, in particular this curve of logarithmic shape where DG KRIGE [163] *presented the experimental variance of samples of fixed size in increasingly large zones of the deposit (panels, quarters, concessions, the whole Rand, etc...).*
The second circumstance was provided by the model with internal self-similarity elaborated by DE WIJS [79], *which allowed to recover this logarithmic function of the variance.*

5 Regionalized Variable and Random Function

"It could be said,
as in *The Emperor of the Moon*,
that all is everywhere and always like here,
up to a degree of magnitude and perfection."

GW LEIBNIZ

The data provide information about regionalized variables, which are simply functions $z(\mathbf{x})$ whose behavior we would like to characterize in a given region of a spatial or time continuum. In applications the regionalized variables are usually not identical with simple deterministic functions and it is therefore of advantage to place them into a probabilistic framework.

In the probabilistic model, a regionalized variable $z(\mathbf{x})$ is considered to be a realization of a random function $Z(\mathbf{x})$ (i.e. an infinite family of random variables constructed at all points $\mathbf{x}$ of a given region $\mathcal{D}$). The advantage of this approach is that we shall only try to characterize simple features of the random function $Z(\mathbf{x})$ and not those of particular realizations $z(\mathbf{x})$.

In such a setting the data values are samples from one particular realization $z(\mathbf{x})$ of $Z(\mathbf{x})$. The epistemological implications of an investigation of a single realization of a phenomenon by probabilistic methods have been discussed in full detail by MATHERON [220] and shall not be reported in the present book. In particular, questions of ergodicity shall not be debated here. We provide an intuitive description of the probabilistic formalization, but we shall not attempt to justify it rigorously. The main reason for the success of the probabilistic approach is that it offers a convenient way of formulating methods which could also be viewed in a deterministic setting, albeit with less elegance.

Multivariate time/space data

In many fields of science data arises which is either time or space dependent—or both. Such data is often multivariate, i.e. several quantities have been measured and need to be examined. The data array, in its most general form, may have the following shape:

$$\begin{array}{c} \\ Samples \end{array} \quad \overset{\textit{Coordinates} \qquad\qquad\qquad \textit{Variables}}{\begin{bmatrix} t_1 & x_1^1 & x_1^2 & x_1^3 & z_1^1 & \dots & z_1^i & \dots & z_1^N \\ \vdots & \vdots & \vdots & \vdots & \vdots & & & & \vdots \\ t_\alpha & x_\alpha^1 & x_\alpha^2 & x_\alpha^3 & z_\alpha^1 & \dots & z_\alpha^i & \dots & z_\alpha^N \\ \vdots & \vdots & \vdots & \vdots & \vdots & & & & \vdots \\ t_n & x_n^1 & x_n^2 & x_n^3 & z_n^1 & \dots & z_n^i & \dots & z_n^N \end{bmatrix}} \tag{5.1}$$

In this array the samples are numbered using an index α and the total number of samples is symbolized by the letter n, thus $\alpha = 1, 2, 3, \dots, n$.

The different variables are labeled by an index i and the number of variables is N, so $i = 1, 2, 3, \dots, N$.

The samples could be pieces of rock of same volume taken by a geologist or they could be plots containing different plants a biologist is interested in. The variables would then be chemical or physical measurements performed on the rocks, or counts of the abundance of different species of plants within the plots.

The samples may have been taken at different moments t_α and at different points $\mathbf{x}_\alpha$, where $\mathbf{x}_\alpha$ is typically[1] the vector of up to three spatial coordinates $(x_\alpha^1, x_\alpha^2, x_\alpha^3) \in \mathbb{R}^3$.

So far the numbers contained in the data set have been arranged into a rectangular table and symbols have been used to describe this table. There is a need for viewing the data in a more general setting.

Regionalized variable

Let us consider that only one property has been measured on different spatial objects (so N= 1) and that the time at which the measurements have been made has not been recorded. In this case the index i and the time coordinate t_α are of no use and can be omitted. The analysis of multivariate phenomena will be taken up in other parts of the book, while examples of time series analysis using geostatistics will be discussed in later chapters.

We have n observations symbolized by

$$z(\mathbf{x}_\alpha), \qquad \text{with} \quad \alpha = 1, \dots, n, \tag{5.2}$$

taken at locations $\mathbf{x}_\alpha$. The sampled objects in a region $\mathcal{D}$ can be considered as a fragment of a larger collection of objects. Many more observations than the few collected could be made, but perhaps because of the cost and effort involved this has not (or cannot) been done. If the objects are points, even infinitely many observations are possible in the region. This possibility of infinitely many observations of the same

[1]There are now applications of geostatistics where $\mathbf{x} \in \mathbb{R}^k$ with an integer $k > 3$, e.g. see [285, 181, 28], and where $\mathbb{R}^k$ is not anymore a geographical space, but an arbitrary parameter or empirical orthogonal functions space.

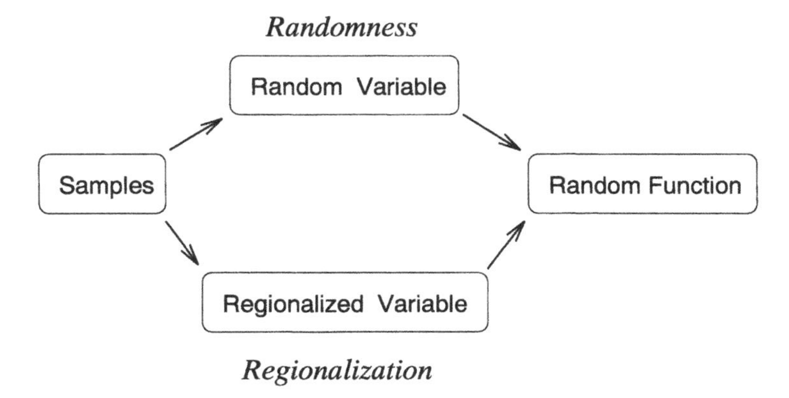

Figure 5.1: The random function model.

kind is introduced by dropping the index α and defining the regionalized variable (for short: ReV) as

$$z(\mathbf{x}) \qquad \text{for all} \quad \mathbf{x} \in \mathcal{D}. \tag{5.3}$$

The data set $\{z(\mathbf{x}_\alpha), \alpha = 1, \ldots, n\}$ is now viewed as a collection of a few values of the ReV.

The study of a regionalized variable will usually involve at least two geometrical aspects: the domain $\mathcal{D}$ in which the ReV is defined and the geometrical *support* for which each sample of the ReV is defined. The latter can for example be a portion of soil that has been analyzed or the integration time of a time-dependent measurement. Let us for now assume point support and defer the question of non-pointwise support to Chapter 10.

Random variable and regionalized value

Each measured value in the data set has a location in the domain $\mathcal{D}$ and we call it a *regionalized value*. A new viewpoint is introduced by considering a regionalized value as the outcome of some random mechanism. Formally this mechanism will be called a *random variable* and a sampled value $z(\mathbf{x}_\alpha)$ represents one draw from the random variable $Z(\mathbf{x}_\alpha)$. At each point $\mathbf{x}_\alpha$ the mechanism responsible for generating a value $z(\mathbf{x}_\alpha)$ may be different, thus $Z(\mathbf{x}_\alpha)$ could a priori have different properties at each point of a region.

Random function

Considering the regionalized values at all points in a given region, the associated function $z(\mathbf{x})$ for $\mathbf{x} \in \mathcal{D}$ is a regionalized variable. The set of values

$$\{z(\mathbf{x}),\, \mathbf{x} \in \mathcal{D}\} \tag{5.4}$$

RAF	$Z(\mathbf{x})$	$\longrightarrow$	$Z(\mathbf{x}_0)$	*The Random Variable's*
$\Downarrow$	$\downarrow$		$\downarrow$	*realization is a*
REV	$z(\mathbf{x})$	$\longrightarrow$	$z(\mathbf{x}_0)$	*Regionalized Value*

Figure 5.2: The regionalized variable as a realization of a random function.

can be viewed as one draw from an infinite set of random variables (one random variable at each point of the domain). The family of random variables

$$\{Z(\mathbf{x}),\ \mathbf{x} \in \mathcal{D}\} \tag{5.5}$$

is called the *random function* (RAF).

Figure 5.1 shows how the random function model has been set up by viewing the data under two different angles. One aspect is that the data values stem from a physical environment (time, 1–2–3D space) and are in some way dependent on their location in the region: they are *regionalized.* The second aspect is that the regionalized sample values $z(\mathbf{x}_\alpha)$ cannot, generally, be modeled with a simple deterministic function $z(\mathbf{x})$. Looking at the values that were sampled, the behavior of $z(\mathbf{x})$ appears to be very complex. Like in many situations where the parameters of a data generating mechanism cannot be determined, a probabilistic approach is chosen, i.e. the mechanism is considered as *random.* The data values are viewed as outcomes of the random mechanism.

Joining together these two aspects of *regionalization* and *randomness* yields the concept of a RAF.

In probabilistic jargon, see Figure 5.2 , the REV $z(\mathbf{x})$ is one realization of the RAF $Z(\mathbf{x})$. A regionalized value $z(\mathbf{x}_0)$ at a specific location $\mathbf{x}_0$ is a realization of a random variable $Z(\mathbf{x}_0)$ which is itself a member of an infinite family of random variables, the RAF $Z(\mathbf{x})$. Capital 'Z' is used to denote random variables while small 'z' is used for their realizations. The point $\mathbf{x}_0$ is an arbitrary point of the region which may or may not have been sampled.

Probability distributions

In this model the random mechanism $Z(\mathbf{x}_0)$ acting at a given point $\mathbf{x}_0$ of the region generates realizations following a probability distribution F

$$P(Z(\mathbf{x}_0) \leq z) \quad = \quad F_{\mathbf{x}_0}(z), \tag{5.6}$$

where P is the probability that an outcome of Z at the point $\mathbf{x}_0$ is lower than a fixed value z.

A bivariate distribution function for two random variables $Z(\mathbf{x}_1)$ and $Z(\mathbf{x}_2)$ at two different locations is

$$P(Z(\mathbf{x}_1) \leq z_1,\ Z(\mathbf{x}_2) \leq z_2) \quad = \quad F_{\mathbf{x}_1,\mathbf{x}_2}(z_1, z_2), \tag{5.7}$$

where P is the probability that simultaneously an outcome of $Z(\mathbf{x}_1)$ is lower than z_1 and an outcome of $Z(\mathbf{x}_2)$ is lower than z_2.

In the same way a multiple distribution function for n random variables located at n different points can be defined

$$F_{\mathbf{x}_1,\ldots,\mathbf{x}_n}(z_1,\ldots,z_n) \quad = \quad P(Z(\mathbf{x}_1) \leq z_1,\ldots,Z(\mathbf{x}_n) \leq z_n). \tag{5.8}$$

Built up in this manner we have an extraordinarily general model which is able to describe any process in nature or technology. In practice, however, we possess only few data from one or several realizations of the RAF and it will be impossible to infer all the mono- and multivariate distribution functions for any set of points. Simplification is needed and it is provided by the idea of stationarity.

Strict stationarity

Stationarity means that characteristics of a RAF *stay the same* when shifting a given set of n points from one part of the region to another. This is called *translation invariance*.

To be more specific, a RAF $Z(\mathbf{x})$ is said to be *strictly stationary* if for any set of n points $\mathbf{x}_1,\ldots,\mathbf{x}_n$ (where n is an arbitrary positive integer) and for any vector $\mathbf{h}$

$$F_{\mathbf{x}_1,\ldots,\mathbf{x}_n}(z_1,\ldots,z_n) \quad = \quad F_{\mathbf{x}_1+\mathbf{h},\ldots,\mathbf{x}_n+\mathbf{h}}(z_1,\ldots,z_n), \tag{5.9}$$

i.e. a translation of a point configuration in a given direction does not change the multiple distribution.

The new situation created by restraining $Z(\mathbf{x})$ to be strictly stationary with all its n-variate distribution functions being shift invariant, can be summarized by a statement of Arlequino (coming back from a trip to the moon): "Everything is everywhere and always the same as here... up to a certain degree of magnitude and perfection"[2]. This position of Arlequino explains best the idea of strict stationarity: things do not change fundamentally when one moves from one part of the universe to another part of it! Naturally, we cannot fully agree with Arlequino's blasé indifference to everything. But there is a restriction "up to a certain degree" in Arlequino's statement and we shall, in an analogous way, loosen the concept of strict stationarity and define several types and degrees of stationarity. These lie in the wide range between the concept

[2] "C'est partout et toujours comme ici, c'est partout et toujours comme chez nous, aux degrés de grandeur et de perfection près." [M SERRES (1968) *Le Système de Leibniz et ses Modèles Mathématiques.* Presses Universitaires de France, Paris.]

of a non-stationary (and non-homogeneous) RAF, whose characteristics change at any time and at any location, and the concept of a strictly stationary random function whose distribution functions are everywhere and always the same.

It should be noted that stationarity is a property of the random function model and not of the regionalized variable. In practice we might say that a given "REV is stationary", but this is of course always meant as a shorthand for "this REV can be considered as a realization of a stationary RAF".

Stationarity of first two moments

Strict stationarity requires the specification of the multipoint distribution (5.9) for any set of points $\{\mathbf{x}_1, \ldots, \mathbf{x}_n\}$. A lighter strategy will be to consider only pairs of points $\{\mathbf{x}_1, \mathbf{x}_2\}$ in the domain and try to characterize only the first two moments, not a full distribution. Naturally such a strategy is ideal in the case of the Gaussian distribution where the first two moments entirely characterize the distribution. In general, however, this approach still works well in practice when the data histogram does not have too heavy tails.

One possibility is to assume the stationarity of the first two moments of the variable: this is *second-order stationarity*. A second possibility is to assume the stationarity of the first two moments of the difference of a pair of values at two points: this receives the name of *intrinsic stationarity* and leads to the notion of variogram.

6 Variogram Cloud

Pairs of sample values are evaluated by computing the squared difference between the values. The resulting dissimilarities are plotted against the separation of sample pairs in geographical space and form the variogram cloud. The cloud is sliced into classes according to separation in space and the average dissimilarities in each class form the sequence of values of the experimental variogram.

Dissimilarity versus separation

We measure the variability of a regionalized variable $z(\mathbf{x})$ at different scales by computing the dissimilarity between pairs of data values, z_α and z_β say, located at points $\mathbf{x}_\alpha$ and $\mathbf{x}_\beta$ in a spatial domain $\mathcal{D}$. The measure for the dissimilarity of two values, labeled $\gamma^\star$, is

$$\gamma^\star_{\alpha\beta} \quad = \quad \frac{(z_\alpha - z_\beta)^2}{2}, \tag{6.1}$$

i.e. half of the square of the difference between the two values.

The two points $\mathbf{x}_\alpha$, $\mathbf{x}_\beta$ in geographical space can be linked by a vector $\mathbf{h} = \mathbf{x}_\beta - \mathbf{x}_a$ as shown on Figure 6.1.

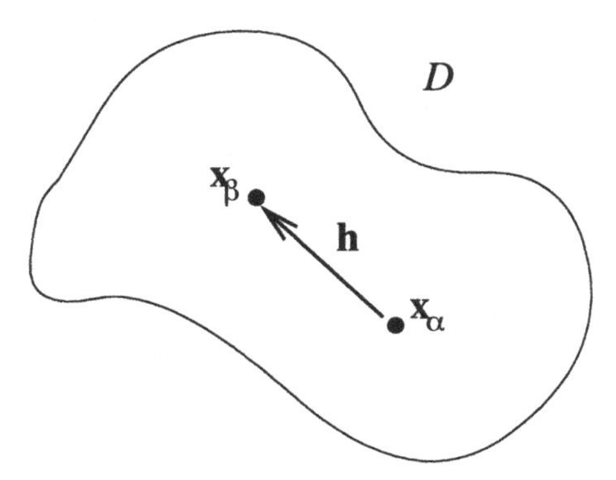

Figure 6.1: A vector linking two points in 2D space.

We let the dissimilarity $\gamma^\star$ depend on the spacing and on the orientation of the point pair described by the vector $\mathbf{h}$,

$$\gamma^\star(\mathbf{h}) \quad = \quad \frac{1}{2}\Big(z(\mathbf{x}_\alpha+\mathbf{h}) - z(\mathbf{x}_\alpha)\Big)^2. \tag{6.2}$$

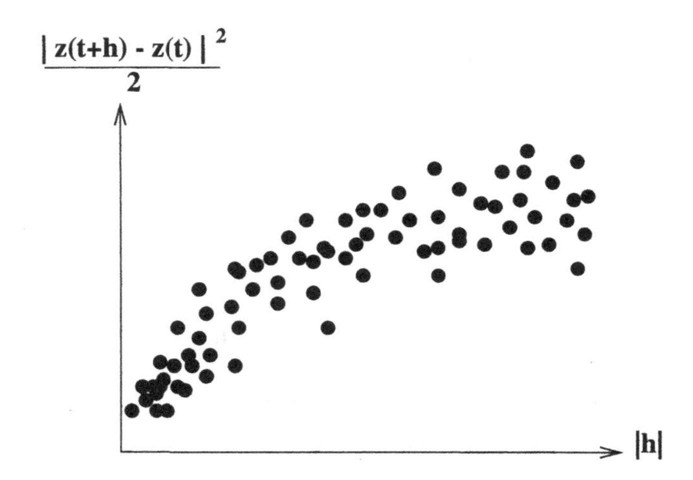

Figure 6.2: Plot of the dissimilarities $\gamma^\star$ against the spatial separation $\mathbf{h}$ of sample pairs: a variogram cloud.

As the dissimilarity is a squared quantity, the sign of the vector $\mathbf{h}$, i.e. the order in which the points $\mathbf{x}_\alpha$ and $\mathbf{x}_\beta$ are considered does not enter into play. The dissimilarity is symmetric with respect to $\mathbf{h}$:

$$\gamma^\star(-\mathbf{h}) \quad = \quad \gamma^\star(+\mathbf{h}). \tag{6.3}$$

Thus on graphical representations the dissimilarities will be plotted against the absolute values of the vectors $\mathbf{h}$. Using all sample pairs in a data set (up to a distance of half the diameter of the region), a plot of the dissimilarities $\gamma^\star$ against the spatial separation $\mathbf{h}$ is produced which is called the *variogram cloud*. A schematic example is given on Figure 6.2.

The dissimilarity often increases with distance as near samples tend to be alike.

The variogram cloud by itself is a powerful tool for exploring features of spatial data. On a graphical computer screen the values of the variogram cloud can be linked to the position of sample pairs on a map representation. The analysis of subsets of the variogram cloud can help in understanding the distribution of the sample values in geographical space. Anomalies, inhomogeneities can be detected by looking at high dissimilarities at short distances. In some cases the variogram cloud consists of two distinct clouds due to the presence of outliers. HASLETT et al. [139] have first developed the use of the variogram cloud in combination with other views on the data using linked windows on a graphic computer screen and they provide many examples showing the power of this exploratory tool.

Actually a variogram cloud seldom looks like what is suggested on Figure 6.2: the variogram cloud usually is dominated by many pairs with low dissimilarity at all scales $\mathbf{h}$ (DIGGLE et al. [90] p51 discuss this question).

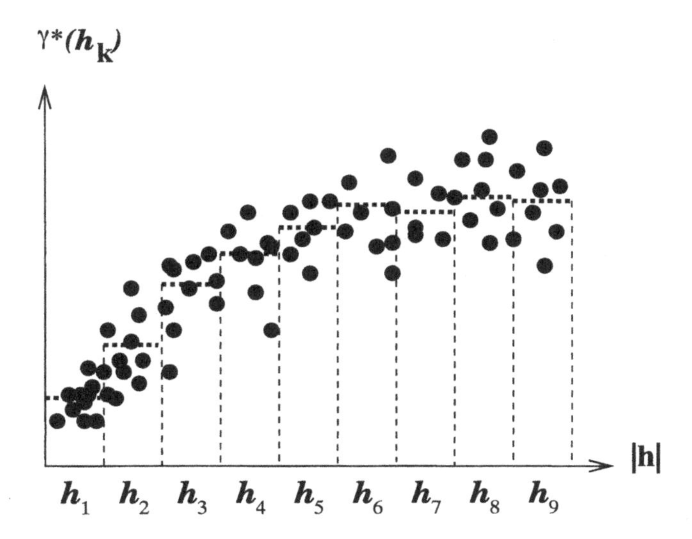

Figure 6.3: The experimental variogram is obtained by averaging the dissimilarities $\gamma^\star$ for given classes $\mathfrak{H}_k$.

Experimental variogram

An average of dissimilarities $\gamma^\star(\mathbf{h})$ can be formed for a given class of vectors $\mathfrak{H}$ by grouping all n_c point pairs that can be linked by a vector $\mathbf{h}$ belonging to $\mathfrak{H}$. Such a class $\mathfrak{H}$ groups vectors whose lengths are within a specified interval of lengths and whose orientation is the same up to a given tolerance on angle. Generally non overlapping vector classes $\mathfrak{H}$ are chosen. The average dissimilarity with respect to a vector class $\mathfrak{H}_k$ is a value of what is termed the *experimental variogram*

$$\gamma^\star(\mathfrak{H}_k) \quad = \quad \frac{1}{2\,n_c}\sum_{\alpha=1}^{n_c}\Big(z(\mathbf{x}_\alpha{+}\mathbf{h}) - z(\mathbf{x}_\alpha)\Big)^2 \qquad \text{with} \quad \mathbf{h} \in \mathfrak{H}_k. \tag{6.4}$$

In practice the experimental variogram is usually computed using vectors $\mathbf{h}$ of a length inferior to half the diameter of the region. For pairs of samples with vectors $\mathbf{h}$ of a length almost equal to the diameter of the region, the corresponding samples are located near the border. Vector classes $\mathfrak{H}$ formed with such pairs will have no contribution from samples at the center of the region and are thus not representative of the whole data set.

An example of an experimental variogram obtained for a sequence of classes $\mathfrak{H}_k$ is sketched on Figure 6.3. The experimental variogram is obtained from the variogram cloud by subdiving it into classes and computing an average for each class.

Usually we can observe that the average dissimilarity between values increases when the spacing between the pairs of sample points is increased. For large spacings the experimental variogram sometimes reaches a sill which can be equal to the variance of the data.

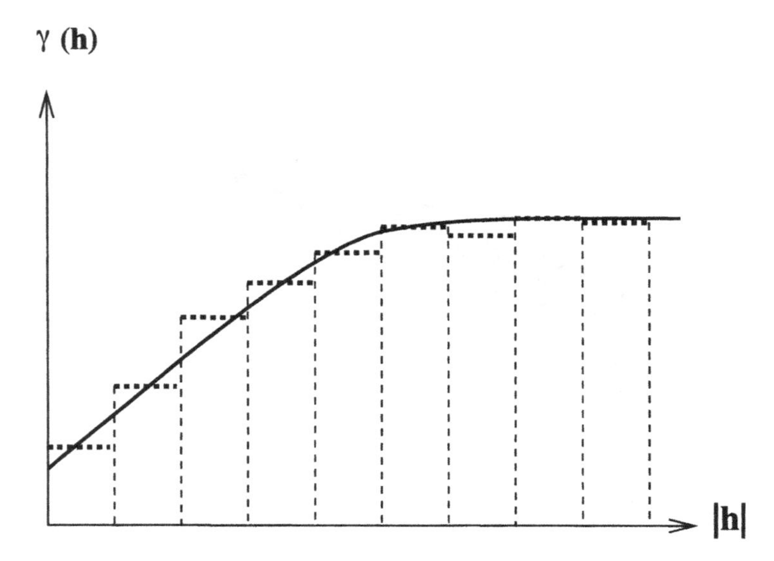

Figure 6.4: The sequence of average dissimilarities is fitted with a theoretical variogram function.

If the slope of the dissimilarity function changes abruptly at a specific scale, this suggests that an intermediate level of the variation has been reached at this scale.

The behavior at very small scales, near the origin of the variogram, is of importance, as it indicates the type of continuity of the regionalized variable: differentiable, continuous but not differentiable, or discontinuous. In this last case, when the variogram is discontinuous at the origin, this is a symptom of a *nugget-effect*, which means that the values of the variable change abruptly at a very small scale, like gold grades when a few gold nuggets are contained in some samples.

When the average dissimilarity of the values is constant for all spacings $\mathbf{h}$, there is no spatial structure in the data. Conversely, a non zero slope of the variogram near the origin indicates structure. An abrupt change in slope indicates the passage to a different structuration of the values in space. We shall learn how to model such transitions with nested theoretical variograms and how to visualize the different types of spatial associations of the values separately as maps by kriging spatial components.

Replacing the experimental by a theoretical variogram

The experimental variogram is replaced by a theoretical variogram function essentially for the reason that the variogram model should have a physical meaning (a random function with the given type of variogram can exist). The use of a theoretical variogram guarantees (using weights subject to a certain constraint) that the variance of any linear combination of sample values is positive.

If the values of the experimental variogram were taken to set up a kriging system, this could lead to negative kriging variances, in a way similar to the Example 3.1 (on page 22) where a problem of multiple regression with missing values was discussed.

A theoretical variogram function is fitted to the sequence of average dissimilarities, as suggested on Figure 6.4. It is important to understand that this fit implies an interpretation of both the behavior at the origin and the behavior at large distances, beyond the range of the experimental variogram. The fit is done by eye because it is generally not so relevant how well the variogram function fits the sequence of points. What counts is the type of continuity assumed for the regionalized variable and the stationarity hypothesis associated to the random function. These assumptions will guide the selection of an appropriate variogram function and this has many more implications than the way the theoretical function is fitted to the experimental variogram. A thorough discussion is found in MATHERON [220].

The different types of behavior at the origin of the variogram model and their impact on kriging are discussed in Chapter 16 and in the Ebro estuary case study (Chapter 27). A weighted least squares fitting algorithm for variograms (and cross variograms) will be presented in Chapter 26. References discussing the automatic inference of generalized covariance functions (of which the variogram functions are a subclass) will be given in Chapter 39.

7 Variogram and Covariance Function

The experimental variogram is a convenient tool for the analysis of spatial data as it is based on a simple measure of dissimilarity. Its theoretical counterpart reveals that a broad class of phenomena are adequately described by it, including phenomena of unbounded variation. When the variation is bounded, the variogram is equivalent to a covariance function.

Regional variogram

The experimental variogram of samples $z(\mathbf{x}_\alpha)$ is the sequence of averages of dissimilarities for different distance classes $\mathfrak{H}_k$. If we had samples for the whole domain $\mathcal{D}$ we could compute the variogram for every possible pair in the domain. The set $\mathcal{D}(\mathbf{h})$, defined as the intersection of the domain $\mathcal{D}$ with a translation $\mathcal{D}_{-\mathbf{h}}$ of itself, describes all the points $\mathbf{x}$ having a counterpart $\mathbf{x}+\mathbf{h}$ in $\mathcal{D}$. The *regional variogram* $g_R(\mathbf{h})$ is the integral over the squared differences of a regionalized variable $z(\mathbf{x})$ for a given lag $\mathbf{h}$

$$g_R(\mathbf{h}) \quad = \quad \frac{1}{2\,|\mathcal{D}(\mathbf{h})|} \int\limits_{\mathcal{D}(\mathbf{h})} \left(z(\mathbf{x}+\mathbf{h}) - z(\mathbf{x})\right)^2 d\mathbf{x}, \tag{7.1}$$

where $|\mathcal{D}(\mathbf{h})|$ is the area (or volume) of the intersection $\mathcal{D}(\mathbf{h})$.

We know the regionalized variable $z(\mathbf{x})$ only at a few locations and it is generally not possible to approximate $z(\mathbf{x})$ by a simple deterministic function. Thus it is convenient to consider $z(\mathbf{x})$ as a realization of a random function $Z(\mathbf{x})$. The associated regional variogram

$$G_R(\mathbf{h}) \quad = \quad \frac{1}{2\,|\mathcal{D}(\mathbf{h})|} \int\limits_{\mathcal{D}(\mathbf{h})} \left(Z(\mathbf{x}+\mathbf{h}) - Z(\mathbf{x})\right)^2 d\mathbf{x} \tag{7.2}$$

is a randomized version of $g_R(\mathbf{h})$. Its expectation defines the theoretical variogram $\gamma(\mathbf{h})$ of the random function model $Z(\mathbf{x})$ over the domain $\mathcal{D}$

$$\gamma(\mathbf{h}) \quad = \quad \mathrm{E}[\,G_R(\mathbf{h})\,]. \tag{7.3}$$

Theoretical variogram

The variation in space of a random function $Z(\mathbf{x})$ can be described by taking the differences between values at pairs of points $\mathbf{x}$ and $\mathbf{x}+\mathbf{h}$:

$$Z(\mathbf{x}+\mathbf{h}) - Z(\mathbf{x}),$$

which are called *increments*.

The theoretical variogram $\gamma(\mathbf{h})$ is defined by the *intrinsic hypothesis*, which is a short form for "a hypothesis of intrinsic stationarity of order two". This hypothesis, which is merely a statement about the type of stationarity characterizing the random function, is formed by two assumptions about the increments:

- the mean $m(\mathbf{h})$ of the increments, called the *drift*, is invariant for any translation of a given vector $\mathbf{h}$ within the domain. Moreover, the drift is supposed to be zero whatsoever the position of $\mathbf{h}$ in the domain.
- the variance of the increments

  has a finite value $2\gamma(\mathbf{h})$ depending on the length and the orientation of a given vector $\mathbf{h}$, but not on the position of $\mathbf{h}$ in the domain.

That is to say, for any pair of points $\mathbf{x}, \mathbf{x}+\mathbf{h} \in \mathcal{D}$ we have

$$\mathrm{E}[\,Z(\mathbf{x}+\mathbf{h}) - Z(\mathbf{x})\,] = m(\mathbf{h}) \;=\; 0, \tag{7.4}$$

$$\mathrm{var}[Z(\mathbf{x}+\mathbf{h}) - Z(\mathbf{x})] \;=\; 2\,\gamma(\mathbf{h}). \tag{7.5}$$

These two properties of an intrinsically stationary random function yield the definition for the theoretical variogram

$$\gamma(\mathbf{h}) \;=\; \frac{1}{2}\,\mathrm{E}\Big[\,(Z(\mathbf{x}+\mathbf{h}) - Z(\mathbf{x}))^2\,\Big]. \tag{7.6}$$

The existence of expectation and variance of the increments does not imply the existence of the first two moments of the random function itself: an intrinsic random function can have an infinite variance although the variance of its increments is finite for any vector $\mathbf{h}$. An intrinsically stationary random function does not need to have a constant mean or a constant variance.

The value of the variogram at the origin is zero by definition

$$\gamma(\mathbf{0}) \;=\; 0. \tag{7.7}$$

The values of the variogram are positive

$$\gamma(\mathbf{h}) \;\geq\; 0, \tag{7.8}$$

and the variogram is an even function

$$\gamma(-\mathbf{h}) \;=\; \gamma(\mathbf{h}). \tag{7.9}$$

The variogram grows slower than $|\mathbf{h}|^2$, i.e.

$$\lim_{|\mathbf{h}|\mapsto\infty} \frac{\gamma(\mathbf{h})}{|\mathbf{h}|^2} \;=\; 0, \tag{7.10}$$

(as otherwise the drift $m(\mathbf{h})$ could not be assumed zero).

Covariance function

The covariance function $C(\mathbf{h})$ is defined on the basis of a hypothesis of stationarity of the first two moments (mean and covariance) of the random function

$$\begin{cases} \mathrm{E}[\,Z(\mathbf{x})\,] = m & \text{for all } \mathbf{x} \in \mathcal{D} \\ \mathrm{E}[\,Z(\mathbf{x}) \cdot Z(\mathbf{x}{+}\mathbf{h})\,] - m^2 = C(\mathbf{h}) & \text{for all } \mathbf{x},\ \mathbf{x}{+}\mathbf{h} \in \mathcal{D}. \end{cases} \tag{7.11}$$

The covariance function is bounded and its absolute value does not exceed the variance

$$|C(\mathbf{h})| \;\leq\; C(\mathbf{0}) = \mathrm{var}(Z(\mathbf{x})). \tag{7.12}$$

Like the variogram, it is an even function: $C(-\mathbf{h}) = C(+\mathbf{h})$. But unlike the variogram, it can take negative values.

The covariance function divided by the variance is called the *correlation function*

$$\rho(\mathbf{h}) \;=\; \frac{C(\mathbf{h})}{C(\mathbf{0})}, \tag{7.13}$$

which is obviously bounded by

$$-1 \;\leq\; \rho(\mathbf{h}) \;\leq\; 1. \tag{7.14}$$

A variogram function can be deduced from a covariance function by the formula

$$\gamma(\mathbf{h}) \;=\; C(\mathbf{0}) - C(\mathbf{h}), \tag{7.15}$$

but in general the reverse is not true, because the variogram is not necessarily bounded. Thus the hypothesis of second-order stationarity is less general than the intrinsic hypothesis (in the monovariate case) and unbounded variogram models do not have a covariance function counterpart.

EXAMPLE 7.1 *For example, the power variogram shown on Figure 7.1*

$$\gamma(\mathbf{h}) \;=\; b\,|\mathbf{h}|^p \qquad \textit{with } 0 < p < 2 \textit{ and } b > 0 \tag{7.16}$$

cannot be obtained from a covariance function as it grows without bounds. Clearly it outgrows the framework fixed by second-order stationarity.

Actually the class of self-similar processes with stationary increments called fractional Brownian motion *has a variogram of the type (7.16) (see [362], p406, for details).*

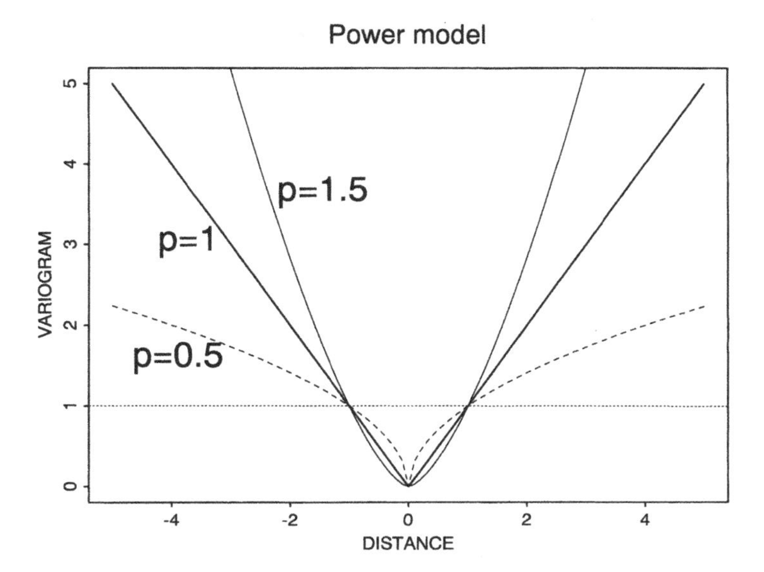

Figure 7.1: Power variograms for different values of the power p (with $b = 1$).

Positive definite function

A covariance function is a positive definite function[1]. This means that the use of a covariance function $C(\mathbf{h})$ for the computation of the variance of a linear combination of n+1 random variables $Z(\mathbf{x}_\alpha)$ (any subset sampled from a second-order stationary random function) must be positive. It is necessarily linked to a positive semi-definite matrix $\mathbf{C}$ of covariances

$$\operatorname{var}\left(\sum_{\alpha=0}^{n} w_\alpha\, Z(\mathbf{x}_\alpha)\right) = \sum_{\alpha=0}^{n}\sum_{\beta=0}^{n} w_\alpha\, w_\beta\, C(\mathbf{x}_\alpha - \mathbf{x}_\beta) = \mathbf{w}^\top \mathbf{C}\,\mathbf{w} \geq 0 \qquad (7.17)$$

for any set of points $\mathbf{x}_\alpha$ and any set of weights w_α (assembled into a vector $\mathbf{w}$).

The continuous covariance functions are characterized by Bochner's theorem as the Fourier transforms of positive measures. This topic is treated in more detail in Chapter 20.

Conditionally negative definite function

The variogram is a *conditionally* negative definite function. The condition to guarantee the positivity of the variance of any linear combination of n+1 random variables,

[1]For functions, in opposition to matrices, no distinction is usually made between "definite" and "semi-definite".

subset of an intrinsic random function, is that the n+1 weights w_α sum up to zero. The variance of a linear combination of intrinsically stationary random variables is defined as

$$\operatorname{var}\Big(\sum_{\alpha=0}^{n} w_\alpha\, Z(\mathbf{x}_\alpha)\Big) = -\sum_{\alpha=0}^{n}\sum_{\beta=0}^{n} w_\alpha\, w_\beta\, \gamma(\mathbf{x}_\alpha - \mathbf{x}_\beta)$$

$$\geq 0, \qquad \text{if } \sum_{\alpha=0}^{n} w_\alpha = 0, \tag{7.18}$$

i.e. any matrix $\mathbf{\Gamma}$ of variogram values is conditionally negative definite

$$[w_\alpha]^\top [\gamma(\mathbf{x}_\alpha - \mathbf{x}_\beta)][w_\alpha] = \mathbf{w}^\top \mathbf{\Gamma}\, \mathbf{w} \leq 0 \qquad \text{for } \sum_{\alpha=0}^{n} w_\alpha = 0. \tag{7.19}$$

To understand the meaning (sufficiency) of the condition on the sum of the weights, it is necessary to compute explicitly the variance of the linear combination.

As the random function $Z(\mathbf{x})$ is intrinsically stationary, only the expectation of its increments has a meaning. A trick enabling to construct increments is to insert an additional random variable $Z(\mathbf{0})$, placed arbitrarily at the origin $\mathbf{0}$ (assuming $\mathbf{0} \in \mathcal{D}$), multiplied by the zero sum weights

$$\operatorname{var}\Big(\sum_{\alpha=0}^{n} w_\alpha\, Z(\mathbf{x}_\alpha)\Big) = \operatorname{var}\Big(-Z(\mathbf{0}) \cdot \underbrace{\sum_{\alpha=0}^{n} w_\alpha}_{0} + \sum_{\alpha=0}^{n} w_\alpha\, Z(\mathbf{x}_\alpha)\Big)$$

$$= \operatorname{var}\Big(\sum_{\alpha=0}^{n} w_\alpha\, (Z(\mathbf{x}_\alpha) - Z(\mathbf{0}))\Big)$$

$$= \sum_{\alpha=0}^{n}\sum_{\beta=0}^{n} w_\alpha\, w_\beta\, \mathrm{E}[\,(Z(\mathbf{x}_\alpha) - Z(\mathbf{0})) \cdot (Z(\mathbf{x}_\beta) - Z(\mathbf{0}))\,]$$

$$= \sum_{\alpha=0}^{n}\sum_{\beta=0}^{n} w_\alpha\, w_\beta\, C_I(\mathbf{x}_\alpha, \mathbf{x}_\beta), \tag{7.20}$$

where $C_I(\mathbf{x}_\alpha, \mathbf{x}_\beta)$ is the covariance of increments formed using the additional variable $Z(\mathbf{0})$.

We also introduce the additional variable $Z(\mathbf{0})$ into the expression of the variogram

$$\gamma(\mathbf{x}_\alpha - \mathbf{x}_\beta) = \frac{1}{2}\,\mathrm{E}[\,(Z(\mathbf{x}_\alpha) + Z(\mathbf{0}) - Z(\mathbf{x}_\beta) - Z(\mathbf{0}))^2\,]$$

$$= \frac{1}{2}\,(2\,\gamma(\mathbf{x}_\alpha) + 2\,\gamma(\mathbf{x}_\beta) - 2\,C_I(\mathbf{x}_\alpha, \mathbf{x}_\beta)), \tag{7.21}$$

so that

$$C_I(\mathbf{x}_\alpha, \mathbf{x}_\beta) = \gamma(\mathbf{x}_\alpha) + \gamma(\mathbf{x}_\beta) - \gamma(\mathbf{x}_\alpha - \mathbf{x}_\beta), \tag{7.22}$$

which incorporates two non-stationary terms $\gamma(\mathbf{x}_\alpha)$ and $\gamma(\mathbf{x}_\beta)$.

Coming back to the computation of the variance of the linear combination, we see that the two non stationary terms are cancelled by the condition on the weights

$$\begin{aligned}
\operatorname{var}\left(\sum_{\alpha=0}^{n} w_\alpha Z(\mathbf{x}_\alpha)\right) &= \sum_{\alpha=0}^{n}\sum_{\beta=0}^{n} w_\alpha w_\beta C_I(\mathbf{x}_\alpha,\mathbf{x}_\beta)\\
= \underbrace{\sum_{\beta=0}^{n} w_\beta}_{0} \sum_{\alpha=0}^{n} w_\alpha \gamma(\mathbf{x}_\alpha) &+ \underbrace{\sum_{\alpha=0}^{n} w_\alpha}_{0} \sum_{\beta=0}^{n} w_\beta \gamma(\mathbf{x}_\beta) - \sum_{\alpha=0}^{n}\sum_{\beta=0}^{n} w_\alpha w_\beta \gamma(\mathbf{x}_\alpha - \mathbf{x}_\beta)\\
= -\sum_{\alpha=0}^{n}\sum_{\beta=0}^{n} w_\alpha w_\beta \gamma(\mathbf{x}_\alpha - \mathbf{x}_\beta) &\qquad \text{for } \sum_{\alpha=0}^{n} w_\alpha = \sum_{\beta=0}^{n} w_\beta = 0. \qquad (7.23)
\end{aligned}$$

When handling linear combinations of random variables, the variogram can only be used together with a condition on the weights guaranteeing its existence. In particular, the variogram cannot in general be used in a simple kriging. Other forms of kriging, with constraints on the weights, are required to use this tool that covers a wider range of phenomena than the covariance function.

Variograms can be characterized on the basis of continuous covariance functions: a function $\gamma(\mathbf{h})$ is a variogram, if the exponential of $-\gamma(\mathbf{h})$ is a covariance function

$$C(\mathbf{h}) = \mathrm{e}^{-\gamma(\mathbf{h})}. \qquad (7.24)$$

This remarquable relation, based on a theorem by SCHOENBERG [292], links (through the defining kernel of the Laplace transform) the conditionally negative functions with the positive definite functions (see also CHOQUET [53]).

COMMENT 7.2 *The covariance function*

$$C(\mathbf{h}) = \exp\left(-\frac{|\mathbf{h}|^p}{a}\right) \qquad \textit{with } 0 < p \leq 2 \textit{ and } a > 0, \qquad (7.25)$$

which is related by Schoenberg's relation to the power variogram model, defines the family of stable *covariance functions. The case $p = 2$ (Gaussian covariance function) is pathological: it corresponds to a deterministic random function (*MATHERON *[202]), which is contradictory with randomness. The case $p = 1$ defines the* exponential *covariance function.*

Fitting the variogram with a covariance function

If the variogram is bounded by a finite value $\gamma(\infty)$, a covariance function can be found such as

$$C(\mathbf{h}) = \gamma(\infty) - \gamma(\mathbf{h}). \qquad (7.26)$$

The experimental variogram can have a shape which suggests to use a bounded variogram function to fit it. The lowest upper bound of the variogram function is described as the *sill*.

When the experimental variogram exhibits a sill, it is possible to fit it with a theoretical variogram that is actually a covariance function $C(\mathbf{h})$ on the basis of the formula for bounded variograms

$$\gamma(\mathbf{h}) \quad = \quad b - C(\mathbf{h}), \tag{7.27}$$

where $b = C(\mathbf{0})$ is the value at the origin of the covariance function.

8 Examples of Covariance Functions

We present a few models of covariance functions. They are defined for isotropic (i.e. rotation invariant) random functions. On the graphical representations the covariance functions are plotted as variograms using the relation $\gamma(\mathbf{h}) = C(\mathbf{0}) - C(\mathbf{h})$.

Nugget-effect model

The covariance function $C(\mathbf{h})$ that models a discontinuity at the origin is the *nugget-effect model*

$$C_{nug}(\mathbf{h}) = \begin{cases} b & \text{for } |\mathbf{h}| = 0 \\ 0 & \text{for } |\mathbf{h}| > 0, \end{cases} \tag{8.1}$$

where b is a positive value.

Its variogram counterpart is zero at the origin and has the value b for $\mathbf{h} \neq 0$. It is shown on Figure 8.1.

The nugget-effect is used to model a discontinuity at the origin of the variogram, i.e. when

$$\lim_{|\mathbf{h}| \mapsto 0} \gamma(\mathbf{h}) = b. \tag{8.2}$$

The nugget-effect is equivalent to the concept of *white noise* in signal processing.

Exponential covariance function

The exponential covariance function model falls off exponentially with increasing distance

$$C_{exp}(\mathbf{h}) = b \exp\left(-\frac{|\mathbf{h}|}{a}\right) \qquad \text{with } a, b > 0. \tag{8.3}$$

The parameter a determines how quickly the covariance falls off. For a value of $\mathbf{h} = 3a$ the covariance function has decreased by 95% of its value at the origin, so that this distance has been termed the *practical range* of the exponential model.

The exponential model is continuous but not differentiable at the origin. It drops asymptotically towards zero for $|\mathbf{h}| \mapsto \infty$.

The variogram equivalent of the exponential covariance function is shown on Figure 8.2.

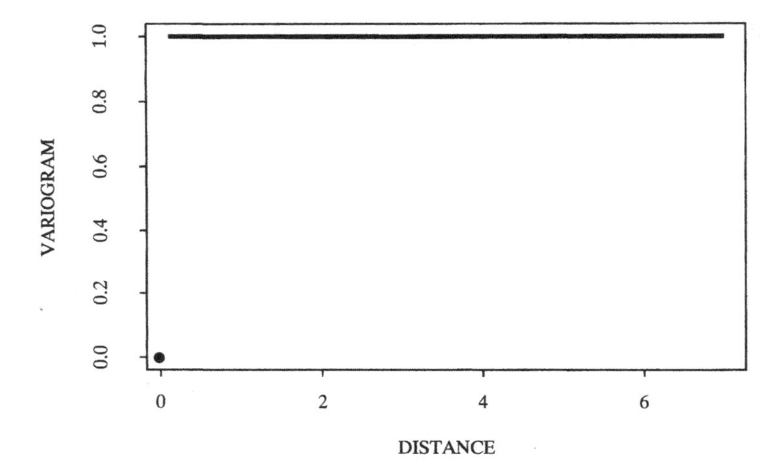

Figure 8.1: A nugget-effect variogram: its value is zero at the origin and $b = 1$ elsewhere.

Spherical model

A commonly used covariance function is the *spherical model*

$$C_{sph}(\mathbf{h}) = \begin{cases} b\left(1 - \dfrac{3}{2}\dfrac{|\mathbf{h}|}{a} + \dfrac{1}{2}\dfrac{|\mathbf{h}|^3}{a^3}\right) & \text{for } 0 \leq |\mathbf{h}| \leq a \\ 0 & \text{for } |\mathbf{h}| > a. \end{cases} \tag{8.4}$$

The parameter a indicates the range of the spherical covariance: the covariance vanishes when the range is reached. The parameter b represents the maximal value of the covariance: the spherical covariance steadily decreases, starting from the maximum b at the origin, until it vanishes when the range is reached.

The nugget-effect model can be considered as a particular case of a spherical covariance function with an infinitely small range. Nevertheless there is an important difference between the two models: $C_{nug}(\mathbf{h})$ describes a discontinuous phenomenon, whose values change abruptly from one location to the other, while $C_{sph}(\mathbf{h})$ represents a phenomenon which is continuous, but not differentiable: it would feel rough, could one touch it.

A corresponding spherical variogram is shown on Figure 8.3. It reaches the sill ($b = 1$) at a range of $a = 3$.

Derivation of the spherical covariance

Imagine a universe with Poisson points, i.e. a 3D-space with points $\mathbf{x}_P$ scattered randomly following a uniform distribution along each coordinate and summing up to

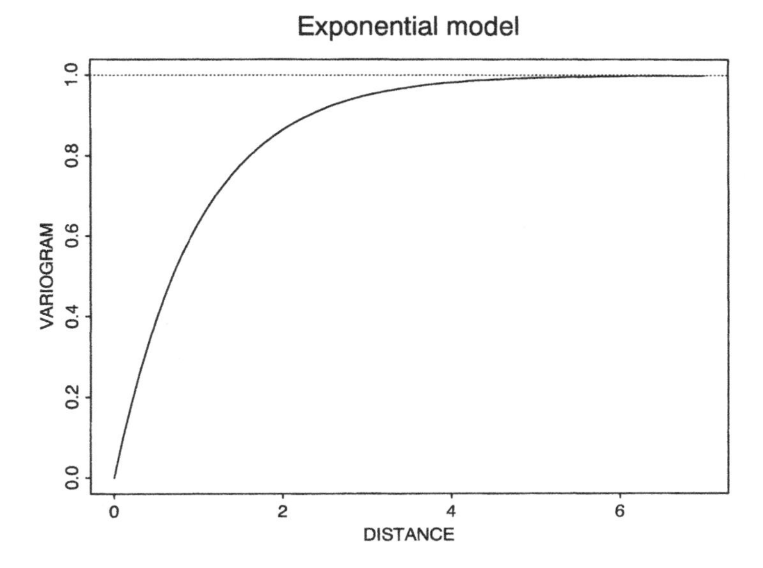

Figure 8.2: An exponential variogram: it rises asymptotically towards a sill $b = 1$. The range parameter is set to $a = 1$. At a practical range of $|\mathbf{h}| = 3$ the exponential model has approached the sill to 95%.

θ points per volume unit on average. A counting function $N(\mathcal{V})$ is defined which counts the number of Poisson points contained in a volume $\mathcal{V}$.

Consider the random function $Z(\mathbf{x}) = N(\mathcal{B}_\mathbf{x})$ which is the count of the number of Poisson points contained in a ball $\mathcal{B}$ centered on a point $\mathbf{x}$. Clearly $\mathcal{B}_\mathbf{x}$ represents the volume of influence of diameter d around a point $\mathbf{x}$ which determines the value of $Z(\mathbf{x})$. The problem at hand is to calculate the covariance function of the random function $Z(\mathbf{x})$.

An indicator function $\boldsymbol{1}_\mathcal{B}(\mathbf{x}')$ is constructed indicating whether a location $\mathbf{x}'$ is inside a ball centered at $\mathbf{x}$

$$\boldsymbol{1}_\mathcal{B}(\mathbf{x}') = \begin{cases} 1, & \text{if } \mathbf{x}' \in \mathcal{B}_\mathbf{x} \\ 0, & \text{if } \mathbf{x}' \notin \mathcal{B}_\mathbf{x}. \end{cases} \tag{8.5}$$

A function $\mathfrak{K}(\mathbf{h})$, the *geometric covariogram,* measures the volume of the intersection of a ball $\mathcal{B}$ with a copy $\mathcal{B}_\mathbf{h}$ of it translated by a vector $\mathbf{h}$

$$\mathfrak{K}(\mathbf{h}) = \int_{-\infty}^{\infty}\int_{-\infty}^{\infty}\int_{-\infty}^{\infty} \boldsymbol{1}_\mathcal{B}(\mathbf{x}')\,\boldsymbol{1}_\mathcal{B}(\mathbf{x}'+\mathbf{h})\,d\mathbf{x}' = \iiint \boldsymbol{1}_\mathcal{B}(\mathbf{x}')\,\boldsymbol{1}_{\mathcal{B}_\mathbf{h}}(\mathbf{x}')\,d\mathbf{x}' \tag{8.6}$$

$$= |\mathcal{B} \cap \mathcal{B}_\mathbf{h}|. \tag{8.7}$$

Conversely, it is worth noting that the intersection $\mathcal{B} \cap \mathcal{B}_{-\mathbf{h}}$ of the ball with a copy of itself translated by $-\mathbf{h}$ represents the set of points $\mathbf{x}' \in \mathcal{B}$ which have a neighbor $\mathbf{x}' + \mathbf{h}$ within the ball, as shown on Figure 8.4,

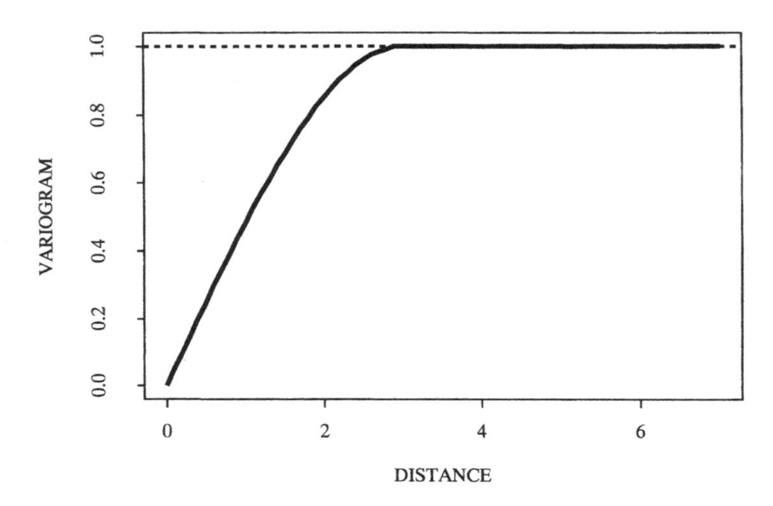

Figure 8.3: A spherical variogram with a sill $b = 1$ and a range $a = 3$.

$$\mathfrak{K}(\mathbf{h}) = \int\limits_{\mathbf{x}' \in \mathcal{B} \cap \mathcal{B}_{-\mathbf{h}}} d\mathbf{x}' = |\mathcal{B} \cap \mathcal{B}_{-\mathbf{h}}|. \tag{8.8}$$

The covariance of $Z(\mathbf{x})$ can now be expressed as

$$C(\mathbf{h}) = \mathrm{E}[\, N(\mathcal{B}) N(\mathcal{B}_{\mathbf{h}}) \,] - \mathrm{E}[\, N(\mathcal{B}) \,] \, \mathrm{E}[\, N(\mathcal{B}_{\mathbf{h}}) \,], \tag{8.9}$$

and as the counts $N(\mathcal{V})$ are independent in any subvolume,

$$\begin{aligned} C(\mathbf{h}) &= \mathrm{E}[\, N(\mathcal{B} \cap \mathcal{B}_{\mathbf{h}})^2 \,] - \mathrm{E}^2[\, N(\mathcal{B} \cap \mathcal{B}_{\mathbf{h}}) \,] \\ &= \theta \, |\mathcal{B} \cap \mathcal{B}_{\mathbf{h}}| \\ &= \theta \, \mathfrak{K}(\mathbf{h}). \end{aligned} \tag{8.10}$$

Calculating explicitly the volume of the intersection of two spheres of equal size whose centers are separated by a vector $\mathbf{h}$ yields the formula for the spherical covariance

$$C(\mathbf{h}) = \begin{cases} \theta |\mathcal{B}| \left(1 - \dfrac{3}{2} \dfrac{|\mathbf{h}|}{d} + \dfrac{1}{2} \dfrac{|\mathbf{h}|^3}{d^3} \right) & \text{for } 0 \leq |\mathbf{h}| \leq d, \\ 0 & \text{for } |\mathbf{h}| > d, \end{cases} \tag{8.11}$$

where $\theta |\mathcal{B}| = \theta \pi d^3 / 6 = C(\mathbf{0})$ represents the variance of $Z(\mathbf{x})$ and $|\mathcal{B}|$ is the volume of the spheres.

The diameter d of the spheres is equal to the range of the covariance function as it indicates the distance at which the covariance vanishes. The range of the spherical

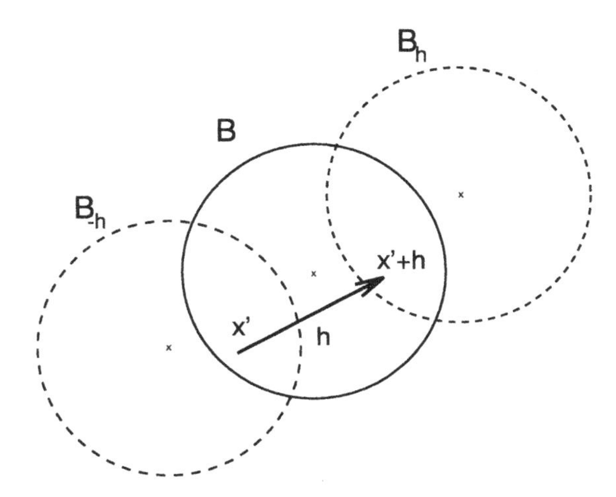

Figure 8.4: The intersection $\mathcal{B} \cap \mathcal{B}_{-\mathbf{h}}$ describes the set of points $\mathbf{x}' \in \mathcal{B}$ which have a neighbor $\mathbf{x}'+\mathbf{h}$ inside $\mathcal{B}$.

covariance function is the maximal distance at which the volumes of influence of two random variables $Z(\mathbf{x})$ and $Z(\mathbf{x}+\mathbf{h})$ can overlap and share information.

In applications large objects (as compared to the scale of the investigation) can condition the spatial structure of the data. The maximal size of these morphological objects in a given direction can often be read from the experimental variogram and interpreted as the range of a spherical model.

The shape of objects conditioning the morphology of a regionalized variable may not be spherical in many applications. This will result in anisotropical behavior of the variogram.

9 Anisotropy

Experimental calculations can reveal a very different behavior of the experimental variogram in different directions. This is called an *anisotropic* behavior. As variogram models are defined for the isotropic case, we need to examine transformations of the coordinates which allow to obtain anisotropic random functions from the isotropic models. In practice anisotropies are detected by inspecting experimental variograms in different directions and are included into the model by tuning predefined anisotropy parameters.

Geometric Anisotropy

In 2D-space a representation of the behavior of the experimental variogram can be made by drawing a map of iso-variogram lines as a function of a vector $\mathbf{h}$. Ideally if the iso-variogram lines are circular around the origin, the variogram obviously only depends on the length of the vector $\mathbf{h}$ and the phenomenon is isotropic.

If not, the iso-variogram lines can in many applications be approximated by concentric ellipses defined along a set of perpendicular main axes of anisotropy. This type of anisotropy, called the *geometric anisotropy,* can be obtained by a linear transformation of the spatial coordinates of a corresponding isotropic model. It allows to relate the class of ellipsoidally anisotropic random functions to a corresponding isotropic random function. This is essential because variogram models are defined for the isotropic case. The linear transformation extends in a simple way a given isotropic variogram to a whole class of ellipsoidally anisotropic variograms.

Rotating and dilating an ellipsoid

We have a coordinate system for $\mathbf{h} = (h_1, \ldots, h_n)$ with n coordinates. In this coordinate system the surfaces of constant variogram describe an ellipsoid and we search a new coordinate system for $\tilde{\mathbf{h}}$ in which the iso-variogram lines are spherical.

As a first step a rotation matrix $\mathbf{Q}$ is sought which rotates the coordinate system $\mathbf{h}$ into a coordinate system $\mathbf{h}' = \mathbf{Q}\mathbf{h}$ that is parallel to the principal axes of the ellipsoid, as shown on Figure 9.1 in the 2D case. The directions of the principal axes should be known from experimental variogram calculations.

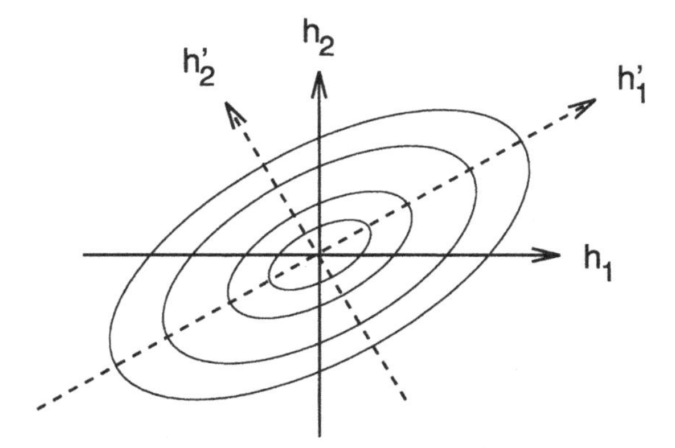

Figure 9.1: The coordinate system for $\mathbf{h} = (h_1, h_2)$ is rotated into the system $\mathbf{h}'$ parallel to the main axes of the concentric ellipses.

In 2D the rotation is given by the matrix

$$\mathbf{Q} = \begin{pmatrix} \cos\theta & \sin\theta \\ -\sin\theta & \cos\theta \end{pmatrix}, \tag{9.1}$$

where θ is the rotation angle.

In 3D the rotation is obtained by a composition of elementary rotations. The convention is to use Euler's angles and the corresponding rotation matrix is

$$\mathbf{Q} = \begin{pmatrix} \cos\theta_3 & \sin\theta_3 & 0 \\ -\sin\theta_3 & \cos\theta_3 & 0 \\ 0 & 0 & 1 \end{pmatrix} \begin{pmatrix} 1 & 0 & 0 \\ \cos\theta_2 & \sin\theta_2 & 0 \\ -\sin\theta_2 & \cos\theta_2 & 0 \end{pmatrix} \begin{pmatrix} \cos\theta_1 & \sin\theta_1 & 0 \\ -\sin\theta_1 & \cos\theta_1 & 0 \\ 0 & 0 & 1 \end{pmatrix}$$

The angle θ_1 defines a rotation of the plane h_1h_2 around h_3 such that h_1 is brought into the plane $h_1'h_2'$. With θ_2 a rotation is performed around the intersection of the planes h_1h_2 and $h_1'h_2'$ bringing h_3 in the position of h_3'. The third rotation with an angle θ_3 rotates everything around h_3' in its final position.

The second step in the transformation is to operate a shrinking or dilation of the principal axes of the ellipsoid using a diagonal matrix

$$\sqrt{\mathbf{\Lambda}} = \begin{pmatrix} \sqrt{\lambda_1} & & 0 \\ & \ddots & \\ 0 & & \sqrt{\lambda_n} \end{pmatrix}, \tag{9.2}$$

which transforms the system $\mathbf{h}'$ into a new system $\widetilde{\mathbf{h}}$ in which the ellipsoids become spheres

$$\widetilde{\mathbf{h}} = \sqrt{\mathbf{\Lambda}}\mathbf{h}'. \tag{9.3}$$

Conversely, if r is the radius of a sphere around the origin in the coordinate system of the isotropic variogram, it is obtained by calculating the length of any vector $\widetilde{\mathbf{h}}$ pointing on the surface of the sphere

$$r = |\widetilde{\mathbf{h}}| = \sqrt{\widetilde{\mathbf{h}}^\top \widetilde{\mathbf{h}}}. \tag{9.4}$$

This yields the equation of an ellipsoid in the $\mathbf{h}'$ coordinate system

$$(\mathbf{h}')^\top \mathbf{\Lambda} \mathbf{h}' = r^2. \tag{9.5}$$

The diameters d_p (principal axes) of the ellipsoid along the principal directions are thus

$$d_p = \frac{2r}{\sqrt{\lambda_p}} \tag{9.6}$$

and the principal directions are the vectors $\mathbf{q}_p$ of the rotation matrix.

Finally once the ellipsoid is determined the anisotropic variogram is specified on the basis of an isotropic variogram by

$$\gamma(r) = \gamma(\sqrt{\mathbf{h}^\top \mathbf{B} \mathbf{h}}), \tag{9.7}$$

where $\mathbf{B} = \mathbf{Q}^\top \mathbf{\Lambda} \mathbf{Q}$.

Exploring 3D space for anisotropy

In 3D applications the anisotropy of the experimental variogram can be explored taking advantage of the geometry of a regular icosahedron (20 faces, 30 edges) centered at the origin. The 15 lines joining opposite edges through the origin are used as leading directions for the experimental calculations. The lines are evenly distributed in space and can be grouped into 5 systems of Cartesian coordinates forming the basis of trirectangular trieders.

The range of a geometrically anisotropic variogram describes an ellipsoid whose principal directions are given by a set of Cartesian coordinates. Five possible ellipsoids for describing the range can now be tested by composing up to four times a rotation $\mathbf{R}$ yielding the rotation matrix

$$\begin{aligned} \mathbf{Q} &= (\mathbf{R})^k \\ &= \left(\frac{1}{2}\right)^k \begin{pmatrix} 1 & -(g+1) & g \\ g+1 & g & -1 \\ g & 1 & g+1 \end{pmatrix}^k \qquad \text{with } k = 1, \ldots, 4, \end{aligned} \tag{9.8}$$

where $g = (\sqrt{5}-1)/2 \cong 0.618$ is the *golden mean*.

Zonal anisotropy

It can happen that experimental variograms calculated in different directions suggest a different value for the sill. This is termed a *zonal anisotropy*.

For example, in 2D the sill along the x_2 coordinate might be much larger than along x_1. In such a situation a common strategy is to fit first to an isotropic model $\gamma_1(\mathbf{h})$ to the experimental variogram along the x_1 direction. Second, to add a geometrically anisotropic variogram $\gamma_2(\mathbf{h})$, which is designed to be without effect along the x_1 coordinate by providing it with a very large range in that direction through an anisotropy coefficient. The final variogram model is then

$$\gamma(\mathbf{h}) \quad = \quad \gamma_1(\mathbf{h}) + \gamma_2(\mathbf{h}), \tag{9.9}$$

in which the main axis of the anisotropy ellipse for $\gamma_2(\mathbf{h})$ is very large in the direction x_1.

The underlying random function model overlays two uncorrelated processes $Z_1(\mathbf{x})$ and $Z_2(\mathbf{x})$

$$Z(\mathbf{x}) \quad = \quad Z_1(\mathbf{x}) + Z_2(\mathbf{x}). \tag{9.10}$$

From the point of view of the regionalized variable, the anisotropy of $\gamma_2(\mathbf{h})$ can be due to morphological objects which are extremely elongated in the direction of x_1, crossing the borders of the domain. These units slice up the domain along x_1 thus creating a *zonation* along x_2, which explains the additional variability to be read on the variogram in that direction.

Nonlinear deformations of space

In air pollution and climatological studies it is frequent that data is available for several replications N_t in time at stations in 2D space. For every pair of locations $(\mathbf{x}_\alpha, \mathbf{x}_\beta)$ in geographical space a variogram value $\overline{\gamma^\star}(\mathbf{x}_\alpha, \mathbf{x}_\beta)$ can be computed by averaging the dissimilarities $\gamma^\star_{\alpha\beta}$ between the two stations for the N_t replications in time. It is often the case for pairs of stations at locations $(\mathbf{x}_\alpha, \mathbf{x}_\beta)$ and $(\mathbf{x}_{\alpha'}, \mathbf{x}_{\beta'})$ with separation vectors $\mathbf{h} \cong \mathbf{h}'$ approximately of the same length and orientation that the values $\overline{\gamma^\star}(\mathbf{x}_\alpha, \mathbf{x}_\beta)$ and $\overline{\gamma^\star}(\mathbf{x}_{\alpha'}, \mathbf{x}_{\beta'})$ are nevertheless very different!

To cope with this problem *spatial correlation mapping* has been developed, inspired by techniques used in morphometrics. SAMPSON & GUTTORP [286] and MONESTIEZ & SWITZER [229] have proposed smooth nonlinear deformations of space $f(\mathbf{x})$ for which the variogram $\gamma(r) = \gamma(|\widetilde{\mathbf{h}}|)$, with $\widetilde{\mathbf{h}} = f(\mathbf{x}) - f(\mathbf{x}')$, is isotropic. The deformation of the geographical space for which the $\overline{\gamma^\star}(\mathbf{x}_\alpha, \mathbf{x}_\beta)$ values best fit a given theoretical model is obtained by multidimensional scaling . The resulting somewhat grotesque looking maps showing the deformed geographical space turn out to be a valuable exploratory tool for understanding the covariance structure of the stations, especially when this can be done for different time periods.

10 Extension and Dispersion Variance

Measurements can represent averages over volumes, surfaces or intervals, called their *support*. The computation of variances depends intimately on the supports that are involved as well as on a theoretical variogram associated to a pointwise support. This is illustrated with an application from industrial hygienics. Furthermore, three simple sampling designs are examined from a geostatistical perspective.

Support

In the investigation of regionalized variables the variances are a function of the size of the domain. On Table 10.1 the results of computations of means and variances in nested 2D domains $\mathcal{D}_n$ are shown.

	Size	Mean $m(\mathcal{D}_n)$	Variance $\sigma^2(\cdot\|\mathcal{D}_n)$
$\mathcal{D}_1$	32×32	20.5	7.4
$\mathcal{D}_2$	64×64	20.1	13.8
$\mathcal{D}_3$	128×128	20.1	23.6
$\mathcal{D}_4$	256×256	20.8	34.6
$\mathcal{D}_5$	512×512	18.8	45.0

Table 10.1: Nested 2D domains $\mathcal{D}_n$ for which the variance increases with the size of the domain (from a simulation of an intrinsic random function by C LAJAUNIE)

In this example the variance $\sigma^2(\cdot|\mathcal{D}_n)$ of point samples in a domain $\mathcal{D}_n$, increases steadily with the size of the domain whereas the mean does not vary following a distinctive pattern. This illustrates the influence that a change in the size of a support (here the domain $\mathcal{D}_n$) can have on a statistic like the variance.

In applications generally two or more supports are involved as illustrated by the Figure 10.1. In mining the samples are collected on a support that can be considered pointwise (only a few cm^3); subsequently small blocs v (m^3) or larger panels V (100m^3) have to be estimated within deposits $\mathcal{D}$. In soil pollution small surface units s are distinguished from larger portions S. In industrial hygiene the problem may be set in terms of time supports: with average measurements on short time intervals Δt the excess over a limit value defined for a work day T should be estimated.

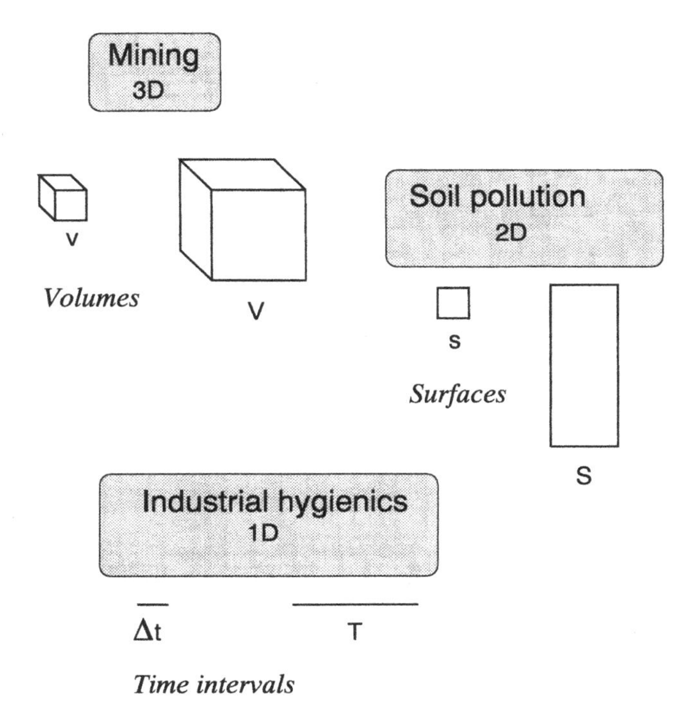

Figure 10.1: Supports in 1, 2, 3D in different applications.

Extension variance

With regionalized variables it is necessary to take account of the spatial disposal of points, surfaces or volumes for which the variance of a quantity should be computed.

The *extension variance* of a point $\mathbf{x}$ with respect to another point $\mathbf{x}'$ is defined as twice the variogram

$$\sigma^2_{\mathrm{E}}(\mathbf{x},\mathbf{x}') \;=\; \operatorname{var}(Z(\mathbf{x})-Z(\mathbf{x}')) \;=\; 2\,\gamma(\mathbf{x}-\mathbf{x}'). \tag{10.1}$$

It represents the square of theoretical error committed when a value at a point $\mathbf{x}$ is "extended" to a point $\mathbf{x}'$.

The extension variance of a small volume v to a larger volume V at a different location (see Figure 10.2) is obtained by averaging the differences between all positions of a point $\mathbf{x}$ in the volume v and a point $\mathbf{x}'$ in V

$$\begin{aligned}\sigma^2_{\mathrm{E}}(v,V) \;&=\; \operatorname{var}(\,Z(v)-Z(V)\,) \qquad (10.2)\\ &=\; 2\,\frac{1}{|v|\,|V|}\int\limits_{\mathbf{x}\,\in v}\int\limits_{\mathbf{x}'\in V}\gamma(\mathbf{x}-\mathbf{x}')\,d\mathbf{x}\,d\mathbf{x}'\end{aligned}$$

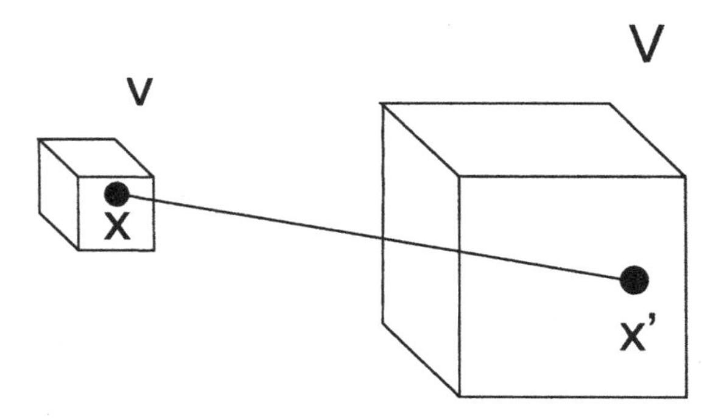

Figure 10.2: Points $\mathbf{x} \in v$ and $\mathbf{x}' \in V$.

$$
-\frac{1}{|v|^2} \int_{\mathbf{x}\in v} \int_{\mathbf{x}'\in v} \gamma(\mathbf{x}-\mathbf{x}')\, d\mathbf{x}\, d\mathbf{x}'
$$

$$
-\frac{1}{|V|^2} \int_{\mathbf{x}\in V} \int_{\mathbf{x}'\in V} \gamma(\mathbf{x}-\mathbf{x}')\, d\mathbf{x}\, d\mathbf{x}'. \qquad (10.3)
$$

Denoting

$$
\overline{\gamma}(v,V) = \frac{1}{|v|\,|V|} \int_{\mathbf{x}\in v} \int_{\mathbf{x}'\in V} \gamma(\mathbf{x}-\mathbf{x}')\, d\mathbf{x}\, d\mathbf{x}', \qquad (10.4)
$$

we have

$$
\sigma_{\mathrm{E}}^2(v,V) = 2\,\overline{\gamma}(v,V) - \overline{\gamma}(v,v) - \overline{\gamma}(V,V). \qquad (10.5)
$$

The extension variance depends on variogram integrals $\overline{\gamma}(v,V)$, whose values can either be read in charts (see JOURNEL & HUIJBREGTS [156], chap. II) or integrated numerically on a computer.

Dispersion variance

Suppose a large volume V is partitioned into n smaller units v of equal size. The experimental *dispersion variance* of the values z_v^α of the small volumes v_α building up V is given by the formula

$$
s^2(v|V) = \frac{1}{n}\sum_{\alpha=1}^{n}(z_v^\alpha - z_V)^2, \qquad (10.6)
$$

where

$$
z_V = \frac{1}{n}\sum_{\alpha=1}^{n} z_v^\alpha. \qquad (10.7)
$$

Considering all possible realizations of a random function we write

$$S^2(v|V) \;=\; \frac{1}{n}\sum_{\alpha=1}^{n}(Z_v^\alpha - Z_V)^2. \tag{10.8}$$

The theoretical formula for the dispersion variance is obtained by taking the expectation

$$\begin{aligned}\sigma^2(v|V) \;&=\; \mathrm{E}[\,S^2(v|V)\,]\\ &=\; \frac{1}{n}\sum_{\alpha=1}^{n}\mathrm{E}\Big[\,(Z_v^\alpha - Z_V)^2\Big],\end{aligned} \tag{10.9}$$

in which we recognize the extension variances

$$\sigma^2(v|V) \;=\; \frac{1}{n}\sum_{\alpha=1}^{n}\sigma_{\mathrm{E}}^2(v_\alpha, V). \tag{10.10}$$

Expressing the extension variances in terms of variogram integrals

$$\begin{aligned}\sigma^2(v|V) \;&=\; \frac{1}{n}\sum_{\alpha=1}^{n}(2\,\overline{\gamma}(v,V) - \overline{\gamma}(v,v) - \overline{\gamma}(V,V))\\ &=\; -\overline{\gamma}(v,v) - \overline{\gamma}(V,V) + \frac{2}{n}\sum_{\alpha=1}^{n}\frac{1}{|v_\alpha|\,|V|}\int\limits_{\mathbf{x}\,\in v_\alpha}\int\limits_{\mathbf{x}'\in V}\gamma(\mathbf{x}-\mathbf{x}')\,d\mathbf{x}\,d\mathbf{x}'\\ &=\; -\overline{\gamma}(v,v) - \overline{\gamma}(V,V) + \frac{2}{n|v|\,|V|}\underbrace{\sum_{\alpha=1}^{n}\int\limits_{\mathbf{x}\,\in v_\alpha}}_{\mathbf{x}\,\in V}\int\limits_{\mathbf{x}'\in V}\gamma(\mathbf{x}-\mathbf{x}')\,d\mathbf{x}\,d\mathbf{x}'\\ &=\; -\overline{\gamma}(v,v) - \overline{\gamma}(V,V) + 2\,\overline{\gamma}(V,V),\end{aligned} \tag{10.11}$$

so that we end up with the simple formula

$$\sigma^2(v|V) \;=\; \overline{\gamma}(V,V) - \overline{\gamma}(v,v). \tag{10.12}$$

The theoretical determination of the dispersion variance reduces to the computation of the variogram integrals $\overline{\gamma}(v,v)$ and $\overline{\gamma}(V,V)$ associated to the two supports v and V.

Krige's relation

Starting from the formula of the dispersion variance, first we see that for the case of the point values (denoted by a dot) the dispersion formula reduces to one term

$$\sigma^2(\cdot|V) \;=\; \overline{\gamma}(V,V). \tag{10.13}$$

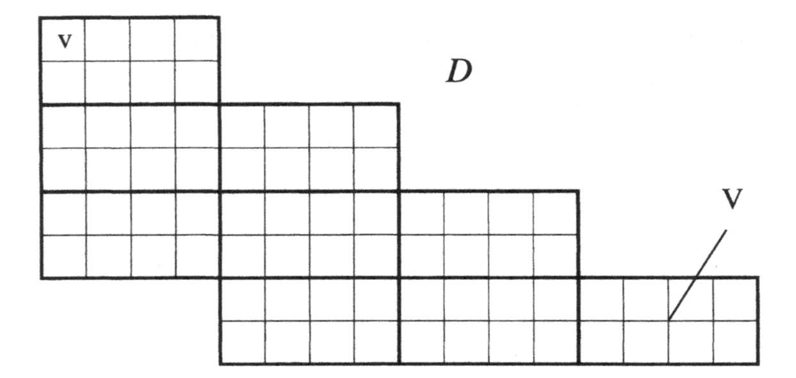

Figure 10.3: A domain $\mathcal{D}$ partitioned into volumes V which are themselves partitioned into smaller volumes v.

Second, we notice that $\sigma^2(v|V)$ is the difference between the dispersion variances of point values in V and in v

$$\sigma^2(v|V) \quad = \quad \sigma^2(\cdot|V) - \sigma^2(\cdot|v). \tag{10.14}$$

Third, it becomes apparent that the dispersion variance of point values in V can be decomposed into

$$\sigma^2(\cdot|V) \quad = \quad \sigma^2(\cdot|v) + \sigma^2(v|V). \tag{10.15}$$

This decomposition can be generalized to non point supports. Let $\mathcal{D}$ be a domain partitioned into large volumes V which are themselves partitioned into small units v as represented on Figure 10.3. Then the relation between the three supports v, V and $\mathcal{D}$ can be expressed theoretically by what is called *Krige's relation*

$$\sigma^2(v|\mathcal{D}) = \sigma^2(v|V) + \sigma^2(V|\mathcal{D}). \tag{10.16}$$

As the dispersion variances are basically differences of variogram averages over given supports, the sole knowledge of the pointwise theoretical variogram model makes dispersion variance computations possible for any supports of interest.

Change of support effect

In the early days of ore reserve estimation, mining engineers used a method called the *polygon method.* It consists in defining a polygon around each sample, representing the area of influence of the sample value, in such a way that the ore deposit is partitioned by the polygons. The reserves are estimated as a linear combination of the grades with the corresponding areas of influence. In the polygon method each sample value is extended to its area of influence, neglecting the fact that the samples are obtained from pointwise measurements while the polygons represent a much larger support.

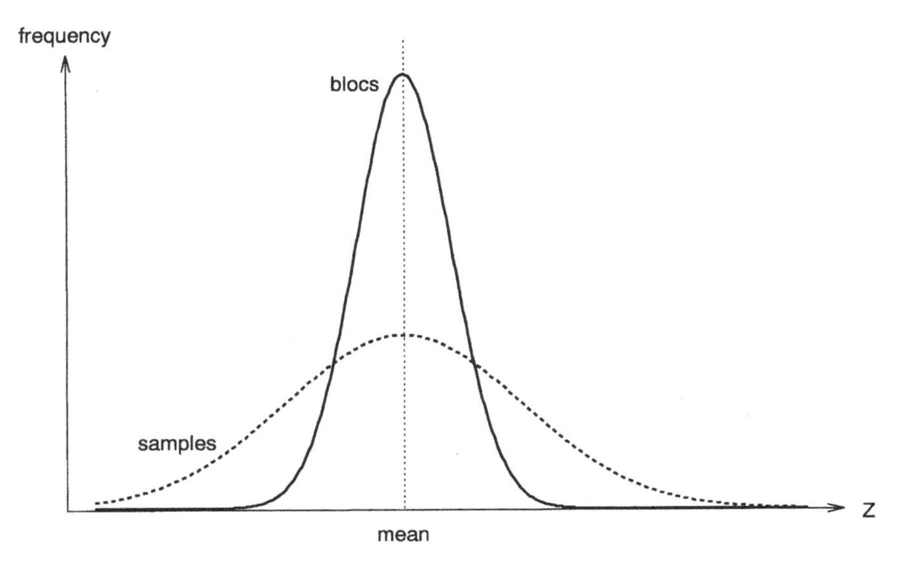

Figure 10.4: The distribution of block values is narrower than the distribution of values at the sample points.

In the case of a square grid the polygons are square blocks v which partition the exploration area. The value at each grid node is extended to each area of influence v. The method implies that the distribution of average values of the blocks is the same as the distribution of the values at the sample points. From Krige's relation we know that this cannot be true: the distribution of the values for a support v is narrower than the distribution of point values (as represented on Figure 10.4) because the variance $\sigma^2(\cdot|v)$ of the points in v generally is not negligible.

In mining, the *cut-off value* defines a grade above which a mining block should be sent to production. Mining engineers are interested in the proportion of the values above the cut-off value which represent the part of a geological body which is of economical interest. If the cut-off grade is a value substantially above the mean, the polygon method will lead to a systematic overestimation of the ore reserves as shown on Figure 10.5. To avoid systematic over- or underestimation the *support effect* needs to be taken into account.

Change of support: affine model

In this section we consider a stationary random function $Z(\mathbf{x})$ with a mean m and a variance σ^2. The mean m is not changed by a change of support and, whatever the distribution, we have the physical fact,

$$\mathrm{E}[\,Z(\mathbf{x})\,] \quad = \quad \mathrm{E}[\,Z_v(\mathbf{x})\,] = m, \tag{10.17}$$

i.e. the mean of the point variable $Z(\mathbf{x})$ is the same as that of the block variable $Z_v(\mathbf{x})$.

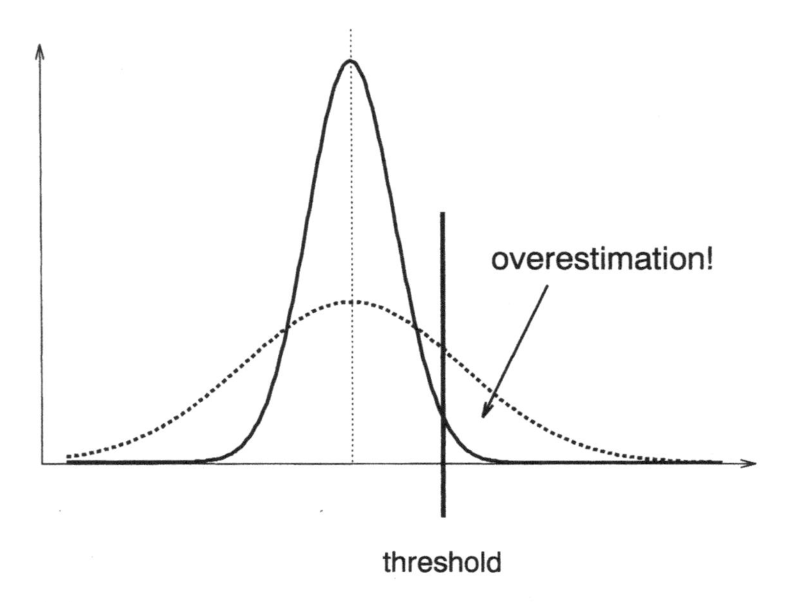

Figure 10.5: The proportion of sample values above the cut-off value is greater than the proportion of block values: the polygon method leads to a systematic overestimation in this case.

The *affine model* is based on the assumption that the standardized point variable follows the same distribution as the standardized block variable. This model is suitable for a Gaussian random function as the distribution of block grades is also Gaussian, i.e. if $Z(\mathbf{x}) \sim \mathcal{N}(m, \sigma^2)$,then $Z_v(\mathbf{x}) \sim \mathcal{N}(m, \sigma_v^2)$ and

$$\frac{Z(\mathbf{x}) - m}{\sigma} \stackrel{\mathcal{L}}{=} \frac{Z_v(\mathbf{x}) - m}{\sigma_v} \sim \mathcal{N}(0, 1), \qquad (10.18)$$

where $\stackrel{\mathcal{L}}{=}$ means that the two quantities are identically distributed.

The distribution of the block values is therefore simply obtained from the distribution of the point values by an affine transformation,

$$Z_v(\mathbf{x}) \stackrel{\mathcal{L}}{=} m + r\,(Z(\mathbf{x}) - m) \sim \mathcal{N}(m, \sigma_v^2), \qquad (10.19)$$

where $r = \sigma_v / \sigma$ is the change of support coefficient.

In practice if the point variance σ^2 and the variogram $\gamma(\mathbf{h})$ are known, the bloc variance is computed by the formula (10.12) of the dispersion variance,

$$\sigma_v^2 = \overline{C}(v, v) = \sigma^2 - \overline{\gamma}(v, v) \qquad (10.20)$$

as, assuming a large domain in comparison to the range of the variogram, $\sigma^2 = \overline{\gamma}(\infty, \infty)$. The change of support coefficient can then readily be computed. The affine change of support model is appropriate only if the data comply with the Gaussian distributional assumption. Otherwise the affine model should only be used for v relatively small as compared to the range.

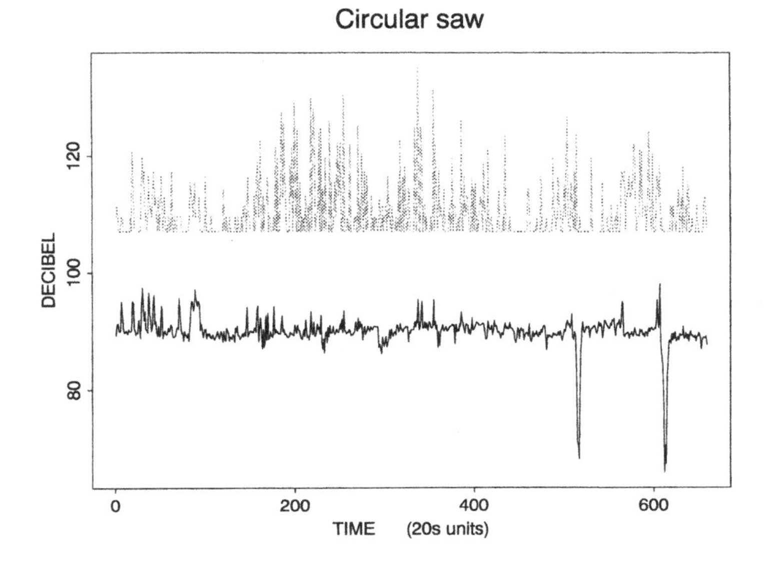

Figure 10.6: Measurements of maximal noise level L_{max} (dots) and average noise L_{eq} (plain) on time intervals of 20 seconds during 3 hours and 40 minutes. They represent the exposure of a worker to the noise of a circular saw.

Application: acoustic data

A series of 659 measurements of equivalent noise levels L_{eq} (expressed in dB_A) averaged over 20 seconds were performed on a worker operating with a circular saw. The problem is to evaluate whether a shorter or larger time integration interval would be of interest.

The $L_{eq}(t)$ are not an additive variable and need to be transformed back to the sound exposure $V_{eq}(t)$. The average sound exposure $V_{eq}(t)$ is defined as the integral (over time interval Δt) of the squared ratio of the instant acoustic pressures $p(x)$ against the reference acoustic pressure p_0

$$V_{eq}(t) = \frac{10^{-9}}{\Delta t} \int_{t-\Delta t/2}^{t+\Delta t/2} \left(\frac{p(x)}{p_0} \right)^2 dx \tag{10.21}$$

$$= \exp(\alpha \, L_{eq}(t) - \beta), \tag{10.22}$$

where $\alpha = (\ln 10)/10$ and $\beta = \ln 10^9$.

The measurements were taken continuously during a period of 3 hours and 40 minutes. The Figure 10.6 shows with a continuous line the time series (in dB_A) of the equivalent noise levels L_{eq} integrated over intervals of 20 seconds. The maximal noise levels L_{max} within these time intervals are plotted with a dotted line (when they are above 107 dB). We observe in passing that the averaging over 20 seconds has

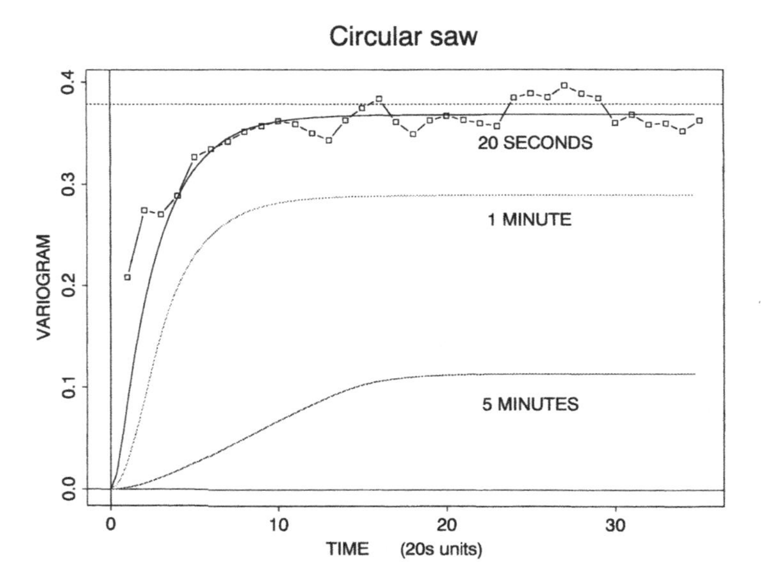

Figure 10.7: Experimental variogram of the sound exposure V_{eq} and a regularized exponential variogram model for time intervals of $\Delta t = 20$s, 1mn and 5mn.

enormously reduced the variation.

The theoretical variogram of the sound exposure was modeled with a pointwise exponential model

$$\gamma(h) \;=\; b\left(1-\mathrm{e}^{-|h|/a}\right) \qquad \text{with } a,b>0. \tag{10.23}$$

The sill is $b = .42$ and the range parameter is $a = 2.4$. It corresponds to a practical range of $3a = 7.2$ time units, i.e. 2.4 minutes, which is the time of a typical repetitive working operation.

The support of the sound exposure, i.e. the integration time, has an impact on the shape of the theoretical variogram: it alters the behavior at the origin, reduces the value of the sill and increases the range. The exponential variogram regularized over time intervals Δt is defined by the formula ([156], p84)

$$\gamma_{\Delta t}(h) \;=\; \begin{cases} \dfrac{b\,a^2}{(\Delta t)^2}\left(2\mathrm{e}^{-\Delta t/a}-2+\dfrac{2h}{a}+\mathrm{e}^{-h/a}\left(2-\mathrm{e}^{-\Delta t/a}\right)-\mathrm{e}^{(h-\Delta t)/a}\right) \\ \qquad\qquad \text{for } 0\le h\le \Delta t, \\ \dfrac{b\,a^2}{(\Delta t)^2}\left(\mathrm{e}^{-\Delta t/a}-\mathrm{e}^{\Delta t/a}+\left(\mathrm{e}^{-\Delta t/a}+\mathrm{e}^{\Delta t/a}-2\right)\cdot\left(1-\mathrm{e}^{-h/a}\right)\right) \\ \qquad\qquad \text{for } h>\Delta t. \end{cases} \tag{10.24}$$

The Figure 10.7 shows the experimental variogram together with the exponential

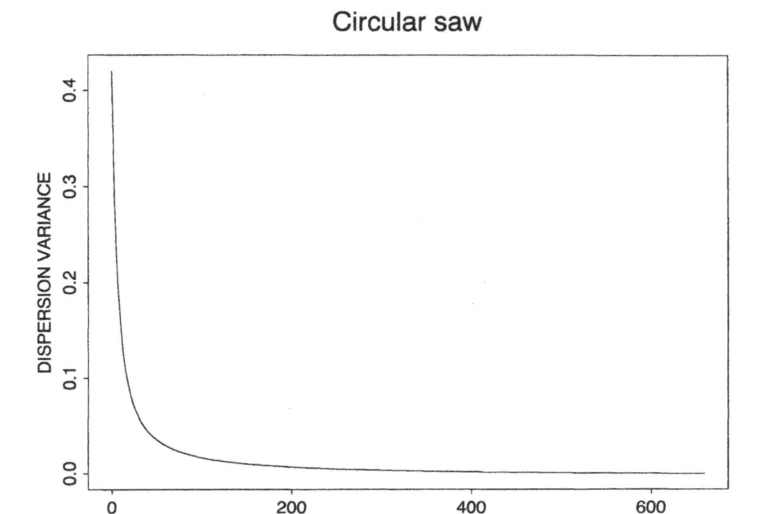

Figure 10.8: Curve of dispersion variances $\sigma^2(\Delta t|\mathcal{D})$ as a function of the integration time Δt with fixed $\mathcal{D}$.

model regularized over time lags of 20 seconds, 1 and 5 minutes illustrating the effect of a modification of the support on the shape of the theoretical variogram.

Finally a curve of the dispersion variance of the sound exposure as a function of the integration time Δt is represented on Figure 10.8. The dispersion variance for an exponential model is calculated with the formula

$$\begin{aligned} \sigma^2(\Delta t|\mathcal{D}) &= \overline{\gamma}(\mathcal{D},\mathcal{D}) - \overline{\gamma}(\Delta t, \Delta t) \\ &= F(\mathcal{D}) - F(\Delta t), \end{aligned} \tag{10.25}$$

where, for $L = \Delta t, \mathcal{D}$,

$$F(L) = b\left(1 + \frac{2a}{L}\left(\frac{a}{L} - 1\right) - \frac{2a^2}{L^2}\exp\left(-\frac{L}{a}\right)\right). \tag{10.26}$$

As the practical range of the variogram is relatively short (2.4 minutes), it can be learned from Figure 10.8 that for a time integration support of less than 1/2 hour (90 time units) a small increase of the support leads to large dropping of the dispersion variance. Conversely it does not seem to make much difference if the integration is changed from 1 hour to 2 hours. With a short practical range the essential part of the variability can only be recovered using an integration time much shorter than 1/2 hour.

Comparison of sampling designs

The concepts of estimation and dispersion variance can be used to compare three sampling designs with n samples

A - regular grid: the domain $\mathcal{D}$ is partitioned into n cubic cells v at the center of which a sample $z(\mathbf{x}_\alpha)$ has been taken;

B - random grid: the n samples are taken at random within the domain $\mathcal{D}$;

C - random stratified grid: the domain $\mathcal{D}$ is partitioned into n cubic cells v inside each of which one sample is taken at random.

For ***design A***, with a regular grid the global estimation variance σ^2_{EG} is computed as

$$\begin{aligned}\sigma^2_{\text{EG}} &= \operatorname{var}\Big(Z^\star_{\mathcal{D}} - Z_{\mathcal{D}}\Big)\\ &= \mathrm{E}\Big[\Big(\frac{1}{n}\sum_{\alpha=1}^{n} Z(\mathbf{x}_\alpha) - \frac{1}{n}\sum_{\alpha=1}^{n} Z(v_\alpha)\Big)^2\Big]\\ &= \mathrm{E}\Big[\Big(\frac{1}{n}\sum_{\alpha=1}^{n} (Z(\mathbf{x}_\alpha) - Z(v_\alpha))\Big)^2\Big]. \end{aligned} \tag{10.27}$$

If we consider that the elementary errors $Z(\mathbf{x}_\alpha){-}Z(v_\alpha)$ are independent from one cell to the other

$$\begin{aligned}\sigma^2_{\text{EG}} &= \frac{1}{n^2}\sum_{\alpha=1}^{n} \mathrm{E}\Big[\Big(Z(\mathbf{x}_\alpha) - Z(v_\alpha)\Big)^2\Big]\\ &= \frac{1}{n^2}\sum_{\alpha=1}^{n} \sigma^2_{\text{E}}(\mathbf{x}_\alpha, v_a). \end{aligned} \tag{10.28}$$

As the points $\mathbf{x}_\alpha$ are at the centers $\mathbf{x}_c$ of cubes of the same size we have for design A

$$\sigma^2_{\text{EG}} = \frac{1}{n}\,\sigma^2_{\text{E}}(\mathbf{x}_c, v). \tag{10.29}$$

For ***design B***, the samples are supposed to be located at random in the domain (Poisson points). We shall consider one realization z with random coordinates X_1, X_2, X_3. The expectation will be taken on the coordinates. The global estimation variance is

$$\begin{aligned}s^2_{\text{EG}} &= \mathrm{E}_X\big[\,(z^\star_{\mathcal{D}} - z_{\mathcal{D}})^2\,\big]\\ &= \mathrm{E}_X\Big[\Big(\frac{1}{n}\sum_{\alpha=1}^{n} z(X_1^\alpha, X_2^\alpha, X_3^\alpha) - z(\mathcal{D})\Big)^2\Big]. \end{aligned} \tag{10.30}$$

Assuming elementary errors to be independent, we are left with

$$s^2_{\mathrm{EG}} \;=\; \frac{1}{n^2}\sum_{\alpha=1}^{n} \mathrm{E}_X[\,(z(X_1^\alpha, X_2^\alpha, X_3^\alpha) - z(\mathcal{D}))^2\,]. \tag{10.31}$$

We now write explicitly the expectation over the random locations distributed with probabilities $1/|\mathcal{D}|$ over the domain

$$\begin{aligned}
s^2_{\mathrm{EG}} &= \frac{1}{n^2}\sum_{\alpha=1}^{n} \int_{x_1}\int_{x_2}\int_{x_3} p(x_1^\alpha, x_2^\alpha, x_3^\alpha)\cdot\Big(z(x_1^\alpha, x_2^\alpha, x_3^\alpha) - z(\mathcal{D})\Big)^2\, dx_1\, dx_2\, dx_3\\
&= \frac{1}{n^2}\sum_{\alpha=1}^{n} \frac{1}{|\mathcal{D}|}\int_{x_1}\int_{x_2}\int_{x_3} \Big(z(x_1^\alpha, x_2^\alpha, x_3^\alpha) - z(\mathcal{D})\Big)^2\, dx_1\, dx_2\, dx_3\\
&= \frac{1}{n^2}\sum_{\alpha=1}^{n} s^2(\cdot|V)\\
&= \frac{1}{n}\, s^2(\cdot|V).
\end{aligned} \tag{10.32}$$

Generalizing the formula from one realization z to the random function Z and taking the expectation (over Z), we have for design B

$$\sigma^2_{\mathrm{EG}} \;=\; \frac{1}{n}\,\sigma^2(\cdot|V). \tag{10.33}$$

For ***design C***, each sample point is located at random within a cube v_α and the global estimation variance for one realization z is

$$\begin{aligned}
s^2_{\mathrm{EG}} &= \mathrm{E}_X[\,(z^\star_{\mathcal{D}} - z_{\mathcal{D}})^2\,]\\
&= \mathrm{E}_X\Big[\,\Big(\frac{1}{n}\sum_{\alpha=1}^{n}(\,z(X_1^\alpha, X_2^\alpha, X_3^\alpha) - z(v_\alpha)\,)\Big)^2\,\Big]\\
&= \frac{1}{n}\, s^2(\cdot|v).
\end{aligned} \tag{10.34}$$

Randomizing z to Z and taking the expectation, we have for design C

$$\sigma^2_{\mathrm{EG}} \;=\; \frac{1}{n}\,\sigma^2(\cdot|v). \tag{10.35}$$

Comparing the random grid B with the random stratified grid C we know from Krige's relation that

$$\sigma^2(\cdot|v) \;\leq\; \sigma^2(\cdot|V), \tag{10.36}$$

and thus design C is a better strategy than design B.

To compare the random stratified grid C with the regular grid A we have to compare the extension variance of the central point in the cube

$$\sigma^2_{\mathrm{E}}(\mathbf{x}_c, v) \quad = \quad 2\,\overline{\gamma}(\mathbf{x}_c, v) - \overline{\gamma}(v, v), \qquad (10.37)$$

with the dispersion variance of a point in the cube

$$\sigma^2(\cdot|v) \quad = \quad \overline{\gamma}(v, v). \qquad (10.38)$$

It turns out that the former is usually lower than the latter (see [51], p136, for a numerical example). The regular grid is superior to the random stratified grid from the point of view of global dispersion variance as the samples cover evenly the region.

However for computing the experimental variogram, an advantage can be seen in using unequally spaced data: they will provide more information about small-scale variability than evenly placed samples. This helps in modeling the variogram near the origin.

11 Ordinary Kriging

Ordinary kriging is the most widely used kriging method. It serves to estimate a value at a point of a region for which a variogram is known, using data in the neighborhood of the estimation location. Ordinary kriging can also be used to estimate a block value. With local second-order stationarity, ordinary kriging implicitly evaluates the mean in a moving neighborhood. To see this, first a kriging estimate of the local mean is set up, then a simple kriging estimator using this kriged mean is examined.

Ordinary kriging problem

We wish to estimate a value at $\mathbf{x}_0$ as represented on Figure 11.1 using the data values

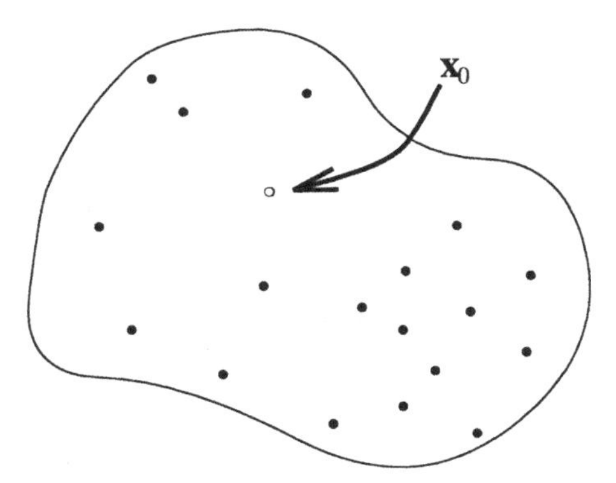

Figure 11.1: A domain with irregularly spaced sample points (black dots) and a location of interest $\mathbf{x}_0$.

from the n neighboring sample points $\mathbf{x}_\alpha$ and combining them linearly with weights w_α

$$Z^{\star}_{\mathrm{OK}}(\mathbf{x}_0) \quad = \quad \sum_{\alpha=1}^{n} w_\alpha \, Z(\mathbf{x}_\alpha). \tag{11.1}$$

Obviously we have to constrain the weights to sum up to one because in the particular case when all data values are equal to a constant, the estimated value should also be equal to this constant.

We assume that the data are part of a realization of an intrinsic random function $Z(\mathbf{x})$ with a variogram $\gamma(\mathbf{h})$.

The unbiasedness is warranted with unit sum weights

$$\begin{aligned}
\mathrm{E}[\, Z^{\star}(\mathbf{x}_0) - Z(\mathbf{x}_0)\,] &= \mathrm{E}\Big[\sum_{\alpha=1}^{n} w_\alpha\, Z(\mathbf{x}_\alpha) - Z(\mathbf{x}_0) \cdot \underbrace{\sum_{\alpha=1}^{n} w_\alpha}_{1}\Big] \\
&= \sum_{\alpha=1}^{n} w_\alpha\, \mathrm{E}[\, Z(\mathbf{x}_\alpha) - Z(\mathbf{x}_0)\,] \\
&= 0, \qquad (11.2)
\end{aligned}$$

because the expectations of the increments are zero.

The estimation variance $\sigma_{\mathrm{E}}^2 = \mathrm{var}(Z^{\star}(\mathbf{x}_0) - Z(\mathbf{x}_0))$ is the variance of the linear combination

$$Z^{\star}(\mathbf{x}_0) - Z(\mathbf{x}_0) = \sum_{\alpha=1}^{n} w_\alpha\, Z(\mathbf{x}_\alpha) - 1 \cdot Z(\mathbf{x}_0) = \sum_{\alpha=0}^{n} w_\alpha\, Z(\mathbf{x}_\alpha), \qquad (11.3)$$

with a weight w_0 equal to -1 and

$$\sum_{\alpha=0}^{n} w_\alpha = 0. \qquad (11.4)$$

Thus the condition that the weights numbered from 1 to n sum up to one also implies that the use of the variogram is *authorized* in the computation of the variance of the estimation error.

COMMENT 11.1 *The variogram is authorized for ordinary kriging, but not for simple kriging, because the latter does not include any constraint on the weights. The condition (11.4) corresponds to the one found in the definition (7.18) of a conditionally negative definite function.*

The estimation variance is

$$\begin{aligned}
\sigma_{\mathrm{E}}^2 &= \mathrm{E}[\, (Z^{\star}(\mathbf{x}_0) - Z(\mathbf{x}_0))^2\,] \\
&= -\gamma(\mathbf{x}_0 - \mathbf{x}_0) - \sum_{\alpha=1}^{n}\sum_{\beta=1}^{n} w_\alpha\, w_\beta\, \gamma(\mathbf{x}_\alpha - \mathbf{x}_\beta) + 2\sum_{\alpha=1}^{n} w_\alpha \gamma(\mathbf{x}_\alpha - \mathbf{x}_0). \qquad (11.5)
\end{aligned}$$

By minimizing the estimation variance with the constraint on the weights, we obtain the *ordinary kriging system* (OK)

$$\begin{pmatrix} \gamma(\mathbf{x}_1-\mathbf{x}_1) & \dots & \gamma(\mathbf{x}_1-\mathbf{x}_n) & 1 \\ \vdots & \ddots & \vdots & \vdots \\ \gamma(\mathbf{x}_n-\mathbf{x}_1) & \dots & \gamma(\mathbf{x}_n-\mathbf{x}_n) & 1 \\ 1 & \dots & 1 & 0 \end{pmatrix} \begin{pmatrix} w_1^{\mathrm{OK}} \\ \vdots \\ w_n^{\mathrm{OK}} \\ \mu_{\mathrm{OK}} \end{pmatrix} = \begin{pmatrix} \gamma(\mathbf{x}_1-\mathbf{x}_0) \\ \vdots \\ \gamma(\mathbf{x}_n-\mathbf{x}_0) \\ 1 \end{pmatrix}, \qquad (11.6)$$

where the $w_\alpha^{\rm OK}$ are weights to be assigned to the data values and where $\mu_{\rm OK}$ is the Lagrange parameter. The left hand side of the system describes the dissimilarities between the data points, while the right hand shows the dissimilarities between each data point and the estimation point $\mathbf{x}_0$.

Performing the matrix multiplication, the ordinary kriging system can be rewritten in the form

$$\begin{cases} \sum_{\beta=1}^{n} w_\beta^{\rm OK}\, \gamma(\mathbf{x}_\alpha - \mathbf{x}_\beta) + \mu_{\rm OK} = \gamma(\mathbf{x}_\alpha - \mathbf{x}_0) \quad \text{for } \alpha = 1, \ldots, n \\ \sum_{\beta=1}^{n} w_\beta^{\rm OK} = 1. \end{cases} \tag{11.7}$$

The estimation variance of ordinary kriging is

$$\sigma_{\rm OK}^2 = \mu_{\rm OK} - \gamma(\mathbf{x}_0 - \mathbf{x}_0) + \sum_{\alpha=1}^{n} w_\alpha^{\rm OK}\, \gamma(\mathbf{x}_\alpha - \mathbf{x}_0). \tag{11.8}$$

Ordinary kriging is an *exact interpolator* in the sense that if $\mathbf{x}_0$ is identical with a data location then the estimated value is identical with the data value at that point

$$Z^\star(\mathbf{x}_0) = Z(\mathbf{x}_\alpha), \qquad \text{if } \mathbf{x}_0 = \mathbf{x}_\alpha. \tag{11.9}$$

This can be easily seen. When $\mathbf{x}_0$ is one of the sample points, the right hand side of the kriging system is equal to one column of the left hand side matrix. A weight vector $\mathbf{w}$ with a weight for that column equal to one and all other weights (including $\mu_{\rm OK}$) equal to zero is a solution of the system. As the left hand matrix is not singular, this is the only solution.

Simple kriging of increments

We deal with an intrinsically stationary random function $Z(\mathbf{x})$ with a variogram $\gamma(\mathbf{h})$. We have data at locations $\mathbf{x}_1, \ldots, \mathbf{x}_n$. Taking arbitrarily a random variable Z at the data location $\mathbf{x}_n$, we can construct an incremental random function

$$Y(\mathbf{x}) = Z(\mathbf{x}) - Z(\mathbf{x}_n), \tag{11.10}$$

whose (non stationary) covariance function is

$$C_Y(\mathbf{x}, \mathbf{y}) = -\gamma(\mathbf{x} - \mathbf{y}) + \gamma(\mathbf{x} - \mathbf{x}_n) + \gamma(\mathbf{y} - \mathbf{x}_n). \tag{11.11}$$

The simple kriging estimate $Y_{\rm SK}^\star(\mathbf{x}_0)$ from increments at locations $\mathbf{x}_1, \ldots, \mathbf{x}_{n-1}$ yields an estimate

$$Z^\star(\mathbf{x}_0) = Y_{\rm SK}^\star(\mathbf{x}_0) + Z(\mathbf{x}_n), \tag{11.12}$$

which is equivalent to the ordinary kriging $Z_{\rm OK}^\star(\mathbf{x})$.

To see this, consider the simple kriging system

$$\sum_{\beta=1}^{n-1} w_\beta^{\mathrm{SK}}\, C_Y(\mathbf{x}_\alpha, \mathbf{x}_\beta) \quad = \quad C_Y(\mathbf{x}_\alpha, \mathbf{x}_0) \qquad \text{for} \quad \alpha = 1, \ldots, n-1. \tag{11.13}$$

Inserting (11.11) we get for each $\alpha = 1, \ldots, n-1$

$$\begin{aligned} &-\sum_{\beta=1}^{n-1} w_\beta^{\mathrm{SK}}\, \gamma(\mathbf{x}_\alpha - \mathbf{x}_\beta) + \sum_{\beta=1}^{n-1} w_\beta^{\mathrm{SK}}\, \gamma(\mathbf{x}_\alpha - \mathbf{x}_n) + \sum_{\beta=1}^{n-1} w_\beta^{\mathrm{SK}}\, \gamma(\mathbf{x}_\beta - \mathbf{x}_n) \\ &\quad = \; -\gamma(\mathbf{x}_\alpha - \mathbf{x}_0) + \gamma(\mathbf{x}_\alpha - \mathbf{x}_n) + \gamma(\mathbf{x}_0 - \mathbf{x}_n), \end{aligned} \tag{11.14}$$

and renaming the terms which do not depend on α

$$\begin{aligned} &-\sum_{\beta=1}^{n-1} w_\beta^{\mathrm{SK}}\, \gamma(\mathbf{x}_\alpha - \mathbf{x}_\beta) - \gamma(\mathbf{x}_\alpha - \mathbf{x}_n) \underbrace{\left(1 - \sum_{\beta=1}^{n-1} w_\beta^{\mathrm{SK}}\right)}_{w_n^{\mathrm{OK}}} \\ &\qquad + \underbrace{\sum_{\beta=1}^{n-1} w_\beta^{\mathrm{SK}}\, \gamma(\mathbf{x}_\beta - \mathbf{x}_n) - \gamma(\mathbf{x}_0 - \mathbf{x}_n)}_{-\mu_{\mathrm{OK}}} \\ &\quad = \; -\gamma(\mathbf{x}_\alpha - \mathbf{x}_0) \qquad \text{for} \quad \alpha = 1, \ldots, n-1. \end{aligned} \tag{11.15}$$

This is the system for the ordinary kriging estimator $Z^\star_{\mathrm{OK}}$ from expression (11.1). We have weights w_α^{OK} summing to one because w_α^{OK} is equal to w_α^{SK} for the first $n-1$ weights, while w_n^{OK} is precisely the difference between one and the sum of those simple kriging weights. The system has $n-1$ equations because there are actually only $n-1$ unknowns in ordinary kriging: both μ_{OK} and w_n^{OK} can be deduced from the $n-1$ other weights.

The equivalence between the ordinary kriging of Z and the simple kriging of its increments Y is no real surprise as both estimators minimize the same estimation variance

$$\mathrm{var}(Z - Z^\star) \quad = \quad \mathrm{var}(Y - Y^\star) \tag{11.16}$$

over the same class of estimators and with the same probabilistic model. As the same minimum is achieved, the resulting kriging variances are equal.

Block kriging

Ordinary kriging can be used to estimate a block value instead of a point value as suggested by the drawing on Figure 11.2. When estimating a block value from point

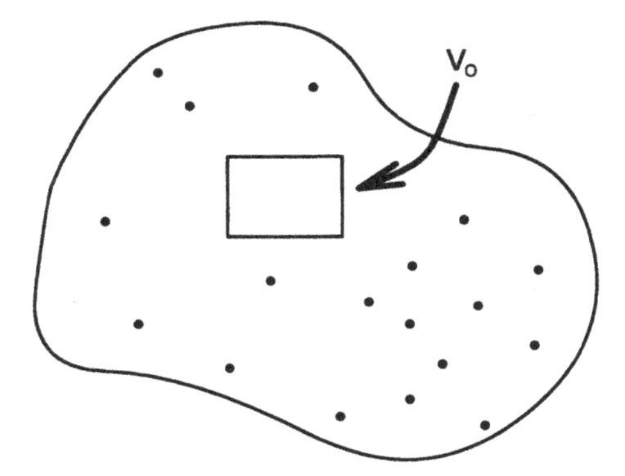

Figure 11.2: A domain with irregularly spaced sample points (black dots) and a block v_0 for which a value should be estimated.

values by

$$Z^{\star}_{v_0} \;=\; \sum_{\alpha=1}^{n} w_\alpha\, Z(\mathbf{x}_\alpha) \tag{11.17}$$

the ordinary kriging is modified in the following way to krige a block

$$\begin{pmatrix} \gamma(\mathbf{x}_1-\mathbf{x}_1) & \dots & \gamma(\mathbf{x}_1-\mathbf{x}_n) & 1 \\ \vdots & \ddots & \vdots & \vdots \\ \gamma(\mathbf{x}_n-\mathbf{x}_1) & \dots & \gamma(\mathbf{x}_n-\mathbf{x}_n) & 1 \\ 1 & \dots & 1 & 0 \end{pmatrix} \begin{pmatrix} w_1^{\mathrm{BK}} \\ \vdots \\ w_n^{\mathrm{BK}} \\ \mu_{\mathrm{BK}} \end{pmatrix} \;=\; \begin{pmatrix} \overline{\gamma}(\mathbf{x}_1, v_0) \\ \vdots \\ \overline{\gamma}(\mathbf{x}_n, v_0)) \\ 1 \end{pmatrix}, \tag{11.18}$$

where the right hand side now contains the average variogram $\overline{\gamma}(\mathbf{x}_\alpha, v_0)$ of each sample point with the block of interest.

The corresponding block kriging variance is

$$\sigma^2_{\mathrm{BK}} \;=\; \mu_{\mathrm{BK}} - \overline{\gamma}(v_0, v_0) + \sum_{\alpha=1}^{n} w_\alpha^{\mathrm{BK}}\, \overline{\gamma}(\mathbf{x}_\alpha, v_0). \tag{11.19}$$

For a second-order stationary random function the block kriging variance can be written in terms of covariances

$$\sigma^2_{\mathrm{BK}} \;=\; \mu_{\mathrm{BK}} + \overline{C}(v_0, v_0) - \sum_{\alpha=1}^{n} w_\alpha^{\mathrm{BK}}\, \overline{C}(\mathbf{x}_\alpha, v_0). \tag{11.20}$$

Assuming the covariance function falls off to zero (like the spherical or exponential functions), if we let the volume v_0 grow very large, the terms $\overline{C}(v_0, v_0)$ and $\overline{C}(\mathbf{x}_\alpha, v_0)$ vanish, i.e.

$$\sigma^2_{\mathrm{BK}} \;\mapsto\; \sigma^2_{\mathrm{KM}} \qquad \text{for } |v_0| \mapsto \infty. \tag{11.21}$$

The same is true for the kriging systems and the estimates. The block kriging tends to be equivalent to the kriging of the mean for large blocks v_0.

Simple kriging with an estimated mean

We assume a second-order stationary $Z(\mathbf{x})$ with known covariance function $C(\mathbf{h})$. We already have learned how to krige the mean using the linear combination

$$m^\star = \sum_{\alpha=1}^{n} w_\alpha^{\text{KM}} z(\mathbf{x}_\alpha) \qquad \text{with} \quad \sum_{\alpha=1}^{n} w_\alpha^{\text{KM}} = 1, \tag{11.22}$$

where w_α^{KM} are the weights of a kriging of the mean and $z(\mathbf{x}_\alpha)$ are data at locations $\mathbf{x}_\alpha$.

Why not insert this estimated mean value into the simple kriging estimator

$$Z^\star_{\text{SKM}}(\mathbf{x}_0) = m^\star + \sum_{\alpha=1}^{n} w_\alpha^{\text{SK}} (Z(\mathbf{x}_\alpha) - m^\star), \tag{11.23}$$

where w_α^{SK} are the simple kriging weights? Replacing $m^\star$ by the corresponding linear combination gives

$$\begin{aligned} Z^\star_{\text{SKM}}(\mathbf{x}_0) &= \sum_{\alpha=1}^{n} w_\alpha^{\text{KM}} Z(\mathbf{x}_\alpha) + \sum_{\alpha=1}^{n} w_\alpha^{\text{SK}} Z(\mathbf{x}_\alpha) - \sum_{\alpha=1}^{n} w_\alpha^{\text{SK}} \sum_{\beta=1}^{n} w_\beta^{\text{KM}} Z(\mathbf{x}_\beta) \\ &= \sum_{\alpha=1}^{n} w_\alpha^{\text{SK}} Z(\mathbf{x}_\alpha) + \sum_{\alpha=1}^{n} w_\alpha^{\text{KM}} Z(\mathbf{x}_\alpha) - \sum_{\alpha=1}^{n} w_\alpha^{\text{KM}} Z(\mathbf{x}_\alpha) \sum_{\beta=1}^{n} w_\beta^{\text{SK}} \\ &= \sum_{\alpha=1}^{n} \Big[w_\alpha^{\text{SK}} + w_\alpha^{\text{KM}} \Big(1 - \sum_{\beta=1}^{n} w_\beta^{\text{SK}} \Big) \Big] Z(\mathbf{x}_\alpha). \end{aligned} \tag{11.24}$$

Introducing a weight

$$\overline{w} = 1 - \sum_{\alpha=1}^{n} w_\alpha^{\text{SK}}, \tag{11.25}$$

called the *weight of the mean*, we have

$$Z^\star_{\text{SKM}}(\mathbf{x}_0) = \sum_{\alpha=1}^{n} \Big[w_\alpha^{\text{SK}} + \overline{w}\, w_\alpha^{\text{KM}} \Big] Z(\mathbf{x}_\alpha) = \sum_{\alpha=1}^{n} w'_\alpha Z(\mathbf{x}_\alpha). \tag{11.26}$$

This looks like the estimator used for ordinary kriging.

We check whether the weights w'_α in this linear combination sum up to one

$$\sum_{\alpha=1}^{n} w'_\alpha = \sum_{\alpha=1}^{n} w_\alpha^{\text{SK}} + \overline{w} \underbrace{\sum_{\alpha=1}^{n} w_\alpha^{\text{KM}}}_{1} = \sum_{\alpha=1}^{n} w_\alpha^{\text{SK}} + 1 - \sum_{\alpha=1}^{n} w_\alpha^{\text{SK}} = 1. \tag{11.27}$$

We examine the possibility that the weights w'_α might be obtained from an ordinary kriging system

$$\begin{aligned} \sum_{\beta=1}^{n} w'_\beta\, C(\mathbf{x}_\alpha - \mathbf{x}_\beta) &= \sum_{\beta=1}^{n} w_\beta^{\text{SK}} C(\mathbf{x}_\alpha - \mathbf{x}_\beta) + \overline{w} \sum_{\beta=1}^{n} w_\beta^{\text{KM}} C(\mathbf{x}_\alpha - \mathbf{x}_\beta) \\ &= C(\mathbf{x}_\alpha - \mathbf{x}_0) + \overline{w}\, \mu_{\text{KM}}. \end{aligned} \tag{11.28}$$

Choosing to call μ' the product of the weight of the mean with the variance of the kriging of the mean, and putting the equations for w'_α together, we indeed have an ordinary kriging system

$$\begin{cases} \sum_{\beta=1}^{n} w'_\beta \, C(\mathbf{x}_\alpha - \mathbf{x}_\beta) = C(\mathbf{x}_\alpha - \mathbf{x}_0) + \mu' & \text{for } \alpha = 1, \ldots, n \\ \sum_{\beta=1}^{n} w'_\beta = 1, \end{cases} \tag{11.29}$$

which shows that ordinary kriging is identical with simple kriging based on the estimated mean: $Z^{\star}_{\text{SKM}}(\mathbf{x}_0) = Z^{\star}_{\text{OK}}(\mathbf{x}_0)$.

The ordinary kriging variance has the following decomposition

$$\sigma^2_{\text{OK}} \quad = \quad \sigma^2_{\text{SK}} + \boxed{\overline{w}^2 \, \sigma^2_{\text{KM}}}\,. \tag{11.30}$$

The ordinary kriging variance is the sum of the simple kriging variance (assuming a known mean) plus the variance due to the uncertainty about the true value of the mean. When the weight of the mean is small, the sum of the weights of simple kriging is close to one and ordinary kriging is close to the simple kriging solution, provided the variance of the kriged mean is also small.

Kriging the residual

Considering $Z(\mathbf{x})$ to be a second-order stationary random function, we can establish the following model

$$Z(\mathbf{x}) \quad = \quad \underbrace{m}_{\text{mean}} + \underbrace{Y(\mathbf{x})}_{\text{residual}} \qquad \text{with} \quad \mathrm{E}[\,Y(\mathbf{x})\,] = 0. \tag{11.31}$$

The mean is uncorrelated with the residual

$$\mathrm{E}[\,m \cdot Y(\mathbf{x})\,] \quad = \quad m\,\mathrm{E}[\,Y(\mathbf{x})\,] = 0, \tag{11.32}$$

because it is a deterministic quantity.

We know how to estimate $Z(\mathbf{x})$ by ordinary kriging and the mean m by a kriging of the mean. Now we would like to krige $Y(\mathbf{x})$ at some location $\mathbf{x}_0$ with the same type of weighted average

$$Y^{\star}(\mathbf{x}_0) \quad = \quad \sum_{\alpha=1}^{n} w_\alpha \, Z(\mathbf{x}_\alpha). \tag{11.33}$$

The unbiasedness is obtained by using weights summing up to zero (instead of one)

$$\begin{aligned}\mathrm{E}[\,Y(\mathbf{x}_0)-Y^\star(\mathbf{x}_0)\,] &= \underbrace{\mathrm{E}[\,Y(\mathbf{x}_0)\,]}_{0}-\sum_{\alpha=1}^{n} w_\alpha\,\underbrace{\mathrm{E}[\,Z(\mathbf{x}_\alpha)\,]}_{m}\\ &= -\,m\,\underbrace{\sum_{\alpha=1}^{n} w_\alpha}_{0}\;=0.\end{aligned} \tag{11.34}$$

The condition on the weights has the effect of removing the mean.

The estimation variance $\mathrm{var}(Y^\star(\mathbf{x}_0)-Y(\mathbf{x}_0))$ is identical to $\mathrm{var}(Z^\star(\mathbf{x}_0)-Z(\mathbf{x}_0))$ as $Y(\mathbf{x})$ has the same covariance function $C(\mathbf{h})$ as $Z(\mathbf{x})$. The system of the ***kriging of the residual*** (KR) is the same as for ordinary kriging, except for the condition on the sum of weights which is zero instead of one.

It is of interest to note that the ordinary kriging system can be decomposed into the kriging of the mean and the kriging of the residual:

$$\begin{pmatrix} C(\mathbf{x}_1-\mathbf{x}_1) & \dots & C(\mathbf{x}_1-\mathbf{x}_n) & 1\\ \vdots & \ddots & \vdots & \vdots\\ C(\mathbf{x}_n-\mathbf{x}_1) & \dots & C(\mathbf{x}_n-\mathbf{x}_n) & 1\\ 1 & \dots & 1 & 0\end{pmatrix}\cdot\left[\begin{pmatrix} w_1^{\mathrm{KM}}\\ \vdots\\ w_n^{\mathrm{KM}}\\ -\mu_{\mathrm{KM}}\end{pmatrix}+\begin{pmatrix} w_1^{\mathrm{KR}}\\ \vdots\\ w_n^{\mathrm{KR}}\\ -\mu_{\mathrm{KR}}\end{pmatrix}\right] = \begin{pmatrix}0\\ \vdots\\ 0\\ 1\end{pmatrix}+\begin{pmatrix} C(\mathbf{x}_1-\mathbf{x}_0)\\ \vdots\\ C(\mathbf{x}_n-\mathbf{x}_0)\\ 0\end{pmatrix} \tag{11.35}$$

The weights of ordinary kriging are composed of the weights of the krigings of the mean and the residual, and so do the Lagrange multipliers

$$w_\alpha^{\mathrm{OK}} = w_\alpha^{\mathrm{KM}}+w_\alpha^{\mathrm{KR}} \qquad \text{and} \qquad \mu_{\mathrm{OK}} = \mu_{\mathrm{KM}}+\mu_{\mathrm{KR}}. \tag{11.36}$$

The estimators are thus compatible at any location of the domain

$$Z^\star(\mathbf{x}_0) = m^\star+Y^\star(\mathbf{x}_0) \qquad \text{for all} \quad \mathbf{x}_0\in\mathcal{D}. \tag{11.37}$$

The kriging variances

$$\sigma^2_{\mathrm{KM}}=\mu_{\mathrm{KM}} \qquad \text{and} \qquad \sigma^2_{\mathrm{KR}}=C(0)-\sum_{\alpha=1}^{n} w_\alpha^{\mathrm{KR}}\,C(\mathbf{x}_\alpha-\mathbf{x}_0) \tag{11.38}$$

however do not add up in an elementary way.

Cross validation

Cross validation is a simple way to compare various assumptions either about the model (e.g. the type of variogram and its parameters, the size of the kriging neighborhood) or about the data (e.g. values that do not fit their neighborhood like outliers or pointwise anomalies).

In the cross validation procedure each sample value $Z(\mathbf{x}_\alpha)$ is removed in turn from the data set and a value $Z^\star(\mathbf{x}_{[\alpha]})$ at that location is estimated using the $n-1$ other samples. The square brackets around the index α symbolize the fact that the estimation is performed at location $\mathbf{x}_\alpha$ excluding the sampled value Z_α.

The difference between a data value and the estimated value

$$Z(\mathbf{x}_\alpha) \quad - \quad Z^\star(\mathbf{x}_{[\alpha]}) \tag{11.39}$$

gives an indication of how well the data value fits into the neighborhood of the surrounding data values.

If the average of the cross-validation errors is not far from zero,

$$\frac{1}{n}\sum_{\alpha=1}^{n}(Z(\mathbf{x}_\alpha) - Z^\star(\mathbf{x}_{[\alpha]})) \;\cong\; 0, \tag{11.40}$$

we can say that there is no apparent bias, while a significant negative (or positive) average error can represent systematic overestimation (respectively underestimation).

The magnitude of the mean squared cross-validation errors is interesting for comparing different estimations:

$$\frac{1}{n}\sum_{\alpha=1}^{n}(Z(\mathbf{x}_\alpha) - Z^\star(\mathbf{x}_{[\alpha]}))^2. \tag{11.41}$$

The kriging standard deviation $\sigma_{[\alpha]}$ represents the error predicted by the model when kriging at location $\mathbf{x}_\alpha$ (omitting the sample at the location $\mathbf{x}_\alpha$). Dividing the cross-validation error by $\sigma_{[\alpha]}$ allows to compare the magnitudes of both the actual and the predicted error:

$$\frac{Z(\mathbf{x}_\alpha) - Z^\star(\mathbf{x}_{[\alpha]})}{\sigma_{[\alpha]}}. \tag{11.42}$$

If the average of the squared standardized cross-validation errors is about one,

$$\frac{1}{n}\sum_{\alpha=1}^{n}\frac{\Big(Z(\mathbf{x}_\alpha) - Z^\star(\mathbf{x}_{[\alpha]})\Big)^2}{\sigma^2_{[\alpha]}} \;\cong\; 1, \tag{11.43}$$

the actual estimation error is equal on average to the error predicted by the model. This last quantity gives an idea about the adequacy of the model and of its parameters.

Kriging with known measurement error variance

Suppose we measure sample values z_α with a device indicating the precision σ_α of each measurement.

The model for each sample location is $Z(\mathbf{x}_\alpha) = Y(\mathbf{x}_\alpha) + \varepsilon_\alpha$, where ε_α is the measurement error, with a known variance $\text{var}(\varepsilon_\alpha) = \sigma^2_\alpha$, which may be different for each sample. The $Y(\mathbf{x}_\alpha)$ represents the quantity we actually wish to measure. We assume it is second-order stationary and shall suppose that we have access to its covariance function $C(\mathbf{h})$.

We need also to assume that the measurement error is not spatial,

$$\text{cov}(Y(\mathbf{x}_\alpha), \varepsilon_\alpha) = 0, \tag{11.44}$$

and that the errors are independent

$$\text{cov}(\varepsilon_\alpha, \varepsilon_\beta) = 0 \qquad \text{for} \quad \alpha \neq \beta. \tag{11.45}$$

We wish to filter the measurement error by kriging. This amounts to estimate $Y(\mathbf{x})$ from the data $Z(\mathbf{x}_\alpha)$ with the linear combination

$$Y^\star(\mathbf{x}_0) = \sum_{\alpha=1}^{n} w_\alpha Z(\mathbf{x}_\alpha). \tag{11.46}$$

The corresponding ordinary kriging system is

$$\begin{pmatrix} \boxed{C(\mathbf{x}_1-\mathbf{x}_1)+\sigma_1^2} & \dots & C(\mathbf{x}_1-\mathbf{x}_n) & 1 \\ \vdots & \ddots & \vdots & \vdots \\ C(\mathbf{x}_n-\mathbf{x}_1) & \dots & \boxed{C(\mathbf{x}_n-\mathbf{x}_n)+\sigma_n^2} & 1 \\ 1 & \dots & 1 & 0 \end{pmatrix} \cdot \begin{pmatrix} w_1 \\ \vdots \\ w_n \\ -\mu \end{pmatrix} = \begin{pmatrix} C(\mathbf{x}_1-\mathbf{x}_0) \\ \vdots \\ C(\mathbf{x}_n-\mathbf{x}_0) \\ 1 \end{pmatrix}. \tag{11.47}$$

In this system the error variances σ^2_α only show in the diagonal of the left hand matrix.

12 Kriging Weights

The behavior of kriging weights in 2D space is discussed with respect to the geometry of the sample/estimation locations, with isotropy or anisotropy and in view of the choice of the type of covariance function. The examples are borrowed from the thesis of RIVOIRARD [264] which contains many more.

Geometry

The very first aspect to discuss about kriging weights is the geometry of the sample points and the estimation point. Let us take a lozenge at the corners of which 4 samples are located. We wish to estimate a value at the center of the lozenge.

Nugget-effect covariance model

In the case of a nugget-effect model the ordinary kriging weights are shown on Figure 12.1. All weights are equal to $1/n$ and do not depend on the geometry.

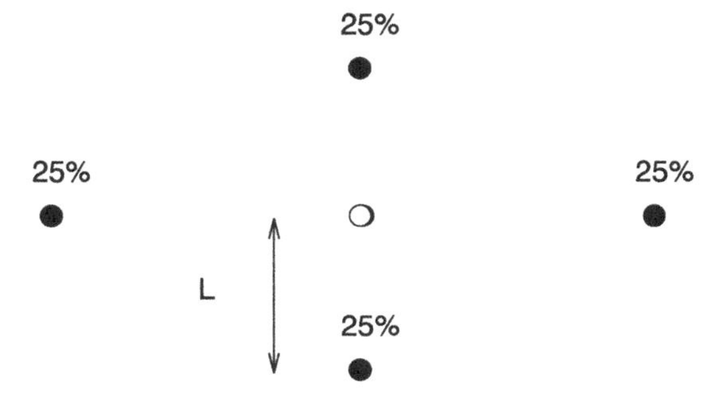

Figure 12.1: Ordinary kriging weights when using a nugget-effect model with samples located at the corners of a lozenge and the estimation point at its center. The kriging variance is $\sigma^2_{\mathrm{OK}} = 1.25\,\sigma^2$.

With a nugget-effect model, when the estimation point does not coincide with a data point, the ordinary kriging variance is equal to

$$\sigma^2_{\mathrm{OK}} \quad = \quad \mu_{\mathrm{OK}} + \sigma^2 = \frac{\sigma^2}{n} + \sigma^2. \tag{12.1}$$

It is larger than the variance σ^2.

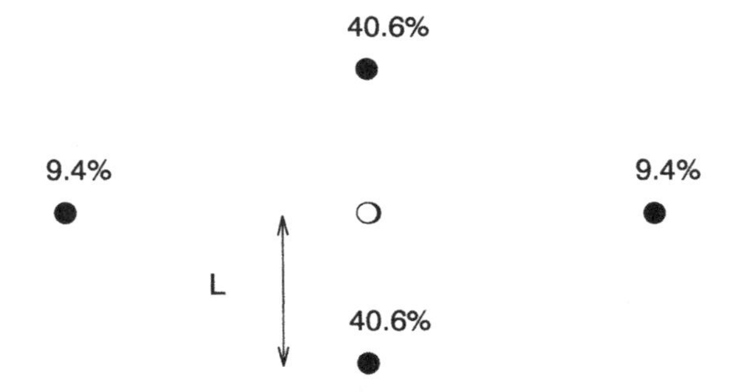

Figure 12.2: With a spherical model of range $a/L = 2$ the remote samples get lower weights. The ordinary kriging variance is $\sigma^2_{\mathrm{OK}} = .84\sigma^2$.

Spherical covariance model

Taking a spherical model of range $a/L = 2$ the weights of the samples at the remote corners of the lozenge get lower weights as to be seen on Figure 12.2. The associated ordinary kriging variance is $\sigma^2_{\mathrm{OK}} = .84\,\sigma^2$.

Gaussian covariance model

Using a Gaussian covariance model with range parameter $a/L = 1.5$ the remote samples have almost no influence on the estimated value at the center of the lozenge as shown on Figure 12.3.

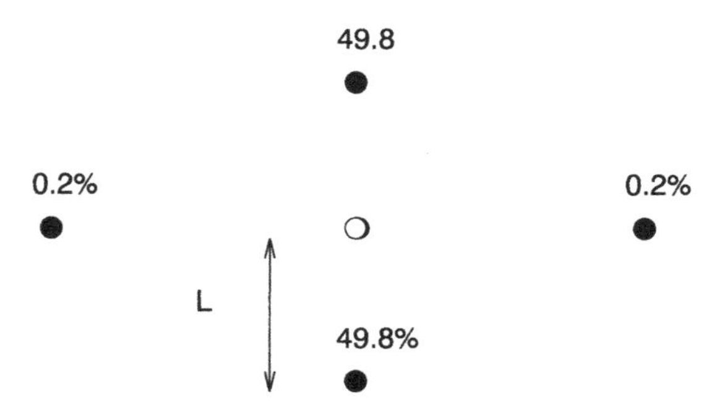

Figure 12.3: Using a Gaussian model with a range parameter $a/L = 1.5$ the remote samples have very little influence on the estimated value. The ordinary kriging variance is $\sigma^2_{\mathrm{OK}} = .30\,\sigma^2$.

The random function associated to the Gaussian model is an analytic function, which implies that the data are assumed to stem from an infinitely differentiable regionalized variable. Such a model is generally not realistic.

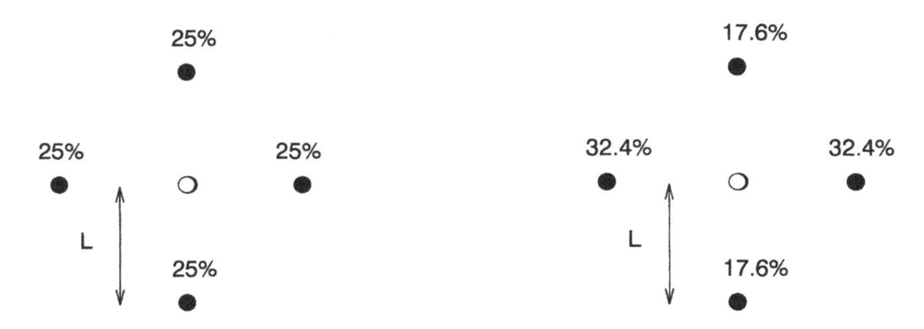

Figure 12.4: Comparing a configuration of four samples at the corners of a square using an isotropic model (on the left) and an anisotropic model (on the right).

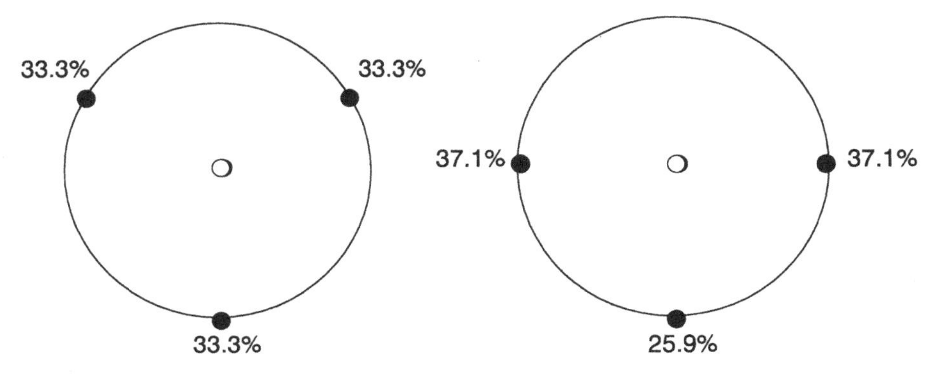

Figure 12.5: Three samples on a circle around the estimation location (on the left $\sigma^2_{\mathrm{OK}} = .45\,\sigma^2$ and on the right $\sigma^2_{\mathrm{OK}} = .48\,\sigma^2$).

Geometric anisotropy

On Figure 12.4 a configuration of four samples at the corners of a square with a central estimation point is used, taking alternately an isotropic (on the left) and an anisotropic model (on the right).

On the left, the isotropic model generates equal weights with this configuration where all sample points are symmetrically located with respect to the estimation point.

On the right, a spherical model with a range of $a/L = 1.5$ in the horizontal direction and a range of $a/L = .75$ in the vertical direction is used as the anisotropic model. The weights are weaker in the direction with the shorter range and this matches intuition.

Relative position of samples

This experiment compares the relative position of samples on a circle of radius L around the estimation point using an isotropic spherical model with a range of $3\,a/L$.

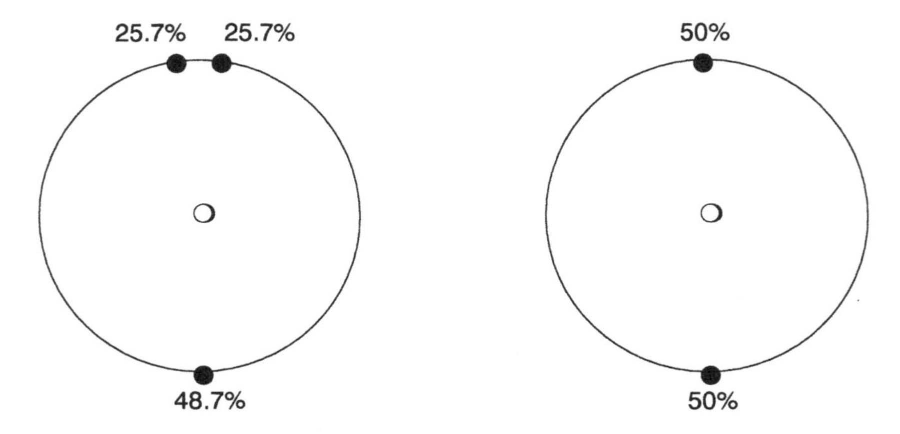

Figure 12.6: On the left, two samples have been set near to each other at the top (σ^2_{OK}= $.526\,\sigma^2$). On the right, the two upper samples have been merged into one single top sample (σ^2_{OK}= $.537\,\sigma^2$).

On the left of Figure 12.5 three samples are arranged in a symmetric way around the estimation point. The weights are identical and the ordinary kriging variance is $\sigma^2_{\text{OK}} = .45\,\sigma^2$. On the right, the two top samples have been moved down around the circle. The weights of these two samples increase because of the gap created at the top of the circle. The ordinary kriging variance is higher than in the symmetric configuration with $\sigma^2_{\text{OK}} = .48\,\sigma^2$.

On the left of Figure 12.6 On the left of Figure 12.6 the two top samples have been set very near to each other and their total weight is slightly higher than the weight of the bottom sample. The ordinary kriging variance is $\sigma^2_{\text{OK}} = .526\,\sigma^2$. On the right of Figure 12.6 the two upper points have actually been merged into a single sample. Symmetry is reestablished: the top and bottom sample locations each share half of the weight. Merging the two upper samples into one top sample has only slightly increased the ordinary kriging variance to $\sigma^2_{\text{OK}} = .537\,\sigma^2$.

Screen effect

A remarkable feature of kriging, which makes it different from other interpolators, is that a sample can screen off other samples located behind it with respect to the estimation location. An example of this phenomenon is given on Figure 12.7 showing two 1D sampling configurations. A spherical model of range parameter $a/L= 2$ was used, where L is the distance from the point A to the estimation location.

At the top of Figure 12.7, two samples A and B are located at different distances of the estimation point and get different ordinary kriging weights of 65.6% and 34.4% according to their proximity to that point. The kriging variance is $\sigma^2_{\text{OK}} = 1.14\,\sigma^2$.

At the bottom of Figure 12.7, a third sample C has been added to the configuration: the weight of B drops down to 2.7% and almost all weight is distributed fairly equally

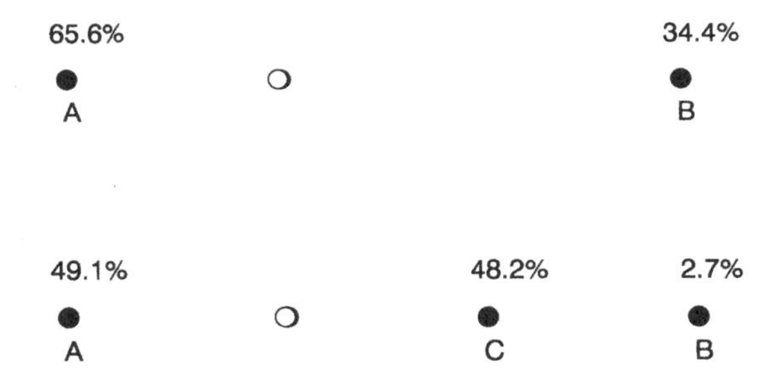

Figure 12.7: At the top, we have two samples at different distances from the estimation point ($\sigma^2_{\mathrm{OK}}= 1.14\,\sigma^2$). At the bottom, a third sample has been added to the configuration ($\sigma^2_{\mathrm{OK}}= .87\,\sigma^2$).

between A and C which are at the same distance from the estimation point. The ordinary kriging variance is brought down to $\sigma^2_{\mathrm{OK}} = .87\,\sigma^2$.

An interpolator based on weights built on the inverse of the distance of a sample to the estimation point does not show screen effects. The weights are shaped without taking any account of the varying amount of samples in different directions around the estimation point.

Factorizable covariance functions

With the 1D configuration at the bottom of Figure 12.7 a total screen off of B by C could be obtained by using an exponential covariance function (instead of the spherical model) and simple kriging (instead of ordinary kriging). This is due to the factorization of the exponential function $C(h) = \exp(-|h|)$:

$$C(x_0, x_{\mathrm{B}}) \quad = \quad C(x_0, x_{\mathrm{C}}) \cdot C(x_{\mathrm{C}}, x_{\mathrm{B}}), \tag{12.2}$$

which implies the conditional independence of $Z(x_0)$ and $Z(x_{\mathrm{B}})$ with respect to $Z(x_{\mathrm{C}})$ for a stationary Gaussian random function $Z(x)$.

In 2D, with $\mathbf{h}= (h_1, h_2)^\top$, the so-called "Gaussian" covariance function (because of the analogy with the distribution of the same name),

$$C(\mathbf{h}) \quad = \quad \mathrm{e}^{-|\mathbf{h}|^2}, \tag{12.3}$$

can be factorized with respect to the two spatial coordinates:

$$C(\mathbf{h}) \quad = \quad \mathrm{e}^{-(h_1)^2 - (h_2)^2} = \mathrm{e}^{-(h_1)^2} \cdot \mathrm{e}^{-(h_2)^2} = C(h_1) \cdot C(h_2). \tag{12.4}$$

An example of simple kriging weights with a Gaussian covariance function is shown on Figure 12.8. The samples D, E, F are completely screened off by the samples A, B, C. The latter are the orthogonal projections of the former on the abscissa of a coordinate system centered on the estimation point. It could be thought that total

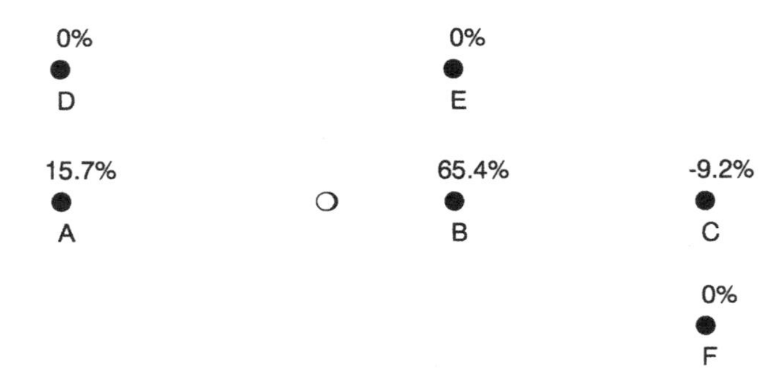

Figure 12.8: Simple kriging with a Gaussian covariance model: samples A, B, C screen off D, E, F (in simple kriging the weights are unconstrained and do not add up to 100%).

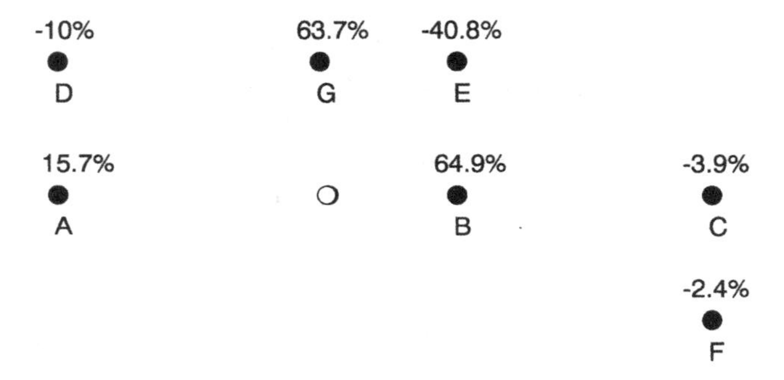

Figure 12.9: Simple kriging with a Gaussian covariance model: adding the point G radically changes the distribution of weights.

screen off will act on samples whose projection on either one of the coordinate axes (centered on the estimation location) is a sample location. There is no such rule: adding a point G above the estimation location as on Figure 12.9 changes radically the overall distribution of weights.

Negative kriging weights

It is interesting to note that many points on Figure 12.9 have received a negative weight. Negative kriging weights are an interesting feature as they make extrapolation possible out of the range of the data values. The counterpart is that when a variable is only defined for positive values (like grades of ore), it may happen that kriged values are slightly negative and need to be readjusted.

COMMENT 12.1 *A formulation of kriging providing only positive weights was given by* BARNES & JOHNSON *[22].* CHAUVET *[43] reviews causes and possible solutions to the presence of significant negative weights, while* MATHERON *[219] examines the question of the existence of covariance, variogram or generalized covariance models that ensure positive kriging weights.*

13 Mapping with Kriging

Kriging can be used as an interpolation method to estimate values on a regular grid using irregularly spaced data. Spatial data may be treated locally by defining a neighborhood of samples around each location of interest in the domain.

Kriging for spatial interpolation

Kriging is certainly not the quickest method for spatial interpolation.

A simpler method, for example, is the *inverse distance interpolation*, which consists in weighting each data by the inverse of the distance to the estimation location (scaling the weights to be unit sum). One could establish an alternate method by using the inverse *squared* distance or, more generally, a power p of the inverse distance. This raises the question of how to choose the parameter p and there is no simple answer. Geostatistics prevents such a problem by starting rightaway with an analysis and an interpretation of the data to determine the parameters entering the interpolation algorithm, i.e. kriging.

The advantages of the geostatistical approach to interpolation are thus:

- kriging is preceded by an analysis of the spatial structure of the data. The representation of the average spatial variability is integrated into the estimation procedure in the form of a variogram model.

- ordinary kriging interpolates exactly: when a sample value is available at the location of interest, the kriging solution is equal to that value.

- kriging, as a statistical method, provides an indication of the estimation error: the kriging standard deviation, which is the square root of the kriging variance.

How is kriging actually used for generating a map?

A regular grid is defined on the computer as shown on Figure 13.1. Each node of the grid becomes the point $\mathbf{x}_0$ in turn and a value is kriged at that location. The result is a map like on Figure 13.2. Here a raster representation of the kriged grid was chosen. Each shaded square is centered on the node of the grid and is shaded with a different grey tone according to the value estimated at that location.

Figure 13.3 shows a raster representation of the corresponding kriging standard deviations. This map of the theoretical kriging estimation errors allows to evaluate the precision of the estimation in any part of the region. It is a useful compendium

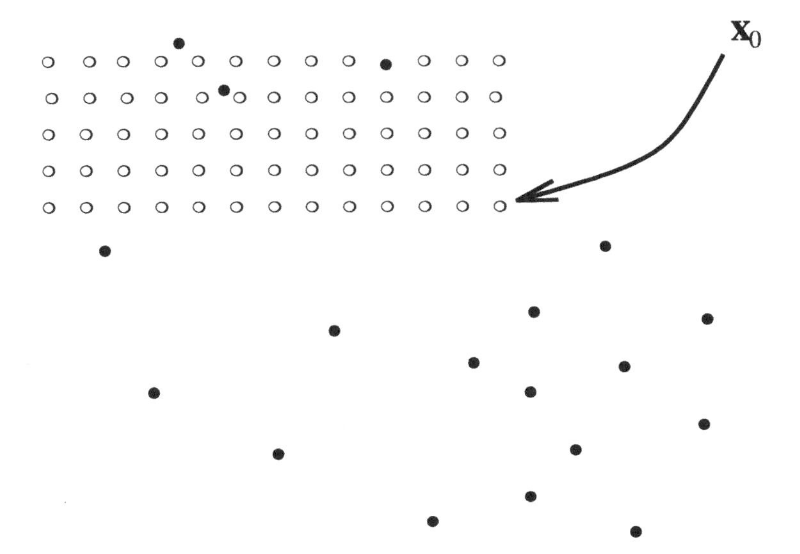

Figure 13.1: To make a map, a regular grid is built for the region of interest and each node becomes the node $\mathbf{x}_0$ in turn.

of the map of the kriged values. It should be understood that the kriging variance is primarily a measure of the density of information around the estimation point.

The white squares on Figure 13.3 correspond to nodes which coincide with a sample location: the kriging variances are zero at these points where exact interpolation takes place.

For a better visual understanding of the spatial structure of the kriged values the grid of regularly spaced values is generally submitted to a second interpolation step. One possibility is to refine the initial grid and to interpolate linearly the kriged values on the finer grid like on the raster representation on Figure 13.4. The result is a smoother version of the raster initially shown on Figure 13.2.

Another way to represent the results of kriging in a smooth way is to contour them with different sets of isolines as shown on Figure 13.5 (the arsenic data has been standardized before kriging: this is why part of the values are negative).

Neighborhood

It may be asked whether it is necessary to involve all samples in the estimation procedure. More precisely, with a spherical covariance model in mind, as the covariance is zero for distances greater than the range, is it necessary to involve samples which are far away from a location of interest $\mathbf{x}_0$?

It seems appealing to draw a circle (sphere) or an ellipse (ellipsoid) around $\mathbf{x}_0$ and to consider only samples which lie within this area (volume) as neighbors of $Z(\mathbf{x}_0)$. Locations whose covariance with $\mathbf{x}_0$ is zero are uncorrelated and the corresponding

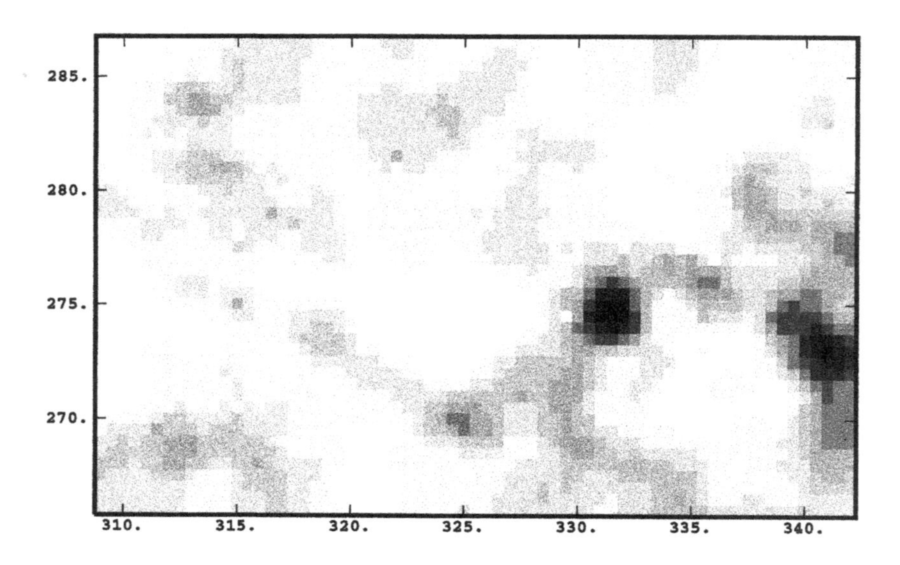

Figure 13.2: Raster representation of kriged values of arsenic: a square is centered on each node of a .5 km × .5 km grid and shaded proportionally to the kriging result.

samples have no direct influence on $Z(\mathbf{x}_0)$. Thus it makes sense to try to define for each location $\mathbf{x}_0$ a local neighborhood of samples nearest to $\mathbf{x}_0$.

At first sight, with the spherical covariance model in mind, the radius of the neighborhood centered on $\mathbf{x}_0$ could be set equal to the range of the covariance function and samples further away could be excluded as represented on Figure 13.6. However, the spherical covariance function expresses only the direct covariances between points, but not the partial covariances of the estimation point with a data point conditionally on the other sample points. The partial correlations are reflected in the kriging weights, which are in general not zero for points beyond the range from the estimation location.

In practice, because of the generally highly irregular spatial arrangement and density of the data, the definition of the size of a local neighborhood is not straightforward. A criterion in common use for data on a regular mesh is to check for a given estimation point if adding more samples leads to a significant decrease of the kriging variance, i.e. an increase in the precision of the estimation. Cross validation criteria can also be used to compare different neighborhoods.

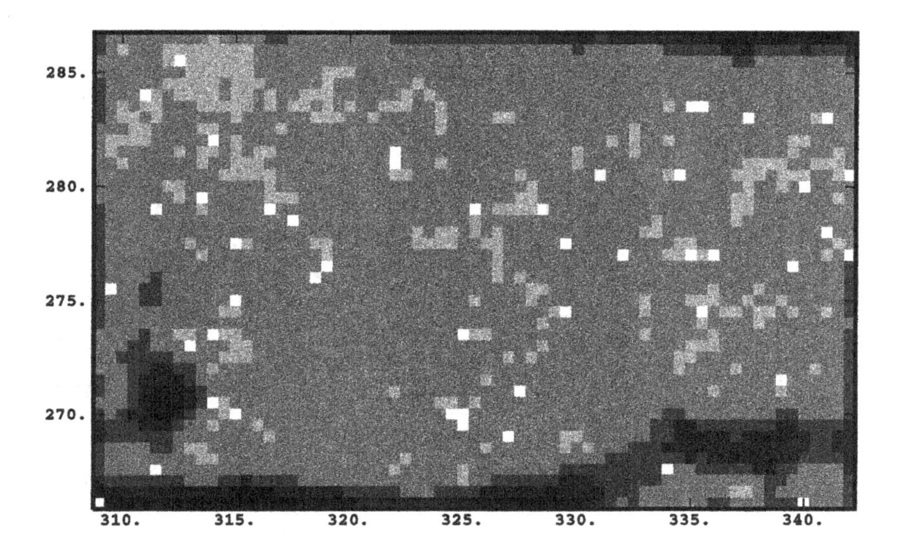

Figure 13.3: Raster representation of the kriging standard deviations.

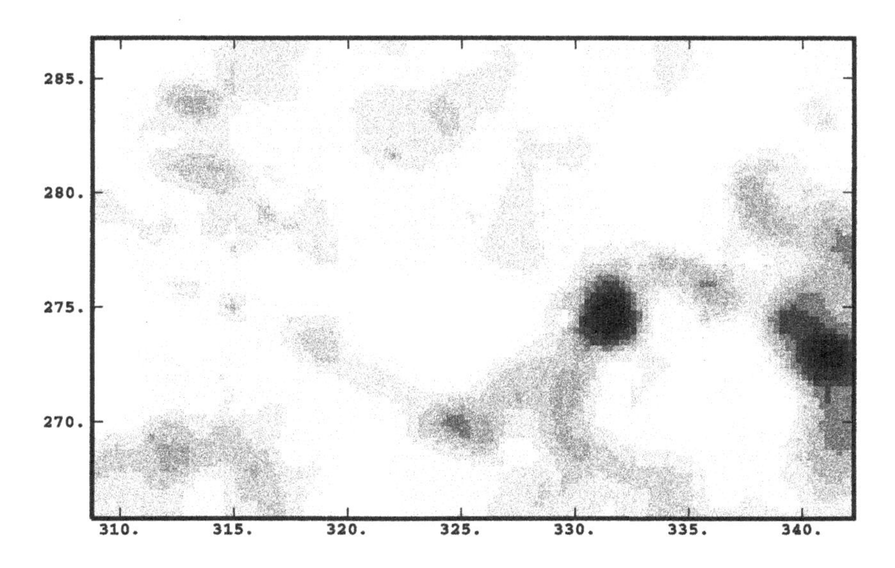

Figure 13.4: Smoothed version of the raster representation of arsenic: a finer grid is used and the kriged values are interpolated linearly on it.

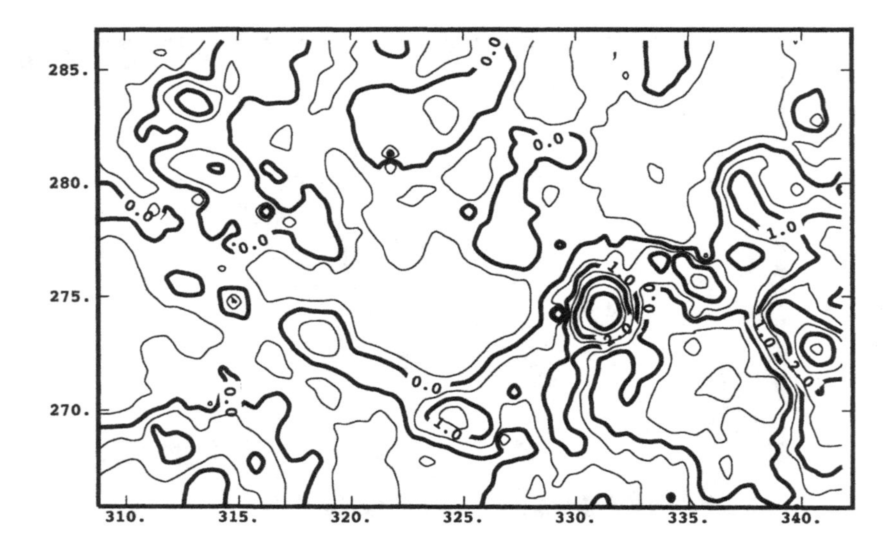

Figure 13.5: Isoline representation of kriged values of arsenic: the kriged values at the grid nodes are contoured with a set of thin and thick isolines (arsenic data was reduced to zero mean and unit variance before kriging).

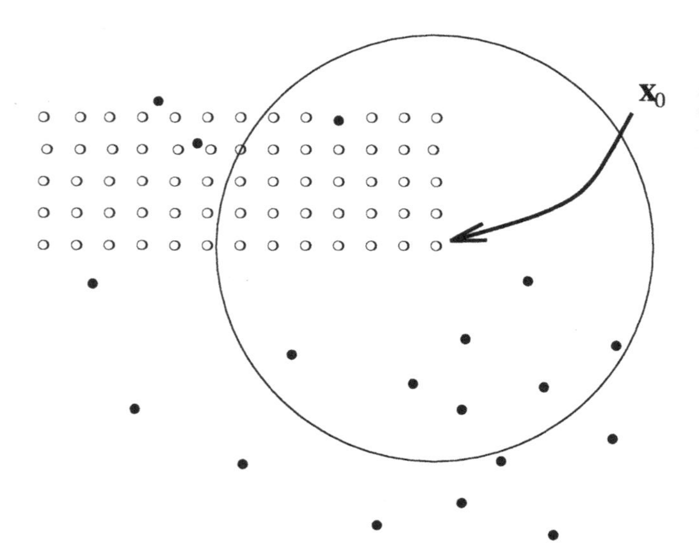

Figure 13.6: Around a location of interest $\mathbf{x}_0$ of the grid a circular neighborhood is defined and the samples within the circle are selected for kriging.

14 Linear Model of Regionalization

A regionalized phenomenon can be thought as being the sum of several independent subphenomena acting at different characteristic scales. A linear model is set up which splits the random function representing the phenomenon into several uncorrelated random functions, each with a different variogram or covariance function. In subsequent chapters we shall see that it is possible to estimate these spatial components by kriging, or to filter them. The linear model of regionalization brings additional insight about kriging when the purpose is to create a map using data irregularly scattered in space.

Spatial anomalies

In geochemical prospecting or in environmental monitoring of soil the aim is to detect zones with a high average grade of mineral or pollutant. A *spatial anomaly* with a significant grade and a relevant spatial extension needs to be detected. We now discuss the case of geochemical exploration.

When prospecting for mineral deposits a distinction can be made between geological, superficial and geochemical anomalies as illustrated on Table 14.1.

A deposit is a *geological anomaly* with a significant average content of a given raw material and enough spatial extension to have economic value. Buried at a specific depth, the geological body may be detectable at the surface by indices, the *superficial anomalies*, which can be isolated points, linear elements or dispersion haloes.

Geological and superficial anomalies correspond to a vision of the geological phenomenon in its full continuity. Yet in prospecting only a discrete perception of the phenomenon is possible through samples spread over the region. A superficial anomaly can be apprehended by one or several samples or it can escape the grip of the prospector when it is located between sample points.

A geochemical anomaly, in the strict sense, only exists at the sample points and we can distinguish between:

- *pointwise anomalies* defined on single samples, and

- *groupwise anomalies* defined on several neighboring samples.

If we think of the pointwise and the groupwise anomalies as having been generated by two different uncorrelated spatial processes, we may attribute a nugget-effect to the presence of pointwise anomalies, a short range structure to groupwise anomalies,

3D		2D		Samples
GEOLOGICAL ANOMALY	$\longmapsto$	SUPERFICIAL ANOMALY	$\longmapsto$	GEOCHEMICAL ANOMALY
		isolated spot	$\rightarrow$	point
	$\nearrow$		$\nearrow$	
deposit	$\rightarrow$	fault		
	$\searrow$		$\searrow$	
		aureola	$\rightarrow$	group of points

Table 14.1: Spatial anomalies in geochemical prospecting.

while a further large scale structure may be explained by the geochemical background variation.

Such associations between components of a variogram model and potential spatial anomalies should be handled in a loose manner, merely as a device to explore the spatial arrangement of high and less high values.

Nested variogram model

Often several sills can be distinguished on the experimental variogram and related to the morphology of the regionalized variable. In a pioneering paper SERRA [299] has investigated spatial variation in the Lorraine iron deposit and found up to seven sills, each with a geological interpretation, in the multiple transitions between the micrometric and the kilometric scales.

Let us first define the correlation function $\rho(\mathbf{h})$, obtained by normalizing the covariance function with its value $b = C(0)$ at the origin,

$$\rho(\mathbf{h}) \quad = \quad \frac{C(\mathbf{h})}{b}, \qquad \text{so that} \quad C(\mathbf{h}) = b\,\rho(\mathbf{h}). \tag{14.1}$$

Different sills b_u observed on the experimental variogram are numbered with an index $u = 0, ..., S$. A *nested variogram* is set up by adding S+1 elementary variograms with different coefficients b_u

$$\gamma(\mathbf{h}) \quad = \quad \sum_{u=0}^{S} \gamma_u(\mathbf{h}) = \sum_{u=0}^{S} b_u\, g_u(\mathbf{h}), \tag{14.2}$$

where the $g_u(\mathbf{h})$ are *normalized variograms*, i.e. elementary variogram models with a slope, sill (or asymptotic sill) normalized to one. The coefficients b_u express explicitly the actual value of the nugget-effect, sill or slope.

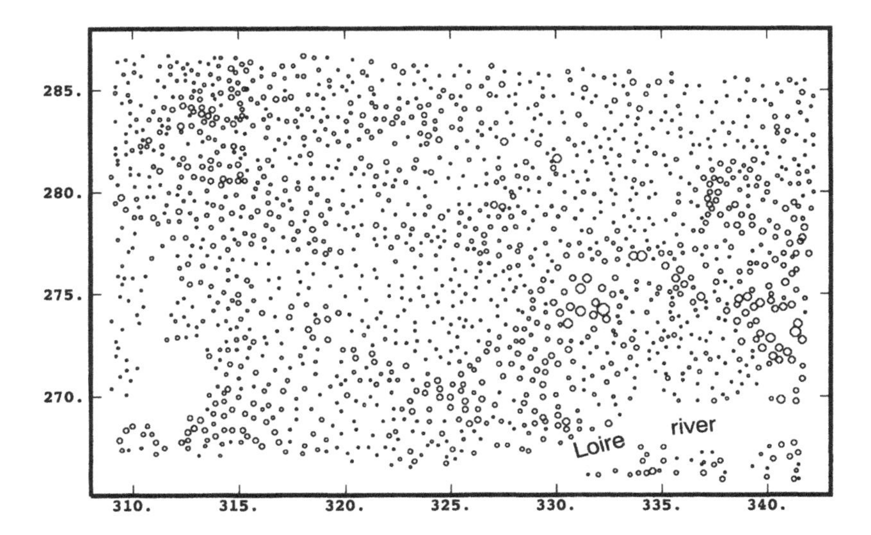

Figure 14.1: Display of the values of arsenic measured on soil samples in a 35 km × 25 km region near the Loire river in France.

If the variograms $\gamma_u(\mathbf{h})$ can be deduced from covariance functions $C_u(\mathbf{h})$, the nested variogram becomes

$$\gamma(\mathbf{h}) \;=\; b - \sum_{u=0}^{S} C_u(\mathbf{h}) = b - \sum_{u=0}^{S} b_u\, \rho_u(\mathbf{h}), \qquad \text{where} \quad b = \sum_{u=0}^{S} b_u, \tag{14.3}$$

with, in a typical case, a nugget-effect covariance function $C_0(\mathbf{h})$ and several spherical covariance functions $C_u(\mathbf{h})$ having different ranges a_u.

On Figure 14.1 a map of arsenic samples taken in soil is shown, covering a region near the Loire river. The arsenic sample locations are represented with circles proportional to the detected values. The mean experimental variogram of this variable is seen on Figure 14.2. The variogram has been fitted with a nested model consisting of a nugget effect plus two spherical functions with a range of 3.5 km and 6.5 km.

Decomposition of the random function

The random function $Z(\mathbf{x})$ associated with a nested variogram model is a sum of spatial components characterizing different spatial scales, i.e. reaching different sills of variation b_u at different scales, except maybe for the last coefficient b_S, which could represent the slope of an unbounded variogram model.

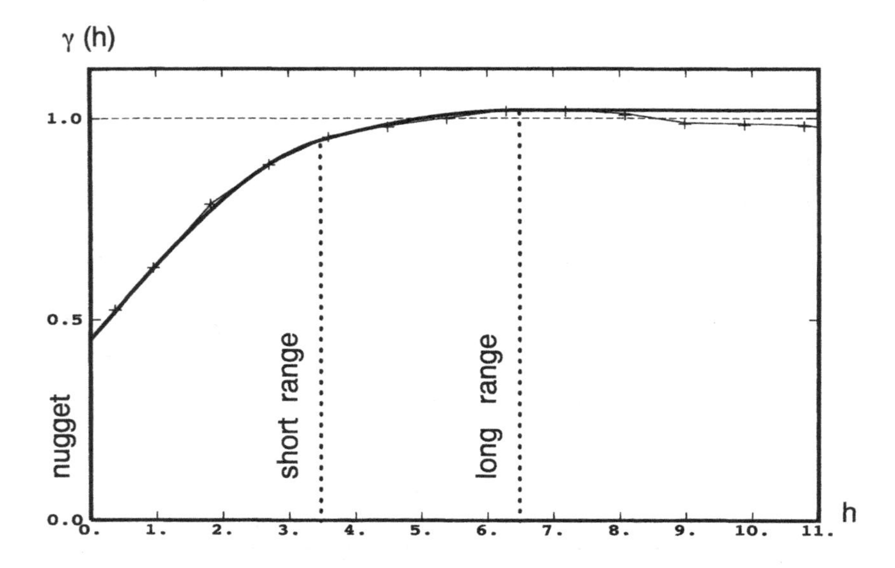

Figure 14.2: Variogram of the arsenic measurements fitted with a nested model consisting of a nugget-effect plus two spherical models of 3.5 km and 6.5 km range.

It is clear that a small-scale component can only be identified if the sampling grid is sufficiently fine. Equally, a large-scale component will only be visible on the variogram if the diameter of the sampled domain is large enough.

A deterministic component (mean m or drift $m(\mathbf{x})$) can also be part of the decomposition.

In the following we present several linear models of regionalization, obtained by assembling together spatial components of different types.

Second-order stationary regionalization

A second-order stationary random function $Z(\mathbf{x})$ can be built by adding uncorrelated zero mean second-order stationary random functions $Z_u(\mathbf{x})$ to a constant m representing the expectation of $Z(\mathbf{x})$

$$Z(\mathbf{x}) \quad = \quad Z_0(\mathbf{x}) + \ldots + Z_u(\mathbf{x}) + \ldots + Z_S(\mathbf{x}) + m, \tag{14.4}$$

where $\operatorname{cov}(Z_u(\mathbf{x}), Z_v(\mathbf{x}+\mathbf{h})) = 0$ for $u \neq v$.

A simple computation shows that the corresponding covariance model is

$$C(\mathbf{h}) \quad = \quad C_0(\mathbf{h}) + \ldots + C_u(\mathbf{h}) + \ldots + C_S(\mathbf{h}). \tag{14.5}$$

Take a random function model with two uncorrelated components

$$Z(\mathbf{x}) \quad = \quad Z_1(\mathbf{x}) + Z_2(\mathbf{x}) + m, \tag{14.6}$$

with $\mathrm{E}[\,Z_1(\mathbf{x})\,Z_2(\mathbf{x}+\mathbf{h})\,] = 0$ and $\mathrm{E}[\,Z_1(\mathbf{x})\,] = \mathrm{E}[\,Z_2(\mathbf{x})\,] = 0$.

Then

$$\begin{aligned}
C(\mathbf{h}) &= \mathrm{E}[\,Z(\mathbf{x}+\mathbf{h})\,Z(\mathbf{x})\,] - m^2 \\
&= \mathrm{E}[\,Z_1(\mathbf{x}+\mathbf{h})\,Z_1(\mathbf{x})\,] + \mathrm{E}[\,Z_2(\mathbf{x}+\mathbf{h})\,Z_2(\mathbf{x})\,] + m^2 \\
&\quad + \mathrm{E}[\,Z_1(\mathbf{x}+\mathbf{h})\,Z_2(\mathbf{x})\,] + \mathrm{E}[\,Z_2(\mathbf{x}+\mathbf{h})\,Z_1(\mathbf{x})\,] \\
&\quad + \mathrm{E}[\,Z_1(\mathbf{x}+\mathbf{h})\,]\,m + \mathrm{E}[\,Z_2(\mathbf{x}+\mathbf{h})\,]\,m \\
&\quad + m\,\mathrm{E}[\,Z_1(\mathbf{x})\,] + m\,\mathrm{E}[\,Z_2(\mathbf{x})\,] - m^2 \\
&= C_1(\mathbf{h}) + C_2(\mathbf{h}). \qquad (14.7)
\end{aligned}$$

Intrinsic regionalization

In the same way an intrinsic regionalization is defined as a sum of $S+1$ components,

$$Z(\mathbf{x}) = Z_0(\mathbf{x}) + \ldots + Z_u(\mathbf{x}) + \ldots + Z_S(\mathbf{x}), \qquad (14.8)$$

for which the increments are zero on average and uncorrelated.

For a two component model,

$$Z(\mathbf{x}) = Z_1(\mathbf{x}) + Z_2(\mathbf{x}), \qquad (14.9)$$

we have the variogram

$$\begin{aligned}
\gamma(\mathbf{h}) &= \mathrm{E}[\,(Z(\mathbf{x}+\mathbf{h}) - Z(\mathbf{x}))^2\,] \\
&= \mathrm{E}\Big[\,\Big((Z_1(\mathbf{x}+\mathbf{h}) - Z_1(\mathbf{x})) + (Z_2(\mathbf{x}+\mathbf{h}) - Z_2(\mathbf{x}))\Big)^2\,\Big] \\
&= \mathrm{E}[\,(Z_1(\mathbf{x}+\mathbf{h}) - Z_1(\mathbf{x}))^2\,] + \mathrm{E}[\,(Z_2(\mathbf{x}+\mathbf{h}) - Z_2(\mathbf{x}))^2\,] \\
&= \gamma_1(\mathbf{h}) + \gamma_2(\mathbf{h}), \qquad (14.10)
\end{aligned}$$

because $\mathrm{E}[\,(Z_1(\mathbf{x}+\mathbf{h}) - Z_1(\mathbf{x}))\cdot(Z_2(\mathbf{x}+\mathbf{h}) - Z_2(\mathbf{x}))\,] = 0$.

Intrinsic regionalization with mostly stationary components

A particular intrinsic random function $Z(\mathbf{x})$ can be constructed by putting together one intrinsic random function $Z_S(\mathbf{x})$ and S second-order stationary functions $Z_u(\mathbf{x}), u = 0, \ldots, S-1$, having means equal to zero and being uncorrelated amongst themselves as well as with the increments of the intrinsic component

$$Z(\mathbf{x}) = Z_0(\mathbf{x}) + \ldots + Z_u(\mathbf{x}) + \ldots + Z_S(\mathbf{x}). \qquad (14.11)$$

The associated variogram model is composed of S elementary structures deduced from covariance functions plus a purely intrinsic structure.

For this mixed linear regionalization model with second order stationary and intrinsically stationary components, we shall develop kriging systems to estimate both types of components in the next chapter.

Locally stationary regionalization

A non-stationary random function $Z(\mathbf{x})$ can be obtained by superposing $S{+}1$ uncorrelated second-order zero mean random functions $Z_u(\mathbf{x})$ and a drift $m(\mathbf{x})$ (the expectation of $Z(\mathbf{x})$ at the location $\mathbf{x}$)

$$Z(\mathbf{x}) \;=\; Z_0(\mathbf{x}) + \ldots + Z_u(\mathbf{x}) + \ldots + Z_S(\mathbf{x}) + m(\mathbf{x}). \qquad (14.12)$$

By assuming a very smooth function $m_l(\mathbf{x})$, which varies slowly from one end of the domain to the other end, the drift is locally almost constant and the random function

$$Z(\mathbf{x}) \;=\; Z_0(\mathbf{x}) + \ldots + Z_u(\mathbf{x}) + \ldots + Z_S(\mathbf{x}) + m_l(\mathbf{x}) \qquad (14.13)$$

can be termed *locally stationary* (i.e. locally second-order stationary), the expectation of $Z(\mathbf{x})$ being approximatively equal to a constant inside any rather small neighborhood of the domain.

The locally stationary regionalization model is the adequate framework for the typical application with the following three steps

1. the experimental variogram is used to describe spatial variation;
2. only covariance functions (i.e. bounded variogram functions) are fitted to the experimental variogram;
3. a moving neighborhood is used for ordinary kriging.

In the first step the experimental variogram filters out $m_l(\mathbf{x})$ at short distances, because it is approximatively equal to a constant in a local neighborhood. For distances up to the radius of the neighborhood the variogram thus estimates well the underlying covariance functions, which are fitted in the second step. In the third step the ordinary kriging assumes implicitly the existence of a local mean within a moving neighborhood. This local mean corresponds to the constant to which $m_l(\mathbf{x})$ is approximatively equal within not too large a neighborhood. The local mean can be estimated explicitly by a kriging of the mean.

15 Kriging Spatial Components

The components of regionalization models can be extracted by kriging. The extraction of a component of the spatial variation is a complementary operation to the filtering out (rejection) of the other components. This is illustrated by an application on geochemical data.

Kriging of the intrinsic component

For $Z(\mathbf{x})$ defined in the framework of the intrinsic regionalization model with mostly stationary components (presented on page 105), we may want to estimate the intrinsic component $Z_S(\mathbf{x})$ from data about $Z(\mathbf{x})$ in a neighborhood

$$Z_S^\star(\mathbf{x}_0) = \sum_{\alpha=1}^{n} w_\alpha^S \, Z(\mathbf{x}_\alpha). \tag{15.1}$$

The expectation of the estimation error is nil using weights that sum up to one,

$$\begin{aligned}
\mathrm{E}[\, Z_S^\star(\mathbf{x}_0) - Z_S(\mathbf{x}_0)\,] &= \mathrm{E}\Big[\sum_{\alpha=1}^{n} w_\alpha^S \, Z(\mathbf{x}_\alpha) - Z_S(\mathbf{x}_0)\cdot \underbrace{\sum_{\alpha=1}^{n} w_\alpha^S}_{1}\Big] \\
&= \sum_{\alpha=1}^{n} w_\alpha^S \, \mathrm{E}[\, Z(\mathbf{x}_\alpha) - Z_S(\mathbf{x}_0)\,] \\
&= \sum_{\alpha=1}^{n} w_\alpha^S \Big(\sum_{u=0}^{S-1} \underbrace{\mathrm{E}[\, Z_u(\mathbf{x}_\alpha)\,]}_{0} + \underbrace{\mathrm{E}[\, Z_S(\mathbf{x}_\alpha) - Z_S(\mathbf{x}_0)\,]}_{0}\Big) \\
&= 0.
\end{aligned} \tag{15.2}$$

Remember that the constraint of unit sum weights is also needed for the existence of a variogram as a conditionally negative definite function as explained in the presentation of ordinary kriging.

The estimation variance σ_{E}^2 is

$$\begin{aligned}
\sigma_{\mathrm{E}}^2 &= \mathrm{var}(Z_S^\star(\mathbf{x}_0) - Z_S(\mathbf{x}_0)) \\
&= \mathrm{E}\Big[\Big(\sum_{u=0}^{S-1}\sum_{\alpha=1}^{n} w_\alpha^S \, Z_u(\mathbf{x}_\alpha) + \sum_{\alpha=1}^{n} w_\alpha^S \,(Z_S(\mathbf{x}_\alpha) - Z_S(\mathbf{x}_0))\Big)^2\Big].
\end{aligned} \tag{15.3}$$

Taking into account the non correlation between components

$$\sigma_{\mathrm{E}}^2 = \sum_{u=0}^{S-1}\sum_{\alpha=1}^{n}\sum_{\beta=1}^{n} w_\alpha^S w_\beta^S C^u(\mathbf{x}_\alpha-\mathbf{x}_\beta) \tag{15.4}$$
$$- \sum_{\alpha=1}^{n}\sum_{\beta=1}^{n} w_\alpha^S w_\beta^S \gamma^S(\mathbf{x}_\alpha-\mathbf{x}_\beta) - \gamma^S(\mathbf{x}_0-\mathbf{x}_0) + 2\sum_{\alpha=1}^{n} w_\alpha^S \gamma^S(\mathbf{x}_\alpha-\mathbf{x}_0).$$

We can replace the covariances by $C^u(\mathbf{x}_0-\mathbf{x}_0) - \gamma^u(\mathbf{x}_\alpha-\mathbf{x}_\beta)$ because it is always possible to construct a variogram from a covariance function and we get

$$\sigma_{\mathrm{E}}^2 = \sum_{u=0}^{S-1} C^u(\mathbf{x}_0-\mathbf{x}_0) \tag{15.5}$$
$$- \gamma^S(\mathbf{x}_0-\mathbf{x}_0) - \sum_{\alpha=1}^{n}\sum_{\beta=1}^{n} w_\alpha^S w_\beta^S \gamma(\mathbf{x}_\alpha-\mathbf{x}_\beta) + 2\sum_{\alpha=1}^{n} w_\alpha^S \gamma^S(\mathbf{x}_\alpha-\mathbf{x}_0).$$

Minimizing σ_{E}^2 with the constraint on the weights, we have the system

$$\begin{cases} \displaystyle\sum_{\beta=1}^{n} w_\beta^S \gamma(\mathbf{x}_\alpha-\mathbf{x}_\beta) + \mu_S = \gamma^S(\mathbf{x}_\alpha-\mathbf{x}_0) & \text{for } \alpha = 1,\ldots,n \\ \displaystyle\sum_{\beta=1}^{n} w_\beta^S = 1. \end{cases} \tag{15.6}$$

The difference with ordinary kriging is that the terms $\gamma^S(\mathbf{x}_\alpha-\mathbf{x}_0)$, which are specific of the component $Z_S(\mathbf{x})$, appear in the right hand side.

COMMENT 15.1 *If the variogram of the intrinsic component $\gamma_S(\mathbf{h})$ is bounded we can replace it with a covariance function $C_S(\mathbf{h})$ using the relation*

$$\gamma_S(\mathbf{h}) = C_S(\mathbf{0}) - C_S(\mathbf{h}). \tag{15.7}$$

The kriging system for extracting the intrinsic component $Z_S(\mathbf{x})$ can now be viewed in the framework of the locally stationary regionalization model: the system estimates a second order stationary component together with the local mean.

Kriging of a second-order stationary component

For the purpose of kriging a particular second-order stationary component $Z_{u_0}(\mathbf{x})$ in the framework of the intrinsic regionalization model with mostly stationary components (see page 105), we start with the linear combination

$$Z_{u_0}^\star(\mathbf{x}_0) = \sum_{\alpha=1}^{n} w_\alpha^{u_0} Z(\mathbf{x}_\alpha). \tag{15.8}$$

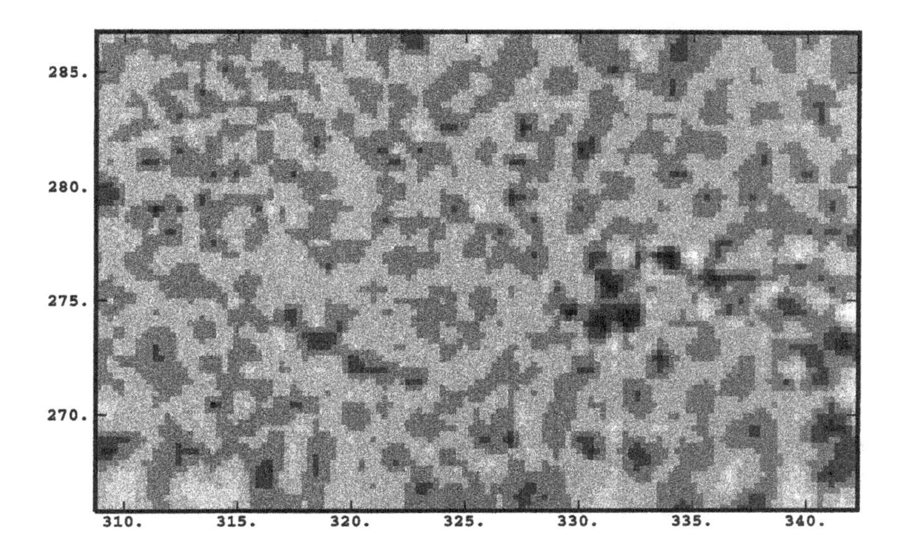

Figure 15.1: Map of the component associated with the short range (3.5 km).

Unbiasedness is achieved on the basis of zero sum weights (introducing a ficticious value $Z(\mathbf{0})$ at the origin, to form increments)

$$\begin{aligned}
\mathrm{E}[\,Z^{\star}_{u_0}(\mathbf{x}_0) - Z_{u_0}(\mathbf{x}_0)\,] &= \mathrm{E}\Big[\sum_{u=0}^{S}\sum_{\alpha=1}^{n} w_{\alpha}^{u_0}\, Z_u(\mathbf{x}_\alpha) - Z_{u_0}(\mathbf{x}_0)\,\Big] \\
&= \mathrm{E}\Big[\sum_{u=0}^{S-1}\sum_{\alpha=1}^{n} w_{\alpha}^{u_0}\, Z_u(\mathbf{x}_\alpha) - Z_{u_0}(\mathbf{x}_0) \\
&\qquad + \sum_{\alpha=1}^{n} w_{\alpha}^{u_0}\, Z_S(\mathbf{x}_\alpha) - Z_S(\mathbf{0}) \cdot \underbrace{\sum_{\alpha=1}^{n} w_{\alpha}^{u_0}}_{0}\,\Big] \\
&= \sum_{u=0}^{S-1}\sum_{\alpha=1}^{n} w_{\alpha}^{u_0}\, \underbrace{\mathrm{E}[\,Z_u(\mathbf{x}_\alpha)\,]}_{0} - \underbrace{\mathrm{E}[\,Z_{u_0}(\mathbf{x}_0)\,]}_{0} \\
&\qquad + \sum_{\alpha=1}^{n} w_{\alpha}^{u_0}\, \underbrace{\mathrm{E}[\,Z_S(\mathbf{x}_\alpha) - Z_S(\mathbf{0})\,]}_{0} \\
&= 0. \qquad (15.9)
\end{aligned}$$

The estimation variance is

$$\begin{aligned}
\sigma_{\mathrm{E}}^2 &= \mathrm{var}(Z^{\star}_{u_0}(\mathbf{x}_0) - Z_{u_0}(\mathbf{x}_0)) \qquad (15.10) \\
&= C^{u_0}(\mathbf{x}_0 - \mathbf{x}_0) - \sum_{\alpha=1}^{n}\sum_{\beta=1}^{n} w_{\alpha}^{u_0}\, w_{\beta}^{u_0}\, \gamma(\mathbf{x}_\alpha - \mathbf{x}_\beta) + 2\sum_{\alpha=1}^{n} w_\alpha\, \gamma^{u_0}(\mathbf{x}_\alpha - \mathbf{x}_0).
\end{aligned}$$

The kriging system for the component $Z_{u_0}(\mathbf{x}_0)$ is then

$$\begin{cases} \sum_{\beta=1}^{n} w_\beta^{u_0}\, \gamma(\mathbf{x}_\alpha - \mathbf{x}_\beta) + \mu_{u_0} = \gamma^{u_0}(\mathbf{x}_\alpha - \mathbf{x}_0) & \text{for } \alpha = 1, \dots, n \\ \sum_{\beta=1}^{n} w_\beta^{u_0} = 0. \end{cases} \tag{15.11}$$

We are thus able to extract from the data different components of the spatial variation which were identified on the experimental variogram with a nested variogram model.

These component kriging techniques bear some analogy to the spectral analysis methods, which are much in use in geophysics. CHILÈS & GUILLEN [49] have compared the two approaches on gravimetric data using covariance functions derived from a model which represents geological bodies as random prisms (see [310], [309]). The variogram obtained is a particular case of a Cauchy model (see Eq. (III.11), p336) with a scale parameter α. The Cauchy model provides variogram models both for gravity ($\alpha = 1/2$) or magnetic data ($\alpha = 3/2$)in physical applications [51], like e.g in problems of electromagnetic compatibility [181]. The results obtained by CHILÈS & GUILLEN by decomposing either the variogram or the spectrum matched well, even though the geological bodies were located at different average depths by each method. An advantage of the geostatistical approach is that the data need not to be interpolated on a regular grid for computing the experimental variogram, while this is necessary for determining the spectrum. The case study is discussed in full detail in CHILÈS & DELFINER [51].

Filtering

Instead of extracting a component of the spatial variation we may wish to reject it. We can filter a spatial component $Z_{u_0}(\mathbf{x})$ by removing the corresponding covariances $C_{u_0}(\mathbf{x}_\alpha - \mathbf{x}_0)$ from the right hand side of the ordinary kriging system.

For example, we might wish to eliminate the component $Z_1(\mathbf{x})$ in the second-order stationary model

$$Z(\mathbf{x}) = Z_1(\mathbf{x}) + \underbrace{Z_2(\mathbf{x}) + m}_{Z_f(\mathbf{x})} \tag{15.12}$$

and to obtain a filtered version $Z_f(\mathbf{x})$ of $Z(\mathbf{x})$. To achieve this by kriging, we use weights w_α^f in the linear combination

$$Z_f^\star(\mathbf{x}_0) = \sum_{\alpha=1}^{n} w_\alpha^f\, Z(\mathbf{x}_\alpha) \tag{15.13}$$

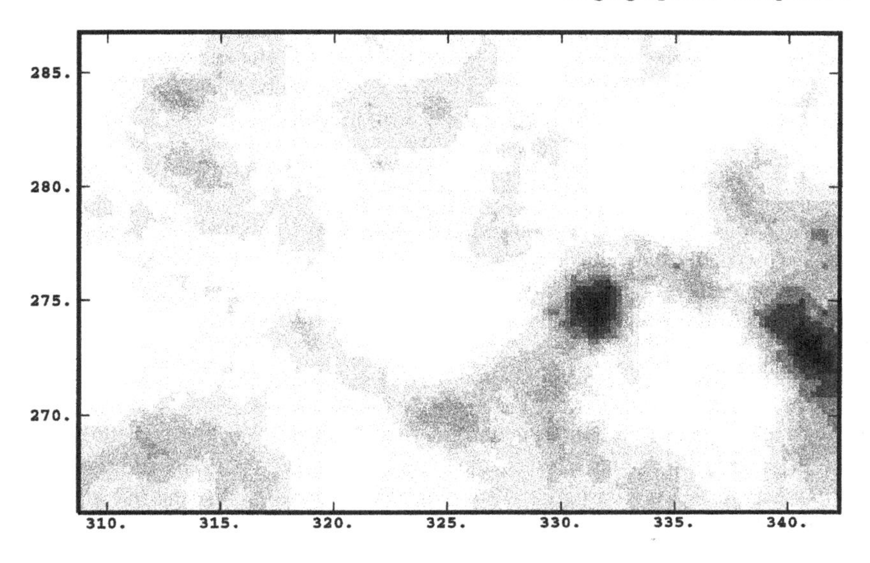

Figure 15.2: Map of the component associated with the long range (6.5 km) together with the estimated mean.

derived from the system

$$\begin{cases} \sum_{\beta=1}^{n} w_{\beta}^{f}\, C(\mathbf{x}_{\alpha}-\mathbf{x}_{\beta}) - \mu_f = C_2(\mathbf{x}_{\alpha}-\mathbf{x}_0) & \text{for } \alpha = 1,\ldots,n \\ \sum_{\beta=1}^{n} w_{\beta}^{f} = 1, & \end{cases} \tag{15.14}$$

where $C_1(\mathbf{h})$ is absent in the right hand side. We obtain thereby an estimation in which the component $Z_1^\star(\mathbf{x})$ is removed.

The filtered estimate can also be computed by subtracting the component from the ordinary kriging estimate

$$Z_f^\star(\mathbf{x}) \quad = \quad Z_{\mathrm{OK}}^\star(\mathbf{x}) - Z_1^\star(\mathbf{x}). \tag{15.15}$$

According to the configuration of data around the estimation point, ordinary kriging implicitly filters out certain components of the spatial variation. This topic will be taken up again in the next chapter when examining why kriging gives a smoothed image of reality.

Application: kriging spatial components of arsenic data

The data was collected during the national geochemical inventory of France and the map of the location of arsenic values in a 35 km $\times$ 25 km region near the Loire river has already been shown on Figure 14.1 (page 103). The values of arsenic were standardized before analysis. The mean experimental variogram of this variable was fitted on Figure 14.2 (page 104) with a nested variogram model,

$$\gamma(\mathbf{h}) \quad = \quad b_0 \operatorname{nug}(\mathbf{h}) + b_1 \operatorname{sph}(\mathbf{h}, 3.5\mathrm{km}) + b_2 \operatorname{sph}(\mathbf{h}, 6.5\mathrm{km}), \qquad (15.16)$$

consisting of a nugget effect variogram $\operatorname{nug}(\mathbf{h})$ and two spherical variograms $\operatorname{sph}(\mathbf{h}, range)$, multiplied by constants b_0, b_1 and b_2.

The corresponding linear regionalization model is made up with three second order stationary components and a local mean

$$Z(\mathbf{x}) \quad = \quad Z_0(\mathbf{x}) + Z_1(\mathbf{x}) + Z_2(\mathbf{x}) + m_l(\mathbf{x}). \qquad (15.17)$$

Kriging was performed using a moving neighborhood consisting of the 50 nearest samples to the estimation point. Several displays of the map of arsenic by ordinary kriging were already presented on Figure 13.2 (page 98), Figure 13.4 (page 99) and Figure 13.5 (page 100). The map of the standard deviations of ordinary kriging was displayed on Figure 13.3 (page 99).

The kriged map of the short range component of the arsenic values is seen on Figure 15.1. It is worthwhile to examine how agglomerates of high values (large circles) on the display of the data about $Z(\mathbf{x})$ on Figure 14.1 (page 103) compare with the darkly shaded areas on the map of the short range component $Z_1(\mathbf{x})$. It is also interesting to inspect how extrapolation acts in areas with no data.

A map of the component associated with the long range variation, together with the estimated mean, is depicted on Figure 15.2. This simultaneous extraction of the long range component $Z_2(\mathbf{x})$ and the local mean $m_l(\mathbf{x})$ is equivalent to filtering out (rejecting) the nugget-effect and the short range components $Z_0(\mathbf{x})$ and $Z_1(\mathbf{x})$. Comparison with Figure 13.4 (page 99) shows that this map is just a smoother version of the ordinary kriging map.

It should be noted that for irregularly spaced data a map of the nugget-effect component $Z_0(\mathbf{x})$ cannot be established. This component can only be estimated at grid nodes which coincide with a data location and is zero elsewhere. It is thus customary to filter out the nugget-effect in order to avoid blobs on the map at locations where extreme data values happen to coincide with estimation nodes (see also a similar discussion on p115).

With regularly spaced data a raster representation of the nugget component can be made using a grid whose nodes are identical with the sample locations.

16 The Smoothness of Kriging

How smooth are estimated values from kriging with irregularly spaced data in a moving neighborhood? By looking at a few typical configurations of data around nodes of the estimation grid and by building on our knowledge of how spatial components are kriged, we can understand the way the estimated values are designed in ordinary kriging.

The sensitivity of kriging to the choice of the variogram model is discussed in connection with an application on topographic data.

Kriging with irregularly spaced data

Let us take a regionalized variable for which three components of spatial variation have been identified. For the following linear model of regionalization we assume local stationarity of order two

$$Z(\mathbf{x}) \quad = \quad Z_0(\mathbf{x}) + Z_1(\mathbf{x}) + Z_2(\mathbf{x}) + m_l(\mathbf{x}). \tag{16.1}$$

The covariance model of the three spatial components is

$$C(\mathbf{h}) \quad = \quad C_0(\mathbf{h}) + C_1(\mathbf{h}) + C_2(\mathbf{h}), \tag{16.2}$$

where we let $C_0(\mathbf{h})$ be a nugget-effect covariance function and $C_1(\mathbf{h}), C_2(\mathbf{h})$ be spherical models with ranges a_1 and a_2, numbered in such a way that $a_1 < a_2$.

Suppose that the sampling grid is highly irregular, entailing a very unequal distribution of data in space. A grid of estimation points is set up, with nodes as close as is required for the construction of a map at a given resolution. At each point the operation of ordinary kriging is repeated, picking up the data in the moving neighborhood.

Assuming a highly irregular arrangement of the data points, very different configurations of sample points around estimation points will arise, four of which we shall examine in the following.

1. $\mathbf{x}_0$ *is more than* a_2 *away from the data*

This configuration of data points around the estimation point can arise when $\mathbf{x}_0$ is located amid a zone without data within the range of the two spherical models, as shown on Figure 16.1. Ordinary kriging is then equivalent to the kriging of the mean. The right hand side covariances of the ordinary kriging system are nil as all distances

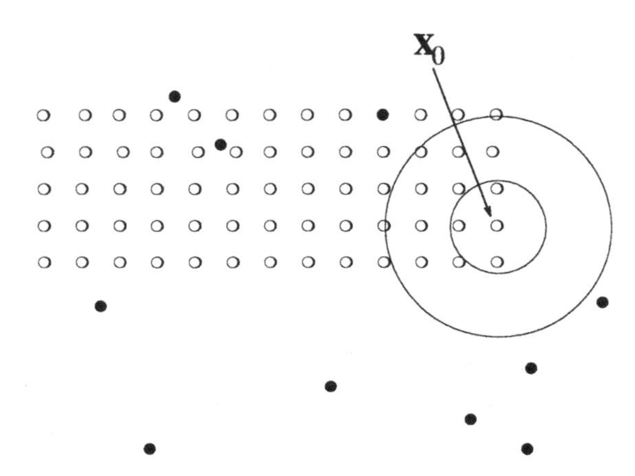

Figure 16.1: No data point is within the range of the two spherical models for this grid node.

involved are greater than a_2. The information transferred to $\mathbf{x}_0$ will be an estimation of the local mean in the neighborhood.

This shows that kriging is very conservative: it will only give an estimate of the mean function when data is far away from the estimation point.

2. $\mathbf{x}_0$ is less than a_2 and more than a_1 away from the nearest data point $\mathbf{x}_\alpha$

With this spatial arrangement, see Figure 16.2, the ordinary kriging is equivalent to a filtering of the components $Z_0(\mathbf{x})$ and $Z_1(\mathbf{x})$, with the equation system

$$\begin{cases} \displaystyle\sum_{\beta=1}^{n} \lambda_\beta \, C(\mathbf{x}_\alpha - \mathbf{x}_\beta) - \mu = \boxed{C_2(\mathbf{x}_\alpha - \mathbf{x}_0)} & \text{for } \alpha = 1, \ldots, n \\ \displaystyle\sum_{\beta=1}^{n} \lambda^\beta = 1. \end{cases} \tag{16.3}$$

The covariances $C_0(\mathbf{h})$ and $C_1(\mathbf{h})$ are not present in the right hand side for kriging at such a grid node.

3. $\mathbf{x}_0$ is less than a_1 away, but does not coincide with the nearest $\mathbf{x}_\alpha$

In this situation, shown on Figure 16.3, ordinary kriging will transfer information not only on the long range component $Z_2(\mathbf{x})$, but also about the short range component $Z_1(\mathbf{x})$, which varies more quickly in space. This creates a more detailed description of the regionalized variable in areas of the map where data is plenty. Ordinary kriging

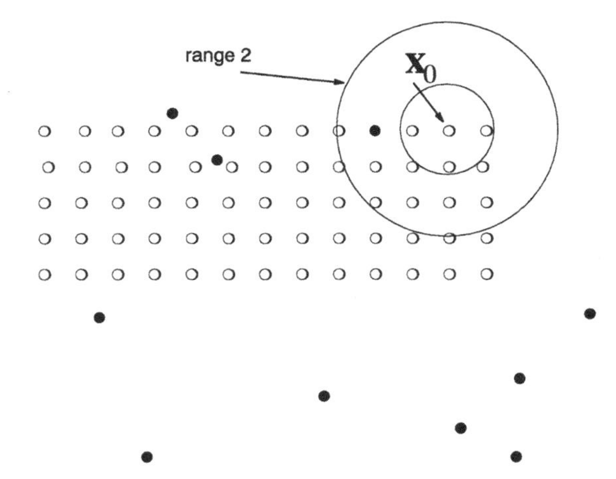

Figure 16.2: One data point is within the range of the second spherical model for this grid node.

is now equivalent to a system filtering the nugget-effect component

$$\begin{cases} \sum_{\beta=1}^{n} \lambda_\beta \, C(\mathbf{x}_\alpha - \mathbf{x}_\beta) - \mu = \boxed{C_1(\mathbf{x}_\alpha - \mathbf{x}_0) + C_2(\mathbf{x}_\alpha - \mathbf{x}_0)} & \text{for } \alpha = 1, \ldots, n \\ \sum_{\beta=1}^{n} \lambda^\beta = 1. \end{cases} \tag{16.4}$$

Only the nugget-effect covariance $C_0(\mathbf{h})$ is absent from the right hand side when kriging at such a location.

4. $\mathbf{x}_0$ *coincides with a data location*

In this last case, when $\mathbf{x}_0 = \mathbf{x}_\alpha$ for a particular value of α, ordinary kriging does not filter anything and restitutes the data value measured at the location $\mathbf{x}_\alpha = \mathbf{x}_0$: it is an exact interpolator!

This property may be perceived as a nuisance when there are only a few locations of coincidence between the grid and the data locations, because then the nugget-effect component will generate spikes in the estimated values at these locations only. To avoid having a map with such blobs at fortuitous locations, it is often advisable to filter systematically the nugget-effect component $Z_0(\mathbf{x})$ in cartographical applications based on irregularly spaced data.

Sensitivity to choice of variogram model

The most important characteristic for the choice of the variogram model is the interpretation of the behavior at the origin. The type of continuity assumed for the

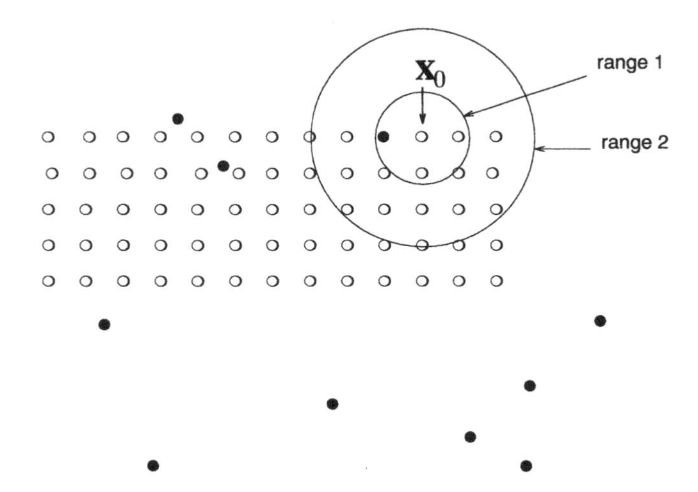

Figure 16.3: One data point is within the range of the two spherical models for this node.

regionalized variable under study has immediate implications for ordinary kriging.

The behavior at the origin can be of three types

1. *discontinuous*, i.e. with a nugget-effect component. The presence of a nugget-effect has the consequence that the kriging estimates will be also discontinuous: each data location will be a point of discontinuity of the estimated surface.

2. *continuous, but not differentiable*. For example a linear behavior near the origin implies that the variogram is not differentiable at the origin. It has the effect that the kriging estimator is not differentiable at the data locations and that the kriged surface is linear in the immediate neighborhood of the data points.

3. *continuous and differentiable*. A quadratic behavior at the origin will generate a quadratic behavior of the kriged surface at the data locations. Models that are differentiable at the origin have the power to extrapolate outside the range of the data in kriging. This implies negative ordinary kriging weights.

Most variogram models are fairly robust with respect to kriging. There however is one pathological model: the so-called "Gaussian" variogram model. This model belongs to the family of *stable variogram models* (that bear this name because of their analogy with the characteristic function of the stable distributions)

$$\gamma(\mathbf{h}) \;=\; b\left(1-\mathrm{e}^{-\dfrac{|\mathbf{h}|^p}{a}}\right) \qquad \text{with } 0<p\leq 2 \text{ and } a,b>0. \tag{16.5}$$

For a power p equal to 2 we have the Gaussian variogram model

$$\gamma(\mathbf{h}) \;=\; b\left(1-\mathrm{e}^{-\dfrac{|\mathbf{h}|^2}{a}}\right), \tag{16.6}$$

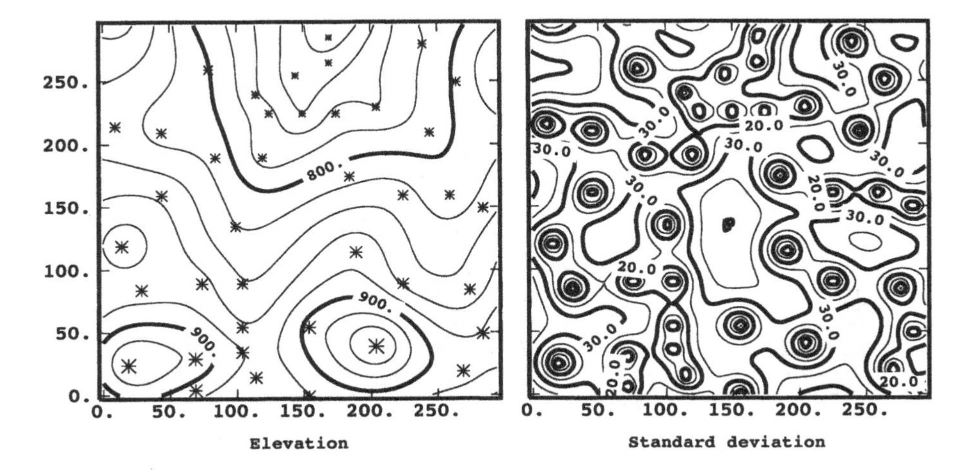

Figure 16.4: On the left: map of elevations obtained by ordinary kriging with unique neighborhood using a stable variogram (power $p = 1.5$) (data points are displayed as stars of size proportional to the elevation). On the right: map of the corresponding standard deviations.

which is infinitely differentiable at the origin. The implication is that the associated random function also has this property. This means that the function is analytic and, if known over a small area, it could be predicted just anywhere in the universe as all derivatives are known. Clearly this model is unrealistic in most applications (besides the fact that it is purely deterministic). Furthermore the use of such an unrealistic model has consequences on kriging which are illustrated in the following example.

Application: kriging topographic data

We use a data set which has become a classic of spatial data analysis since its publication by DAVIS [76]. It consists of 52 measurements of elevation (in feet) in a 300 ft × 300 ft region.

First we fit a stable variogram model with a power of 1.5 and obtain by ordinary kriging with unique neighborhood (all samples used for kriging) the isoline map on the left of Figure 16.4. This is the map most authors get using different methods, models or parameters (see [76], [262], [349], [17], [329]). The map of the kriging standard deviations is seen on the right of Figure 16.4. The standard deviations are comparable to those obtained with a spherical model of range 100 ft and a sill equal to the variance of the data (3770 ft^2). They are of the same order of magnitude as the standard deviation of the data (61.4 ft).

Now, if we increase the power of the stable model from 1.5 to 2 we have a dramatic change in the behavior at the origin and in the extrapolating power of the variogram

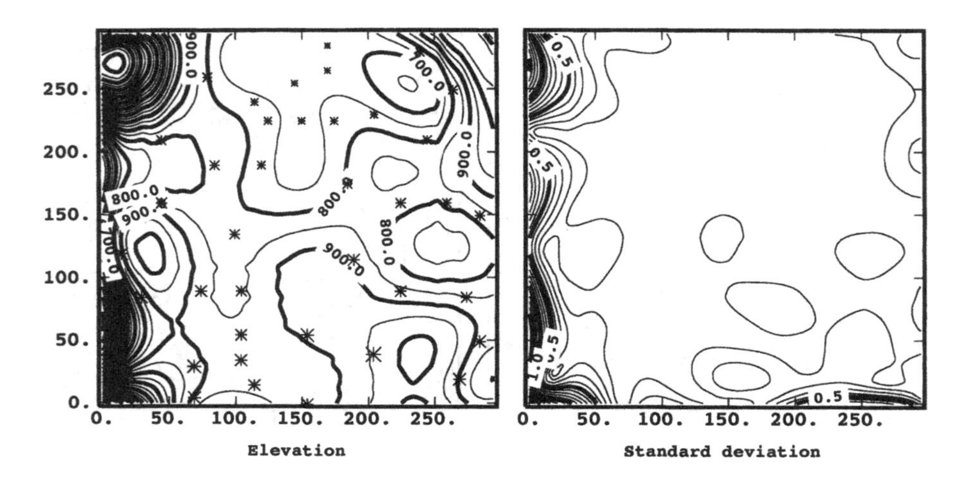

Figure 16.5: Maps kriged using a Gaussian variogram in unique neighborhood.

model. The map of contoured estimates of ordinary kriging using the Gaussian model in unique neighborhood is shown on the left of Figure 16.5. We can observe artificial hills and valleys between data locations. At the borders of the map the estimates extrapolate far outside the range of the data. The kriging standard deviations represented on the map on the right of Figure 16.5 are generally near zero: this complies with the deterministic type of variogram model, which is supposed to describe exhaustively all features of the spatial variation in that region.

Switching from a unique to a moving neighborhood (32 points) using the Gaussian model we get the map to be seen on the left of Figure 16.6. The extrapolation effects are weaker than in unique neighborhood. The isolines seem as drawn by a trembling hand: the neighborhoods are apparently in conflict because the data do not comply with the extreme assumption of continuity built into the model. The standard deviations in moving neighborhood are higher than in unique neighborhood as shown on the right of Figure 16.6, especially on the borders of the map as extrapolation is performed with less data.

The conclusion from this experiment is that the Gaussian model should not be used in practice. A stable variogram model with a power less than 2 will do the job better.

Another instructive exercise is to add a slight nugget-effect (e.g. 1/1000, 1/100, 1/10 of the variance) to a Gaussian variogram and to observe how the map of the estimated values behaves, how the standard deviations increase. The discontinuity added at the origin actually destroys the extreme extrapolative properties of the Gaussian model.

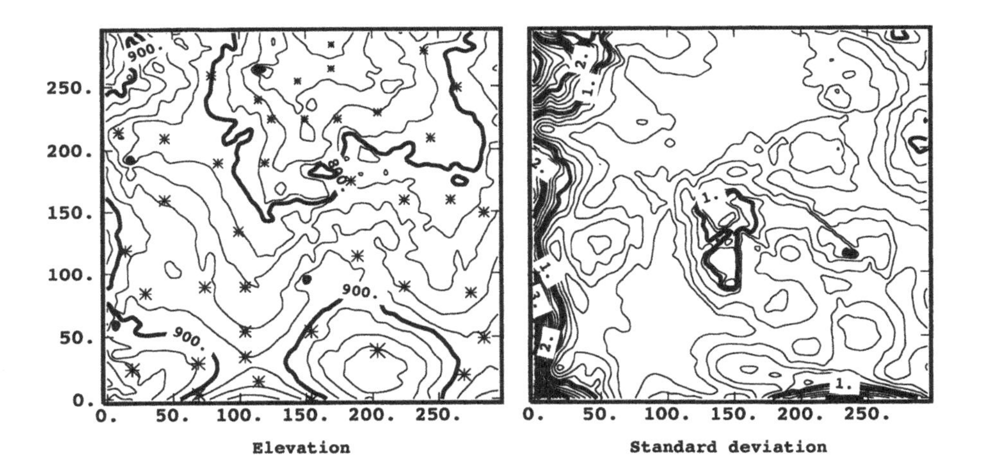

Figure 16.6: Maps kriged using a Gaussian variogram in moving neighborhood.

Part C

Multivariate Analysis

17 Principal Component Analysis

Principal component analysis is the most widely used method of multivariate data analysis owing to the simplicity of its algebra and to its straightforward interpretation.

A linear transformation is defined which transforms a set of correlated variables into uncorrelated factors. These orthogonal factors can be shown to extract successively a maximal part of the total variance of the variables. A graphical display can be produced which shows the position of the variables in the plane spanned by two factors.

Uses of PCA

Principal Component Analysis (PCA) can be used for:

1. data compression,
2. multivariate outlier detection,
3. deciphering a correlation matrix,
4. identifying underlying factors,
5. detecting intrinsic correlation.

The first four topics will be discussed in this chapter, while the problem of intrinsic correlation will be exposed in Chapter 22.

An example of data compression by PCA is given in Example 17.2, the use of PCA in deciphering the structure of a correlation matrix is shown in Example 17.2, multivariate outliers are made visible in Example 17.3 and an application illustrating the interpretation of factors in terms of underlying phenomena is provided in Example 17.6.

Transformation into factors

The basic problem solved by principal component analysis is to transform a set of correlated variables into uncorrelated quantities, which could be interpreted in an ideal (multi-Gaussian) context as independent factors underlying the phenomenon. This is why the uncorrelated quantities are called *factors*, although such an interpretation is not always perfectly adequate.

$\mathbf{Z}$ is the $n \times N$ matrix of data from which the means of the variables have already been subtracted. The corresponding $N \times N$ experimental variance-covariance matrix $\mathbf{V}$ then is

$$\mathbf{V} = [s_{ij}] = \frac{1}{n} \mathbf{Z}^\top \mathbf{Z}. \tag{17.1}$$

Let $\mathbf{Y}$ be an $n \times N$ matrix containing in its rows the n samples of factors Y_p $(p = 1, \ldots, N)$, which are uncorrelated and of zero mean.

The variance-covariance matrix of the factors is diagonal, owing to the fact that the covariances between factors are nil by definition,

$$\mathbf{D} = \frac{1}{n} \mathbf{Y}^\top \mathbf{Y} = \begin{pmatrix} d_{11} & 0 & 0 \\ 0 & \ddots & 0 \\ 0 & 0 & d_{NN} \end{pmatrix}, \tag{17.2}$$

and the diagonal elements d_{pp} are the variances of the factors.

A matrix $\mathbf{A}$ is sought, $N \times N$ orthogonal, which linearly transforms the measured variables into synthetic factors

$$\mathbf{Y} = \mathbf{Z}\,\mathbf{A} \qquad \text{with } \mathbf{A}^\top \mathbf{A} = \mathbf{I}. \tag{17.3}$$

Multiplying this equation from the left by $1/n$ and $\mathbf{Y}^\top$, we have

$$\frac{1}{n} \mathbf{Y}^\top \mathbf{Y} = \frac{1}{n} \mathbf{Y}^\top \mathbf{Z}\,\mathbf{A}, \tag{17.4}$$

and replacing $\mathbf{Y}$ by $\mathbf{Z}\mathbf{A}$ on the right hand side, it follows

$$\frac{1}{n} (\mathbf{Z}\,\mathbf{A})^\top (\mathbf{Z}\,\mathbf{A}) = \frac{1}{n} \mathbf{A}^\top \mathbf{Z}^\top \mathbf{Z}\,\mathbf{A} = \mathbf{A}^\top \frac{1}{n} (\mathbf{Z}^\top \mathbf{Z})\,\mathbf{A}. \tag{17.5}$$

Finally

$$\mathbf{D} = \mathbf{A}^\top \mathbf{V}\,\mathbf{A}, \tag{17.6}$$

that is,

$$\mathbf{V}\,\mathbf{A} = \mathbf{A}\,\mathbf{D}. \tag{17.7}$$

It can immediately be seen that the matrix $\mathbf{Q}$ of orthonormal eigenvectors of $\mathbf{V}$ offers a solution to the problem and that the eigenvalues λ_p are then simply the variances of the factors Y_p. Principal component analysis is nothing else than a statistical interpretation of the eigenvalue problem

$$\mathbf{V}\,\mathbf{Q} = \mathbf{Q}\,\boldsymbol{\Lambda} \qquad \text{with } \mathbf{Q}^\top \mathbf{Q} = \mathbf{I}, \tag{17.8}$$

defining the factors as

$$\mathbf{Y} = \mathbf{Z}\,\mathbf{Q}. \tag{17.9}$$

Maximization of the variance of a factor

Another important aspect of principal component analysis is that it allows to define a sequence of orthogonal factors which successively absorb a maximal amount of the variance of the data.

Take a vector $\mathbf{y}_1$ corresponding to the first factor obtained by transforming the centered data matrix $\mathbf{Z}$ with a vector $\mathbf{a}_1$ calibrated to unit length

$$\mathbf{y}_1 = \mathbf{Z}\mathbf{a}_1 \qquad \text{with } \mathbf{a}_1^\top \mathbf{a}_1 = 1. \tag{17.10}$$

The variance of $\mathbf{y}_1$ is

$$\operatorname{var}(\mathbf{y}_1) = \frac{1}{n}\mathbf{y}_1^\top \mathbf{y}_1 = \frac{1}{n}\mathbf{a}_1^\top \mathbf{Z}^\top \mathbf{Z}\mathbf{a}_1 = \mathbf{a}_1^\top \mathbf{V}\mathbf{a}_1. \tag{17.11}$$

To attribute a maximal part of the variance of the data to $\mathbf{y}_1$, we define an objective function ϕ_1 with a Lagrange parameter λ_1, which multiplies the constraint that the transformation vector $\mathbf{a}_1$ should be of unit norm

$$\phi_1 = \mathbf{a}_1^\top \mathbf{V}\mathbf{a}_1 - \lambda_1(\mathbf{a}_1^\top \mathbf{a}_1 - 1). \tag{17.12}$$

Setting the derivative with respect to $\mathbf{a}_1$ to zero,

$$\frac{\partial \phi_1}{\partial \mathbf{a}_1} = 0 \qquad \Longleftrightarrow \qquad 2\mathbf{V}\mathbf{a}_1 - 2\lambda_1 \mathbf{a}_1 = 0, \tag{17.13}$$

we see that λ_1 is an eigenvalue of the variance-covariance matrix and that $\mathbf{a}_1$ is equal to the eigenvector $\mathbf{q}_1$ associated with this eigenvalue

$$\mathbf{V}\mathbf{q}_1 = \lambda_1 \mathbf{q}_1. \tag{17.14}$$

We are interested in a second vector $\mathbf{y}_2$ orthogonal to the first

$$\operatorname{cov}(\mathbf{y}_2, \mathbf{y}_1) = \operatorname{cov}(\mathbf{Z}\mathbf{a}_2, \mathbf{Z}\mathbf{a}_1) = \mathbf{a}_2^\top \mathbf{V}\mathbf{a}_1 = \mathbf{a}_2^\top \lambda_1 \mathbf{a}_1 = 0. \tag{17.15}$$

The function ϕ_2 to maximize incorporates two constraints: the fact that $\mathbf{a}_2$ should be unit norm and the orthogonality between $\mathbf{a}_2$ and $\mathbf{a}_1$. These constraints bring up two new Lagrange multipliers λ_2 and μ

$$\phi_2 = \mathbf{a}_2^\top \mathbf{V}\mathbf{a}_2 - \lambda_2(\mathbf{a}_2^\top \mathbf{a}_2 - 1) + \mu\, \mathbf{a}_2^\top \mathbf{a}_1. \tag{17.16}$$

Setting the derivative with respect to $\mathbf{a}_2$ to zero

$$\frac{\partial \phi_2}{\partial \mathbf{a}_2} = 0 \qquad \Longleftrightarrow \qquad 2\mathbf{V}\mathbf{a}_2 - 2\lambda_2 \mathbf{a}_2 + \mu\, \mathbf{a}_1 = 0. \tag{17.17}$$

What is the value of μ ? Multiplying the equation by $\mathbf{a}_1^\top$ from the left,

$$2\,\underbrace{\mathbf{a}_1^\top \mathbf{V}\mathbf{a}_2}_{0} - 2\lambda_2\,\underbrace{\mathbf{a}_1^\top \mathbf{a}_2}_{0} + \mu\,\underbrace{\mathbf{a}_1^\top \mathbf{a}_1}_{1} = 0, \tag{17.18}$$

we see that μ is nil (the constraint is not active) and thus

$$\mathbf{V}\mathbf{a}_2 = \lambda_2 \mathbf{a}_2. \tag{17.19}$$

Again λ_2 turns out to be an eigenvalue of the variance-covariance matrix and $\mathbf{a}_2$ is the corresponding eigenvector $\mathbf{q}_2$. Continuing in the same way we find the rest of the N eigenvalues and eigenvectors of $\mathbf{V}$ as an answer to our maximization problem.

EXAMPLE 17.1 (DATA COMPRESSION) *PCA can be seen to as a simple data compression algorithm. This is an example from geophysical exploration following [136].*

Let $\mathbf{Z}$ *be an* $n \times N$ *matrix of* n *seismic profiles and* N *spatially correlated samples. The principal components matrix* $\mathbf{Y}$ *is of size* $n \times N$ *(like* $\mathbf{Z}$*). The variance of each principal component* $\mathbf{y}_p$ *(a column of the matrix* $\mathbf{Y}$*) is given by the corresponding eigenvalue* λ_p*. We assume that the eigenvalues have been ordered by decreasing variance.*

The idea for compressing the data is to retain only the principal components having the largest variances. Keeping M *out of the* N *principal components, which explain a substantial amount of the total variance in an* $n \times M$ *matrix* $\widetilde{\mathbf{Y}}$*, we have the approximation*

$$\widetilde{\mathbf{Z}} \cong \widetilde{\mathbf{Y}}\,\widetilde{\mathbf{Q}}^{\top} \tag{17.20}$$

where $\widetilde{\mathbf{Q}}$ *is an* $N \times M$ *matrix of eigenvectors.*

This approximation is interesting for data compression if M *is substantially smaller than* N*: we can then save considerable storage by keeping the two matrices* $\widetilde{\mathbf{Y}}, \widetilde{\mathbf{Q}}$ *instead of the original data* $\mathbf{Z}$*. The* $n \times N$ *numbers of the matrix* $\mathbf{Z}$ *are then replaced with* $M(n+N)$ *numbers, which will be used to reconstruct an approximate matrix* $\widetilde{Z}$*.*

Following HAGEN *[136], having originally* $n = 200$ *good quality seismic traces in an* $N = 50$ *sample window, if the* $M = 4$ *first principal components express 85% of the total variance, the original data base of* $200 \times 50 = 10,000$ *samples can be reduced to only* $4(200+50) = 1,000$ *samples. This new data base needs only one tenth of the storage space, yet preserving a sufficiently accurate description of the main geological patterns important for reservoir characterization.*

Interpretation of the factor variances

Numbering the eigenvalues of $\mathbf{V}$ from the largest to the lowest, we obtain a sequence of N uncorrelated factors which provide an optimal decomposition (in the least squares sense) of the total variance as

$$\operatorname{tr}(\mathbf{V}) = \sum_{i=1}^{N} s_{ii} = \sum_{p=1}^{N} \lambda_p. \tag{17.21}$$

The eigenvalues indicate the amount of the total variance associated with each factor and the ratio

$$\frac{\text{variance of the factor}}{\text{total variance}} = \frac{\lambda_p}{\mathrm{tr}(\mathbf{V})} \tag{17.22}$$

gives a numerical indication, usually expressed in %, of the importance of the factor.

Generally it is preferable to standardize the variables (subtracting the means and dividing by the standard deviations), so that the principal component analysis is performed on the correlation matrix $\mathbf{R}$. In this framework, when an eigenvalue is lower than 1, we may consider that the associated factor has less explanatory value than any single variable, as its variance is inferior to the unit variance of each variable.

Correlation of the variables with the factors

In general it is preferable to work with standardized variables $\widetilde{z}_{\alpha i}$ to set them on a common scale and make them comparable,

$$\widetilde{z}_{\alpha i} = \frac{z_{\alpha i} - m_i^\star}{\sqrt{s_{ii}}}, \tag{17.23}$$

where $m_i^\star$ and $\sqrt{s_{ii}}$ are the experimental mean and standard deviation of the variable $\mathbf{z}_i$.

The variance-covariance matrix associated to standardized data is the correlation matrix

$$\mathbf{R} = \frac{1}{n}\,\widetilde{\mathbf{Z}}^\top\,\widetilde{\mathbf{Z}}, \tag{17.24}$$

which can be decomposed using its eigensystem as

$$\mathbf{R} = \widetilde{\mathbf{Q}}\,\widetilde{\mathbf{\Lambda}}\,\widetilde{\mathbf{Q}}^\top = \widetilde{\mathbf{Q}}\,\sqrt{\widetilde{\mathbf{\Lambda}}}\left(\widetilde{\mathbf{Q}}\,\sqrt{\widetilde{\mathbf{\Lambda}}}\right)^\top = \widetilde{\mathbf{A}}\,\widetilde{\mathbf{A}}^\top. \tag{17.25}$$

The vectors $\widetilde{\mathbf{a}}_i$, columns of $\widetilde{\mathbf{A}}^\top$, are remarkable in the sense that they contain the correlations r_{ip} between a variable $\widetilde{\mathbf{z}}_i$ and the factors $\mathbf{y}_p$ because

$$r_{ip} = \mathrm{corr}(\widetilde{\mathbf{z}}_i, \mathbf{y}_p) = \sqrt{\widetilde{\lambda}_p}\,\widetilde{q}_{ip} = \widetilde{a}_{ip}. \tag{17.26}$$

The vectors $\widetilde{\mathbf{a}}_i$ are of unit length and their cross product is equal to the correlation coefficient

$$\widetilde{\mathbf{a}}_i^\top\,\widetilde{\mathbf{a}}_j = r_{ij}. \tag{17.27}$$

Owing to their geometry the vectors $\widetilde{\mathbf{a}}_i$ can be used to represent the position of the variables on the surface of the unit hypersphere centered at the origin. The correlation coefficients r_{ij} are the cosines of the angles between the vectors referring to two different variables.

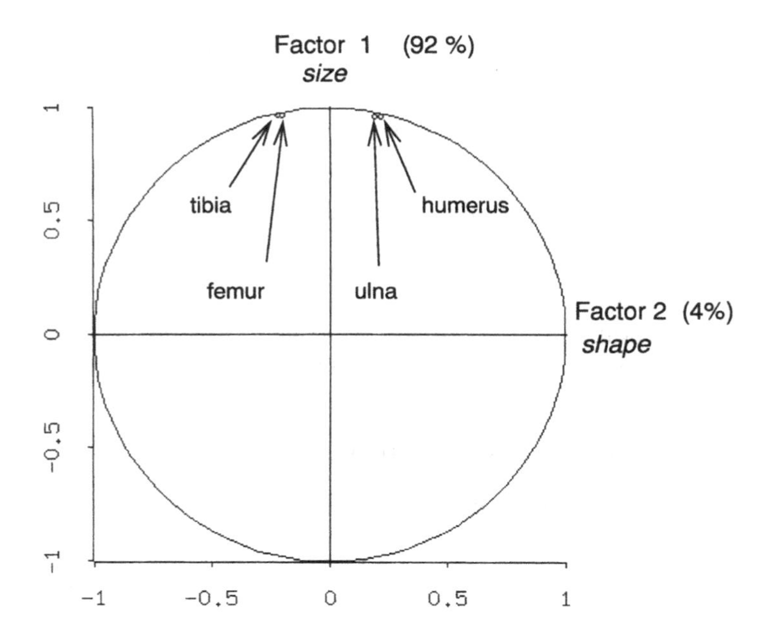

Figure 17.1: The position of the bonelength variables inside the circle of correlations for the first two principal components. Each axis is labeled with the number of the factor and the proportion of the total variance (in %) extracted by the factor.

The position of the variables on the surface of the hypersphere can be projected towards a plane defined by a pair of axes representing factors, which passes through the origin. The coordinates of the projection are given by the correlation coefficients r_{ip} and the corresponding graphical representation is therefore called the *circle of correlations*. The circle of correlations shows the proximity of the variables inside a unit circle and is useful to evaluate the affinities and the antagonisms between the variables. Statements can easily be made about variables which are located near the circumference of the unit circle because the proximities in 2-dimensional space then correspond to proximities in N-dimensional space. For the variables located further away from the circumference of the unit circle it is necessary to check on other principal axes whether the proximities really correspond to proximities on the surface of the hypersphere. Two variables near each other on the projection plane may have been projected, one from the upper hemisphere, the other from the lower hemisphere.

EXAMPLE 17.2 (DECIPHERING A CORRELATION MATRIX) *The correlation matrix of the lengths of 276 leghorn fowl bones (see* MORRISON, *1978, p282) on Table 17.1 seems trivial to interpret at first look: all variables are well correlated. Big birds have large bones and small birds have small bones.*

Inspection of the plane of the first two factors on Figure 17.1 showing the position of the length variables inside the correlation circle, reveals two things. The correla-

Humerus	1			
Ulna	.940	1		
Tibia	.875	.877	1	
Femur	.878	.886	.924	1
	Humerus	Ulna	Tibia	Femur

Table 17.1: Correlation matrix between the lengths of bones of fowl.

tions of all variables with the first factor (92% of the total variance) are all very strong and of the same sign as can be seen on Table 17.2.

	PC1	PC2
Humerus	.96	-.22
Ulna	.96	-.20
Tibia	.96	.22
Femur	.96	.20
Component variance	3.69	.17
Percentage	92 %	4%

Table 17.2: Correlations between bone lengths and the first two principal components.

This factor represents the difference in size *of the fowls, which explains most of the variation, and has an obvious impact on all bone lengths. The second factor (4% of the variance) splits up the group of the four variables into two distinct subgroups. The second factor is due to the difference in* shape *of the birds: the leg bones (tibia and femur) do not grow in the same manner as the wing bones (humerus and ulna). Some birds have shorter legs or longer wings than others (and vice-versa).*

Having a second look at the correlation matrix on Table 17.1, it is readily seen that the correlations among the wing variables (.94) and among the leg variables (.92) are stronger than the rest. Principal component analysis does not extract anything new or hidden from the data. It is merely a help to read and decipher the structure of a correlation matrix. This feature becomes especially useful for large correlation matrices.

EXAMPLE 17.3 (MULTIVARIATE OUTLIERS) *In a study on soil pollution data, the seven variables Pb, Cd, Cr, Cu, Ni, Zn, Mo were logarithmically transformed and standardized. The principal components calculated on the correlation matrix extracted 38%, 27%, 15%, 9%, 6%, 3% and 2% of the total variance. So the first two factors represent about two thirds (65%) of the total variance. The third factor (15%) hardly explains more variance than any of the original variables taken alone (1/7 = 14.3%).*

The Figure 17.2 shows the correlations with the first two factors. Clearly the first factor exhibits, like in the previous biological example, mainly a size effect. This is more appropriately termed a dilution factor *because the measured elements constitute*

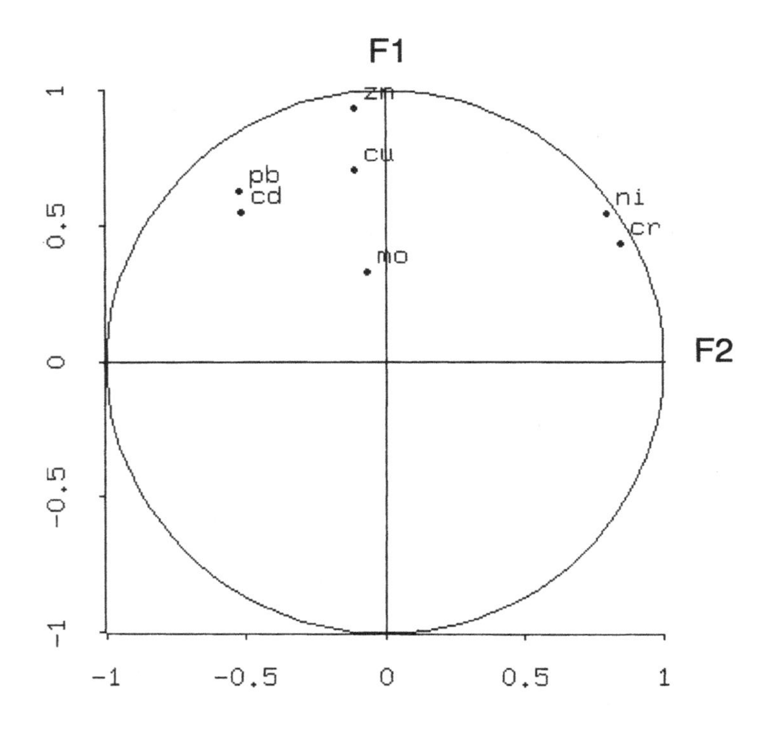

Figure 17.2: Unit circle showing the correlation of seven soil pollution variables with the first two principal components.

a very small proportion of the soil and their proportion in a soil sample is directly proportional to the overall pollution of the soil. The second factor looks interesting: it shows the antagonism between two pairs of elements: Pb, Cd and Ni, Cr.

The correlation coefficients between the seven soil pollution elements are listed on Table 17.3. The ordering of the variables was chosen to be the same as on the second factor: Pb, Cd, Zn, Cu, Mo, Ni and Cr (from left to right on Figure 17.2). This exemplifies the utility of a PCA as a device to help reading a correlation matrix.

In this data set of 100 samples there are three multivariate outliers that could be identified with a 3D representation of the sample cloud in the coordinate system of the first three factors (which concentrate 80% of the variance). On a modern computer screen the cloud can easily be rotated into a position that shows best one of its features like on Figure 17.3.

This need for an additional rotation illustrates the fact that PCA only provides an optimal projection plane for the samples if the cloud is of ellipsoidal shape.

In the general case of non-standardized data it is possible to build a graph showing the correlations between the set of variables and a pair of factors. The variance-covariance matrix $\mathbf{V}$ is multiplied from the left and the right with the matrix $\mathbf{D}_{s^{-1}}$ of

Pb	1						
Cd	.40	1					
Zn	.59	.54	1				
Cu	.42	.27	.62	1			
Mo	.20	.19	.30	-.07	1		
Ni	-.04	-.08	.40	.29	.07	1	
Cr	-.14	-.11	.29	.11	.13	.86	1
	Pb	Cd	Zn	Cu	Mo	Ni	Cr

Table 17.3: Correlation matrix between the seven soil pollution elements.

the inverses of the standard deviations (see Example I.4)

$$\mathbf{D}_{s^{-1}}\,\mathbf{V}\,\mathbf{D}_{s^{-1}} = \mathbf{D}_{s^{-1}}\,\mathbf{Q}\,\boldsymbol{\Lambda}\,\mathbf{Q}^{\top}\,\mathbf{D}_{s^{-1}} \tag{17.28}$$

$$= \mathbf{D}_{s^{-1}}\,\mathbf{Q}\,\sqrt{\boldsymbol{\Lambda}}\,(\mathbf{D}_{s^{-1}}\,\mathbf{Q}\,\sqrt{\boldsymbol{\Lambda}}\,)^{\top} \tag{17.29}$$

from what the general formula to calculate the correlation between a variable and a factor can be deduced

$$\operatorname{corr}(\mathbf{z}_i, \mathbf{y}_p) = \sqrt{\frac{\lambda_p}{s_{ii}}}\, q_{ip}. \tag{17.30}$$

EXERCISE 17.4 *Deduce from*

$$\frac{1}{n}\widetilde{\mathbf{Z}}^{\top}\mathbf{Y} = \mathbf{R}\,\widetilde{\mathbf{Q}} = \widetilde{\mathbf{Q}}\,\widetilde{\boldsymbol{\Lambda}} \qquad \textit{with}\ \widetilde{\mathbf{Q}}^{\top}\,\widetilde{\mathbf{Q}} = \mathbf{I} \tag{17.31}$$

that the correlation coefficient between a vector $\widetilde{\mathbf{z}}_i$ *and a factor* $\mathbf{y}_p$ *is*

$$r_{ip} = \sqrt{\widetilde{\lambda}_p}\,\widetilde{q}_{ip}. \tag{17.32}$$

From this it can be seen that when a standardized variable is orthogonal to all others, it will be identical with a factor of unit variance in the principal component analysis.

EXERCISE 17.5 *Let* $\widetilde{\mathbf{Z}}$ *be a matrix of standardized data and* $\mathbf{Y}$ *the matrix of corresponding factors obtained by a principal component analysis* $\mathbf{Y} = \widetilde{\mathbf{Z}}\,\widetilde{\mathbf{Q}}$ *where* $\widetilde{\mathbf{Q}}$ *is the matrix of orthogonal eigenvectors of the correlation matrix* $\mathbf{R}$.

Show that

$$\mathbf{R} = \operatorname{corr}(\widetilde{\mathbf{Z}}, \mathbf{Y})\,[\operatorname{corr}(\widetilde{\mathbf{Z}}, \mathbf{Y})\,]^{\top}. \tag{17.33}$$

EXAMPLE 17.6 (INTERPRETING THE FACTORS) *We have 1054 soil samples from lateritic terrain in Mali analysed for 14 elements: Fe, Al, V, P, Cr, Cu, Nb, As, Mo, Si, Ti, Ce, Zr, Y; they are described in* ROQUIN ET AL. *[273, 274]. This spatially*

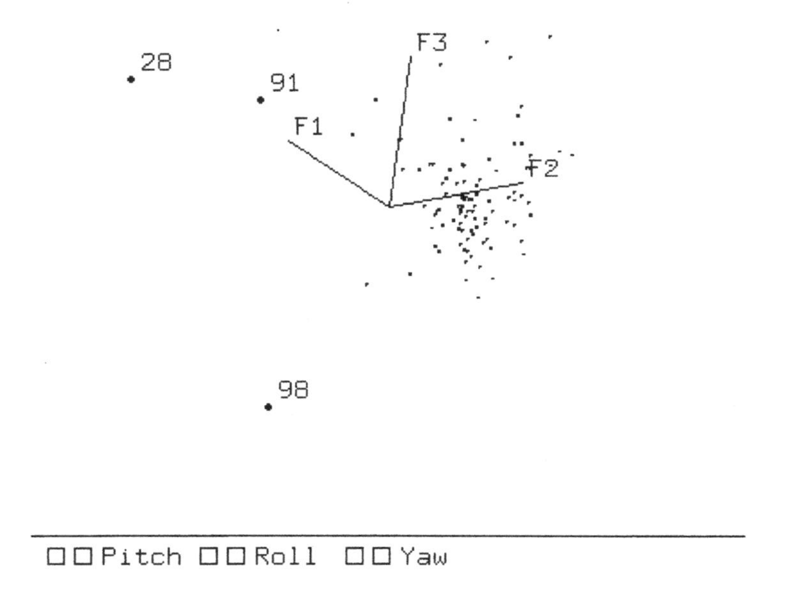

Figure 17.3: The cloud of 100 samples in the coordinate system of the first three factors after an appropriate rotation on a computer screen ("Pitch", "Roll" and "Yaw" are the zones of the computer screen used by the pointer to activate the 3D rotation of the cloud). Three outliers (with the corresponding sample numbers) can be seen.

autocorrelated data by the way turned out to be intrinsically correlated [183], a concept that will be introduced in Chapter 22.

The Table 17.4 shows the very simple structure of the correlation matrix: an opposition between the duricrust variables (Fe, Al, V, P, Cr, Cu, Nb, As, Mo) and the variables of the flats (Si, Ti, Ce, Zr, Y); the variables are positively correlated within each group and negatively correlated between groups.

The Figure 17.4 is called the circle of correlations. It displays the correlations r_{ip} between the original variables $\mathbf{z}_i$ and a pair of principal components (factors). The coordinates of the variables on Figure 17.4 are obtained using the values of correlations with the first (ordinate) and the second (abscissa) principal component. The first principal component can be termed a "duricrust factor" as it displays in an obvious way the opposition between the variables characteristic of the duricrust variables (Fe,...) and the flats (Si,...).

The Figure 17.5 plots the sample cloud in the coordinate system provided by the first (ordinate) and the second (abscissa) principal components. Two subclouds can be seen: white coloured dots represent the samples from the duricrusts and black dots

	Fe	Al	V	P	Cr	Cu	Nb	As	Mo	Si	Ti	Ce	Zr
Al	.69												
V	.97	.69											
P	.89	.53	.87										
Cr	.94	.72	.95	.82									
Cu	.77	.67	.72	.71	.73								
Nb	.72	.43	.81	.71	.73	.50							
As	.87	.60	.87	.84	.83	.69	.76						
Mo	.79	.67	.81	.74	.78	.65	.78	.85					
Si	-.97	-.75	-.94	-.87	-.91	-.76	-.73	-.86	-.80				
Ti	-.93	-.59	-.90	-.83	-.85	-.66	-.69	-.82	-.72	.94			
Ce	-.76	-.44	-.73	-.67	-.73	-.50	-.57	-.64	-.54	.77	.81		
Zr	-.89	-.73	-.86	-.78	-.82	-.70	-.65	-.80	-.74	.94	.91	.70	
Y	-.92	-.68	-.89	-.80	-.86	-.68	-.68	-.80	-.73	.96	.96	.84	.93

Table 17.4: Correlation matrix of the Mali geochemical variables.

represent the samples from the flats.

The Figure 17.6 shows the geographical map of sample locations. The white coloured dots are samples classified as "duricrust" while the black dots are viewed as from "flats". Actually this map matches well the geological map displayed in ROQUIN ET AL. *[273].*

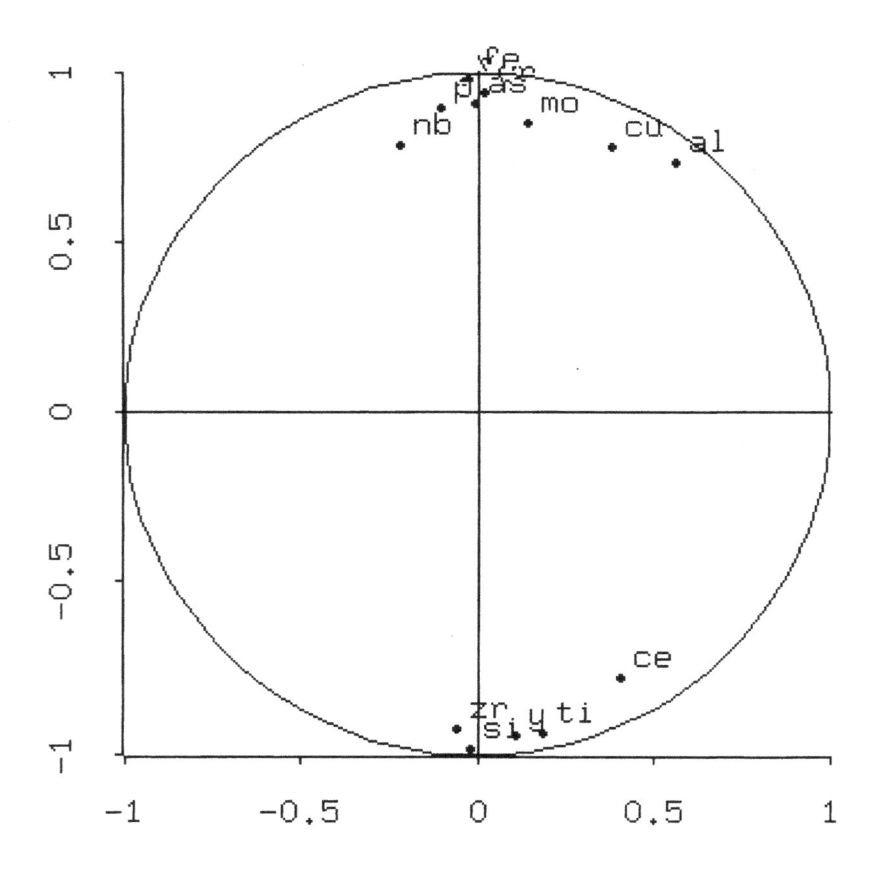

Figure 17.4: Circle of correlations for the first two principal components of the Mali geochemical variables: PC1 (ordinate) against PC2 (abscissa).

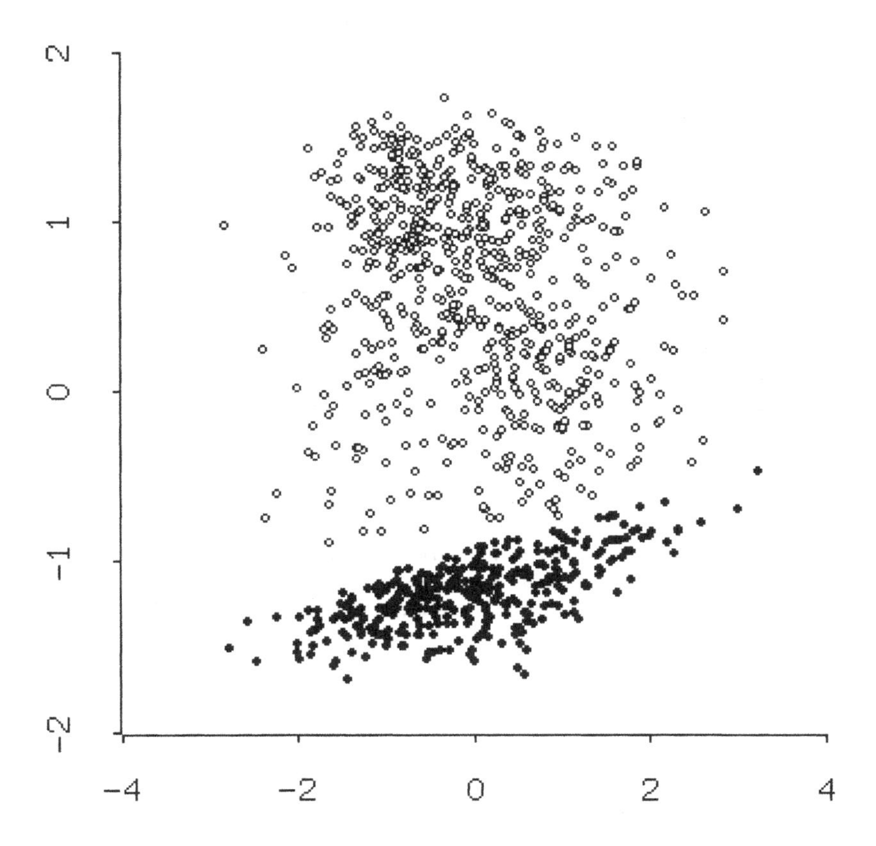

Figure 17.5: Sample cloud for the first two principal components of the Mali geochemical variables. PC1 (ordinate) against PC2 (abscissa). Duricrusts: white (Fe-Al), Flats: black (SiO_2).

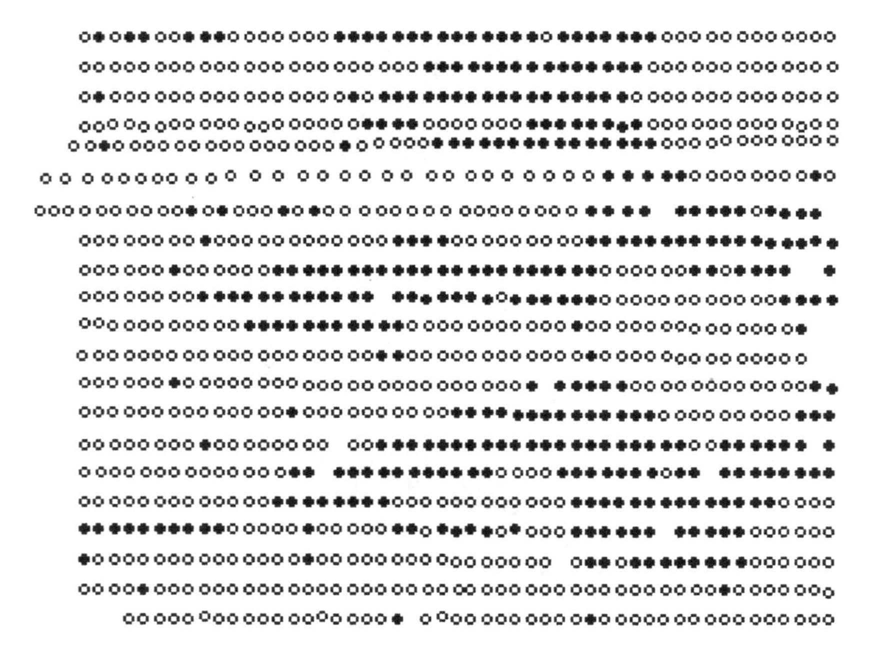

Figure 17.6: Geographical map of sample locations in a 4 × 5 km^2 area. Duricrusts: white (Fe-Al), Flats: black (SiO_2) .

18 Canonical Analysis

In the previous chapter we have seen how principal component analysis was used to determine linear orthogonal factors underlying a set of multivariate measurements. Now we turn to the more ambitious problem to compare two groups of variables by looking for pairs of orthogonal factors, which successively assess the strongest possible links between the two groups.

Factors in two groups of variables

Canonical analysis proposes to study simultaneously two groups of variables measuring various properties on the same set of samples. Successively pairs of factors are established which link best the two groups of variables. The aim being, when interpreting the factors, to see for each factor underlying one group of properties, if it can be connected to a factor underlying another set of properties gained from the same set of samples.

The $n \times N$ centered data table $\mathbf{Z}$ is split into two groups of variables,

$$\mathbf{Z} = [\mathbf{Z}_1 | \mathbf{Z}_2], \tag{18.1}$$

that is, the matrix $\mathbf{Z}$ is formed by putting together the columns of an $n \times M$ matrix $\mathbf{Z}_1$ and an $n \times L$ matrix $\mathbf{Z}_2$ with $N = M + L$.

The variance-covariance matrix $\mathbf{V}$ associated with $\mathbf{Z}$ can be subdivided into four blocs

$$\mathbf{V} = \frac{1}{n} [\mathbf{Z}_1 | \mathbf{Z}_2]^{\top} [\mathbf{Z}_1 | \mathbf{Z}_2] = \begin{pmatrix} \mathbf{V}_{11} & \mathbf{C}_{12} \\ \mathbf{C}_{21} & \mathbf{V}_{22} \end{pmatrix}, \tag{18.2}$$

where the matrices $\mathbf{V}_{11}$ of order $M \times M$ and $\mathbf{V}_{22}$ of order $L \times L$ are the variance-covariance matrices associated with $\mathbf{Z}_1$ and $\mathbf{Z}_2$. The covariance matrices $\mathbf{C}_{12}$ (of order $M \times L$) et $\mathbf{C}_{21}$ (of order $L \times M$) contain the covariances between variables of different groups

$$\mathbf{C}_{12}^{\top} = \mathbf{C}_{21}. \tag{18.3}$$

We are seeking pairs of factors $\{\mathbf{u}_p, \mathbf{v}_p\}$:

$$\mathbf{u}_p = \mathbf{Z}_1 \mathbf{a}_p \qquad \text{and} \qquad \mathbf{v}_p = \mathbf{Z}_2 \mathbf{b}_p, \tag{18.4}$$

which are uncorrelated within their respective groups,

$$\mathbf{u}_p \perp \mathbf{u}_k \qquad \text{and} \qquad \mathbf{v}_p \perp \mathbf{v}_k \qquad \text{for } p \neq k, \tag{18.5}$$

and for which the correlation $\operatorname{corr}(\mathbf{u}_p, \mathbf{v}_p)$ is maximal. This correlation, called the *canonical correlation* between two factors $\mathbf{u}$ and $\mathbf{v}$ is

$$\operatorname{corr}(\mathbf{u}, \mathbf{v}) = \frac{\mathbf{a}^\top \mathbf{Z}_1^\top \mathbf{Z}_2 \mathbf{b}}{\sqrt{\mathbf{a}^\top \mathbf{Z}_1^\top \mathbf{Z}_1 \mathbf{a}} \cdot \sqrt{\mathbf{b}^\top \mathbf{Z}_2^\top \mathbf{Z}_2 \mathbf{b}}}. \tag{18.6}$$

Deciding to norm $\mathbf{a}$ and $\mathbf{b}$ following the metric given by the variance-covariance matrices (supposed to be positive definite) of the two groups,

$$\mathbf{a}^\top \mathbf{V}_{11} \mathbf{a} = \mathbf{b}^\top \mathbf{V}_{22} \mathbf{b} = 1, \tag{18.7}$$

the correlation can be simply written as

$$\operatorname{corr}(\mathbf{u}, \mathbf{v}) = \mathbf{a}^\top \mathbf{C}_{12} \mathbf{b}. \tag{18.8}$$

Intermezzo: singular value decomposition

It is instructive to follow a little algebraic detour. We pose

$$\mathbf{x} = \sqrt{\mathbf{V}_{11}}\, \mathbf{a} \qquad \text{and} \qquad \mathbf{y} = \sqrt{\mathbf{V}_{22}}\, \mathbf{b} \qquad \text{with } \mathbf{x}^\top \mathbf{x} = \mathbf{y}^\top \mathbf{y} = 1, \tag{18.9}$$

in such a way that

$$\operatorname{corr}(\mathbf{u}, \mathbf{v}) = \mathbf{x}^\top \underbrace{\left[\sqrt{\mathbf{V}_{11}}\, \mathbf{C}_{12} \sqrt{\mathbf{V}_{22}}\right]}_{\mathbf{G}} \mathbf{y}. \tag{18.10}$$

For the complete system of vectors $\mathbf{x}$ and $\mathbf{y}$ we have

$$\mathbf{X}^\top \mathbf{G} \mathbf{Y} = \boldsymbol{\Sigma}, \tag{18.11}$$

that is to say the *singular value decomposition* (see appendix on Matrix Algebra for details):

$$\mathbf{G} = \mathbf{X} \boldsymbol{\Sigma} \mathbf{Y}^\top, \tag{18.12}$$

where the canonical correlations, which are the only non zero elements of $\boldsymbol{\Sigma}$, are identical with the singular values of the rectangular matrix $\mathbf{G}$.

Maximization of the correlation

We are looking for a pair of transformation vectors $\{\mathbf{a}, \mathbf{b}\}$ making the correlation between the factors $\{\mathbf{u}, \mathbf{v}\}$ maximal

$$\operatorname{corr}(\mathbf{u}, \mathbf{v}) = \mathbf{a}^\top \mathbf{C}_{12} \mathbf{b} \qquad \text{with } \mathbf{a}^\top \mathbf{V}_{11} \mathbf{a} = \mathbf{b}^\top \mathbf{V}_{22} \mathbf{b} = 1. \tag{18.13}$$

The objective function ϕ is

$$\phi = \mathbf{a}^\top \mathbf{C}_{12}\mathbf{b} - \frac{1}{2}\mu\,(\mathbf{a}^\top \mathbf{V}_{11}\mathbf{a} - 1) - \frac{1}{2}\mu'(\mathbf{b}^\top \mathbf{V}_{22}\mathbf{b} - 1), \tag{18.14}$$

where μ and μ' are Lagrange parameters.

Setting the partial derivatives with respect to $\mathbf{a}$ and $\mathbf{b}$ to zero

$$\frac{\partial\phi}{\partial\mathbf{a}} = 0 \iff \mathbf{C}_{12}\mathbf{b} - \mu\mathbf{V}_{11}\mathbf{a} = 0, \tag{18.15}$$

$$\frac{\partial\phi}{\partial\mathbf{b}} = 0 \iff \mathbf{C}_{21}\mathbf{a} - \mu'\mathbf{V}_{22}\mathbf{b} = 0. \tag{18.16}$$

What is the relation between μ and μ'? Premultiplying the first equation with $\mathbf{a}^\top$ and the second with $\mathbf{b}^\top$,

$$\mathbf{a}^\top \mathbf{C}_{12}\mathbf{b} - \mu\,\mathbf{a}^\top \mathbf{V}_{11}\mathbf{a} = 0 \iff \mu = \mathbf{a}^\top \mathbf{C}_{12}\mathbf{b}, \tag{18.17}$$

$$\mathbf{b}^\top \mathbf{C}_{21}\mathbf{a} - \mu'\mathbf{b}^\top \mathbf{V}_{22}\mathbf{b} = 0 \iff \mu' = \mathbf{b}^\top \mathbf{C}_{21}\mathbf{a}, \tag{18.18}$$

we see that

$$\mu = \mu' = \mathrm{corr}(\mathbf{u}, \mathbf{v}). \tag{18.19}$$

Having found earlier that

$$\mathbf{C}_{12}\mathbf{b} = \mu\mathbf{V}_{11}\mathbf{a} \quad \text{and} \quad \mathbf{C}_{21}\mathbf{a} = \mu\mathbf{V}_{22}\mathbf{b}, \tag{18.20}$$

we premultiply the first equation with $\mathbf{C}_{21}\mathbf{V}_{11}^{-1}$,

$$\mathbf{C}_{21}\mathbf{V}_{11}^{-1}\mathbf{C}_{12}\mathbf{b} = \mu\mathbf{C}_{21}\mathbf{a}, \tag{18.21}$$

and taking into account the second equation we have

$$\mathbf{C}_{21}\mathbf{V}_{11}^{-1}\mathbf{C}_{12}\mathbf{b} = \mu^2\mathbf{V}_{22}\mathbf{b}. \tag{18.22}$$

Finally this results in two eigenvalue problems, defining $\lambda = \mu^2$,

$$\mathbf{V}_{22}^{-1}\mathbf{C}_{21}\mathbf{V}_{11}^{-1}\mathbf{C}_{12}\mathbf{b} = \lambda\mathbf{b}, \tag{18.23}$$

$$\mathbf{V}_{11}^{-1}\mathbf{C}_{12}\mathbf{V}_{22}^{-1}\mathbf{C}_{21}\mathbf{a} = \lambda\mathbf{a}. \tag{18.24}$$

Only the smaller one of the two systems needs to be solved. The solution of the other one can then easily be obtained on the basis of the following transition formulae between the transformation vectors

$$\mathbf{a} = \frac{1}{\sqrt{\lambda}}\mathbf{V}_{11}^{-1}\mathbf{C}_{12}\mathbf{b} \quad \text{and} \quad \mathbf{b} = \frac{1}{\sqrt{\lambda}}\mathbf{V}_{22}^{-1}\mathbf{C}_{21}\mathbf{a}. \tag{18.25}$$

EXERCISE 18.1 *Show that two factors in the same group are orthogonal:*

$$\mathrm{cov}(\mathbf{u}_p, \mathbf{u}_k) = \delta_{pk}. \tag{18.26}$$

EXERCISE 18.2 *Show that two factors in different groups are orthogonal:*

$$\mathrm{cov}(\mathbf{u}_p, \mathbf{v}_k) = \mu\,\delta_{pk}. \tag{18.27}$$

EXERCISE 18.3 *Show that the vector* $\mathbf{a}$ *in the following equation*

$$\mathbf{V}_{11}\mathbf{a} = \lambda\,\mathbf{C}_{12}\mathbf{V}_{22}^{-1}\mathbf{C}_{21}\mathbf{a} \tag{18.28}$$

solves a canonical analysis problem.

19 Correspondence Analysis

Canonical analysis investigates, so to speak, the correspondence between the factors of two groups of quantitative variables. The same approach applied to two qualitative variables, each of which represents a group of mutually exclusive categories, is known under the evocative name of "correspondence analysis".

Disjunctive table

A qualitative variable z is a system of categories (classes) which are mutually exclusive: every sample belongs to exactly one category. The membership of a sample z_α to a category C_i can be represented numerically by an indicator function

$$\mathbf{1}_{C_i}(z_\alpha) = \begin{cases} 1 & \text{if } z_\alpha \in C_i \\ 0 & \text{otherwise.} \end{cases} \tag{19.1}$$

The matrix recording the memberships of the n samples to the N categories of the qualitative variable is the disjunctive table $\mathbf{H}$ of order $n \times N$

$$\mathbf{H} = \Big[\mathbf{1}_{C_i}(z_\alpha)\Big]. \tag{19.2}$$

Each row of the disjunctive table contains only one element equal to 1 and has zeroes elsewhere, as the possibility that a sample belongs simultaneously to more than one category is excluded.

The product of $\mathbf{H}^\top$ with $\mathbf{H}$ results in a diagonal matrix whose diagonal elements n_{ii} indicate the number of samples in the category number i. The division of the n_{ii} by n yields the proportion F_{ii} of samples contained in a category and we have

$$\mathbf{V} = \frac{1}{n}\mathbf{H}^\top\mathbf{H} = \begin{pmatrix} F_{11} & 0 & 0 \\ 0 & \ddots & 0 \\ 0 & 0 & F_{NN} \end{pmatrix}. \tag{19.3}$$

Contingency table

Two qualitative variables measured on the same set of samples are represented, respectively, by a table $\mathbf{H}_1$ of order $n \times M$ and a table $\mathbf{H}_2$ of order $n \times L$.

The product of these two disjunctive tables has as elements the number of samples n_{ij} belonging simultaneously to a category i of the first qualitative variable and a category j of the second qualitative variable. The elements of this table, which is a contingency table of order $M \times L$, can be divided by the sample total n, thus measuring the proportion of samples at the crossing of the categories of two qualitative variables

$$\mathbf{C}_{12} = \frac{1}{n} \mathbf{H}_1^\top \mathbf{H}_2 = [F_{ij}]. \tag{19.4}$$

Canonical analysis of disjunctive tables

Correspondence analysis consists in applying the formalism of canonical analysis to the disjunctive tables $\mathbf{H}_1$ and $\mathbf{H}_2$ by forming the matrices

$$\mathbf{C}_{21}^\top = \mathbf{C}_{12}, \qquad \mathbf{V}_{11} = \frac{1}{n} \mathbf{H}_1^\top \mathbf{H}_1, \qquad \mathbf{V}_{22} = \frac{1}{n} \mathbf{H}_2^\top \mathbf{H}_2. \tag{19.5}$$

Because of the diagonal structure of the matrices $\mathbf{V}_{11}$ and $\mathbf{V}_{22}$, which only intervene in the calibration of the transformation vectors, we can say that correspondence analysis boils down to the investigation of the links between the rows and the columns of the contingency table $\mathbf{C}_{12}$.

Coding of a quantitative variable

A quantitative variable z can be transformed into a qualitative variable by partitioning it with a set of non overlapping (disjunct) intervals C_i. Indicator functions of these intervals allow to constitute a disjunctive table associated with z

$$\mathbf{1}_{C_i}(z_\alpha) = \begin{cases} 1 & \text{if } z_\alpha \text{ belongs to the interval } C_i \\ 0 & \text{otherwise.} \end{cases} \tag{19.6}$$

A subdivision is sometimes given a priori, as for example with granulometric fractions of a soil sample.

The coding of a quantitative variable enables a correspondence analysis which explores the links between the intervals of this quantitative variable and the categories of a qualitative variable.

Contingencies between two quantitative variables

The diagonal matrix $\mathbf{V} = (1/n)\, \mathbf{H}^\top \mathbf{H}$ associated with the partition of a quantitative variable contains the values of the histogram of the samples z_α, because its diagonal elements show the frequencies in the classes of values of z.

The contingency table $\mathbf{C}_{12} = (1/n)\, \mathbf{H}_1^\top \mathbf{H}_2$ of two quantitative variables z_1 and z_2 holds the information about a bivariate histogram, for we find in it the frequencies of samples at the crossing of classes of z_1 and z_2.

Continuous correspondence analysis

The bivariate histogram, that is, the contingency table of two quantitative variables, can be modeled with a bivariate distribution function $F(dZ_1, dZ_2)$. If the corresponding bivariate density function $\varphi(Z_1, Z_2)$ exists, we have

$$F(dZ_1, dZ_2) \quad = \quad F_1(dZ_1)\, F_2(dZ_2)\, \varphi(Z_1, Z_2), \tag{19.7}$$

where $F_1(dZ_1),\ F_2(dZ_2)$ are the marginal distribution functions of Z_1 and Z_2, used to model their respective histograms. Supposing, moreover, the density function φ is square integrable, it can be decomposed into a system of eigenvalues $\lambda_p = \mu_p^2$ and eigenfunctions $f_p(Z_1),\ g_p(Z_2)$

$$\varphi(Z_1, Z_2) \quad = \quad \sum_{p=0}^{\infty} \mu_p\, f_p(Z_1)\, g_p(Z_2). \tag{19.8}$$

This decomposition is the continuous analogue of the singular value decomposition.

EXAMPLE 19.1 *When the bivariate law is bi-Gaussian, the decomposition into eigenvalues yields coefficients μ_p equal to a coefficient ρ power p,*

$$\mu_p \quad = \quad \rho^p \qquad \text{with } |\rho| < 1, \tag{19.9}$$

as well as eigenfunctions equal to normalized Hermite polynomials η_p,

$$\eta_p(Z) \quad = \quad \frac{H_p(Z)}{\sqrt{p!}}. \tag{19.10}$$

The bivariate Gaussian distribution $G(dZ_1, dZ_2)$ can then be decomposed into factors η_p of decreasing importance ρ^p,

$$G(dZ_1, dZ_2) \quad = \quad \sum_{p=0}^{\infty} \rho^p\, \eta_p(Z_1)\, \eta_p(Z_2)\, G(dZ_1)\, G(dZ_2). \tag{19.11}$$

EXAMPLE 19.2 *In non linear geostatistics, interest is focused on the bivariate distribution of two random variables $Z(\mathbf{x}_\alpha)$ and $Z(\mathbf{x}_\beta)$ located at two points $\mathbf{x}_\alpha$ and $\mathbf{x}_\beta$ of a spatial domain. The decomposition of this bivariate distribution results, under specific assumptions, in a so called* isofactorial model. *Isofactorial models are used in* disjunctive kriging *for estimating a non linear function of the data in a spatial domain.*

In particular, in the case of the bi-Gaussian isofactorial model, the coefficient ρ will be equal to the value of the correlation function $\rho(\mathbf{x}_\alpha - \mathbf{x}_\beta)$ between two points $\mathbf{x}_\alpha$ and $\mathbf{x}_\beta$. This topic is pursued in Chapters 32 to 36.

Part D

Multivariate Geostatistics

20 Direct and Cross Covariances

The cross covariance between two random functions can be computed not only at locations $\mathbf{x}$ but also for pairs of locations separated by a vector $\mathbf{h}$. On the basis of an assumption of joint second-order stationarity a cross covariance function between two random functions is defined which only depends on the separation vector $\mathbf{h}$.

The interesting feature of the cross covariance function is that it is generally not an even function, i.e. its values for $+\mathbf{h}$ and $-\mathbf{h}$ may be different. This occurs in time series when the effect of one variable on another variable is delayed.

The cross variogram is an even function, defined in the framework of an intrinsic hypothesis. When delay-effects are an important aspect of the coregionalization, the cross variogram is not an appropriate tool to describe the data.

Cross covariance function

The direct and cross covariance functions $C_{ij}(\mathbf{h})$ of a set of N random functions $Z_i(\mathbf{x})$ are defined in the framework of a joint second order stationarity hypothesis

$$\begin{cases} \mathrm{E}[Z_i(\mathbf{x})] = m_i & \text{for all } \mathbf{x} \in \mathcal{D};\ i = 1, \ldots, N, \\ \mathrm{E}[\,(Z_i(\mathbf{x}) - m_i) \cdot (Z_j(\mathbf{x}+\mathbf{h}) - m_j)\,] & = C_{ij}(\mathbf{h}) \\ & \text{for all } \mathbf{x}, \mathbf{x}+\mathbf{h} \in \mathcal{D};\ i, j = 1, \ldots, N. \end{cases} \tag{20.1}$$

The mean of each variable $Z_i(\mathbf{x})$ at any point of the domain is equal to a constant m_i. The covariance of a variable pair depends only on the vector $\mathbf{h}$ linking a point pair and is invariant for any translation of the point pair in the domain.

A set of cross covariance functions is a positive definite function, i.e. the variance of any linear combination of N variables at $n+1$ points with a set of weights w_α^i needs to be positive. For any set of points $\mathbf{x}_\alpha \in \mathcal{D}$ and any set of weights $w_\alpha^i \in \mathbb{R}$

$$\mathrm{var}\Big(\sum_{i=1}^{N} \sum_{\alpha=0}^{n} w_\alpha^i\, Z_i(\mathbf{x}_\alpha)\Big) = \sum_{i=1}^{N} \sum_{j=1}^{N} \sum_{\alpha=0}^{n} \sum_{\beta=0}^{n} w_\alpha^i\, w_\beta^j\, C_{ij}(\mathbf{x}_\alpha - \mathbf{x}_\beta) \;\geq\; 0. \tag{20.2}$$

Delay effect

The cross covariance function is not a priori an even or an odd function. Generally for $i \neq j$, a change in the order of the variables or a change in the sign of the separation vector $\mathbf{h}$ changes the value of the cross covariance function

$$C_{ij}(\mathbf{h}) \neq C_{ji}(\mathbf{h}) \qquad \text{and} \qquad C_{ij}(-\mathbf{h}) \neq C_{ij}(\mathbf{h}). \tag{20.3}$$

If both the sequence and the sign are changed, we are back to the same value

$$C_{ij}(\mathbf{h}) = C_{ji}(-\mathbf{h}). \tag{20.4}$$

In particular, the maximum of the cross covariance function (assuming positive correlation between a given variable pair) may be shifted away from the origin of the abscissa in a certain direction by a vector $\mathbf{r}$. This shift of the maximal correlation is frequent with time series, where one variable can have an effect on another variable which is not instantaneous. The time for the second variable to react to fluctuations of the first variable causes a *delay* in the correlation between the time series.

EXAMPLE 20.1 *Let $Z_1(\mathbf{x})$ be a random function obtained by shifting the random function $Z_2(\mathbf{x})$ with a vector $\mathbf{r}$, multiplying it with a constant a and adding $\varepsilon(\mathbf{x})$*

$$Z_1(\mathbf{x}) = a\, Z_2(\mathbf{x}+\mathbf{r}) + \varepsilon(\mathbf{x}), \tag{20.5}$$

where $\varepsilon(\mathbf{x})$ is an independent measurement error without spatial correlation, i.e. a quantity with a zero mean and a nugget-effect covariance function $C_{nug}(\mathbf{h})$.

The direct covariance function of $Z_1(\mathbf{x})$ is proportional to that of $Z_2(\mathbf{x})$,

$$C_{11}(\mathbf{h}) = a^2\, C_{22}(\mathbf{h}) + C_{nug}(\mathbf{h}), \tag{20.6}$$

and the cross covariance function is obtained from the even function $C_{22}(\mathbf{h})$ translated by $\mathbf{r}$,

$$C_{12}(\mathbf{h}) = a\, C_{22}(\mathbf{h}+\mathbf{r}). \tag{20.7}$$

It is worth noting that the nugget-effect term is absent from the cross covariance function.

EXERCISE 20.2 *Compute the cross covariance function between $Z_1(\mathbf{x})$ and $Z_2(\mathbf{x})$ for*

$$Z_1(\mathbf{x}) = a_1\, Z_2(\mathbf{x}+\mathbf{r}_1) + a_2\, Z_2(\mathbf{x}+\mathbf{r}_2), \tag{20.8}$$

which incorporates two shifts $\mathbf{r}_1$ and $\mathbf{r}_2$.

Cross variogram

The direct and cross variograms $\gamma_{ij}(\mathbf{h})$ are defined in the context of a joint intrinsic hypothesis for N random functions, when for any $\mathbf{x}, \mathbf{x}+\mathbf{h} \in \mathcal{D}$ and all pairs $i, j = 1, \ldots, N$

$$\begin{cases} \mathrm{E}[Z_i(\mathbf{x}+\mathbf{h}) - Z_i(\mathbf{x})] = 0, \\ \mathrm{cov}\Big[(Z_i(\mathbf{x}+\mathbf{h}) - Z_i(\mathbf{x})), (Z_j(\mathbf{x}+\mathbf{h}) - Z_j(\mathbf{x}))\Big] = 2\,\gamma_{ij}(\mathbf{h}). \end{cases} \tag{20.9}$$

The *cross variogram* is thus defined as half the expectation of the product of the increments of two variables

$$\gamma_{ij}(\mathbf{h}) = \frac{1}{2}\,\mathrm{E}\Big[(Z_i(\mathbf{x}+\mathbf{h}) - Z_i(\mathbf{x}))\cdot(Z_j(\mathbf{x}+\mathbf{h}) - Z_j(\mathbf{x}))\Big]. \tag{20.10}$$

The cross variogram is obviously an even function and it satisfies the following inequality

$$\gamma_{ii}(\mathbf{h})\,\gamma_{jj}(\mathbf{h}) \geq |\gamma_{ij}(\mathbf{h})|^2, \tag{20.11}$$

because the square of the covariance of increments from two variables is bounded by the product of the corresponding increment variances. Actually a matrix $\mathbf{\Gamma}(\mathbf{h}_0)$ of direct and cross variogram values is a positive semi-definite matrix for any fixed $\mathbf{h}_0$ because it is a variance-covariance matrix of increments.

It is appealing to investigate its relation to the cross covariance function in the framework of joint second-order stationarity. The following formula is easily obtained

$$\gamma_{ij}(\mathbf{h}) = C_{ij}(0) - \frac{1}{2}(C_{ij}(-\mathbf{h}) + C_{ij}(+\mathbf{h})), \tag{20.12}$$

which shows that the cross variogram takes the average of the values for $-\mathbf{h}$ and for $+\mathbf{h}$ of the corresponding cross covariance function. Decomposing the cross covariance function into an even and an odd function

$$C_{ij}(\mathbf{h}) = \underbrace{\frac{1}{2}(C_{ij}(+\mathbf{h}) + C_{ij}(-\mathbf{h}))}_{\text{even term}} + \underbrace{\frac{1}{2}(C_{ij}(+\mathbf{h}) - C_{ij}(-\mathbf{h}))}_{\text{odd term}}, \tag{20.13}$$

we see that the cross variogram only recovers the even term of the cross covariance function. It is not adequate for modeling data in which the odd term of the cross covariance function plays a significant role.

The cross covariance function between a random function and its derivative is an odd function. Such antisymmetric behavior of the cross covariance function is found in hydrogeology, when computing the theoretical cross covariance function between water head and transmissivity, where the latter variable is seen as the derivative of the former (see for example [68], p68). Several cross covariance function models between a random function and its derivative are given in [93], [323].

EXAMPLE 20.3 (GAS FURNACE DATA FROM [30]) *Two time series, one corresponding to the fluctuation of a gas input into a furnace and the other being the output of CO_2 from the furnace, are measured at the same time. The chemical reaction between the input variable and the output variable takes several tens of seconds and we can expect a delay effect on measurements taken every 9 seconds.*

The experimental cross covariance function for different distance classes $\mathfrak{H}$ gathering n_c pairs of locations $\mathbf{x}_\alpha, \mathbf{x}_\beta$ according to their separation vectors $\mathbf{x}_\alpha - \mathbf{x}_\beta = \mathbf{h} \in \mathfrak{H}$ is computed as

$$C^\star_{ij}(\mathfrak{H}) = \frac{1}{n_c}\sum_{\alpha=1}^{n_c}(z_i(\mathbf{x}_\alpha) - m_i)\cdot(z_j(\mathbf{x}_\alpha+\mathbf{h}) - m_j). \tag{20.14}$$

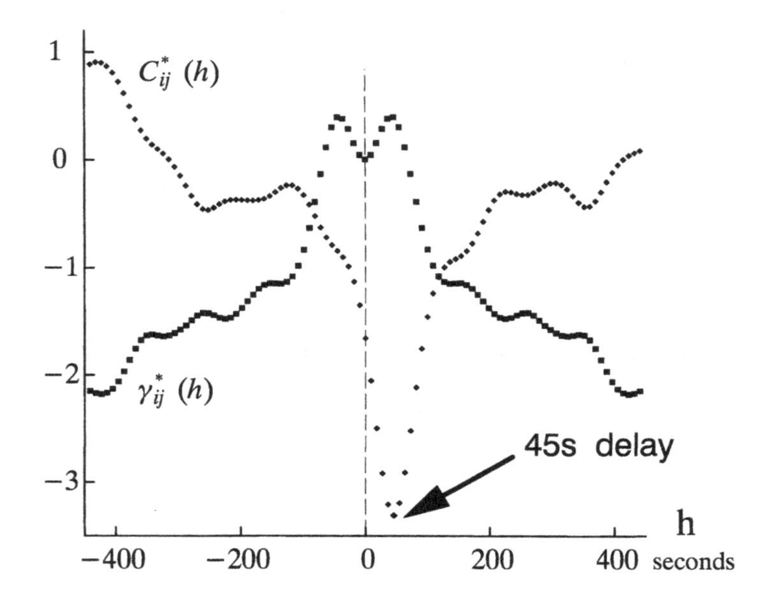

Figure 20.1: Experimental cross covariance $C^{\star}_{ij}(\mathbf{h})$ and experimental cross variogram $\gamma^{\star}_{ij}(\mathbf{h})$.

The experimental cross covariance function shown on Figure 20.1 reveals a delay of 45 seconds between fluctuations in the gas input and a subsequent effect on the rate of carbon dioxide measured at the output of the system.

Figure 20.1 also exhibits the corresponding experimental cross variogram for different classes $\mathfrak{H}$ *with* $\mathbf{h} \in \mathfrak{H}$

$$\gamma^{\star}_{ij}(\mathfrak{H}) = \frac{1}{2n_c} \sum_{\alpha=1}^{n_c} (z_i(\mathbf{x}_\alpha+\mathbf{h})-z_i(\mathbf{x}_\alpha)) \cdot (z_j(\mathbf{x}_\alpha+\mathbf{h})-z_j(\mathbf{x}_\alpha)). \tag{20.15}$$

The experimental cross variogram is symmetric with respect to the ordinate and is not suitable for detecting a delay.

Figure 20.2 displays a plot of the decomposition of the experimental cross covariance into an even and an odd term. The even term has the same shape as the cross variogram when viewed upside down. The odd term measures the degree of asymmetry of the cross covariance: in case of symmetry the odd term would be identically zero.

In practice the experimental cross covariance function should always be plotted (together with its decomposition into even and odd terms) before attempting to use a cross variogram. It has to be checked whether there are any important asymmetries in the cross covariance functions. For example, if one of the variables in the pair is the derivative of the other, the cross covariance will be antisymmetric, while its even term as well as the cross variogram will be identically zero. In such a case, using directly

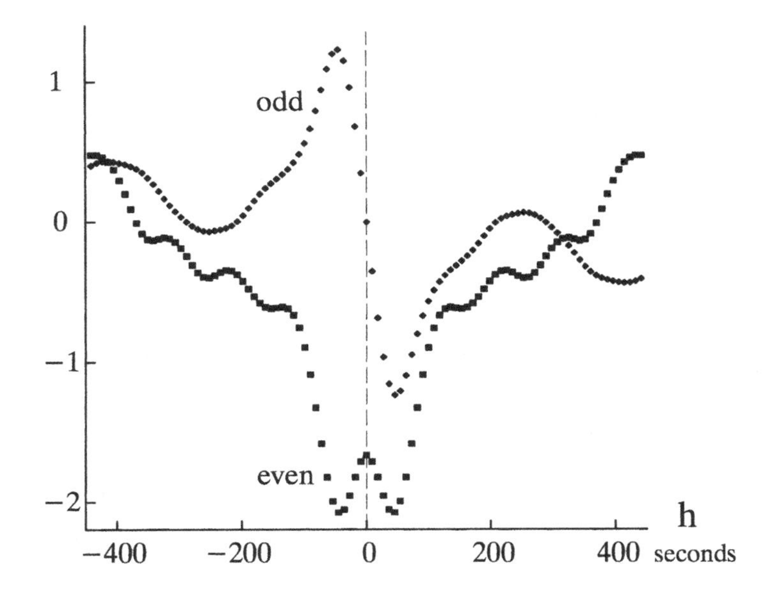

Figure 20.2: Even and odd terms of the experimental cross covariance.

the experimental cross variogram would lead the data analyst to miss an important non-linear relation between the variables.

Pseudo cross variogram

An alternate generalization of the variogram, the pseudo cross variogram $\pi_{ij}(\mathbf{h})$, has been proposed by MYERS [233] and CRESSIE [64] by considering the variance of cross increments,

$$\operatorname{var}\Big[Z_i(\mathbf{x}+\mathbf{h})-Z_j(\mathbf{x})\Big] \quad = \quad 2\,\pi_{ij}(\mathbf{h}), \tag{20.16}$$

instead of the covariance of the direct increments as in the definition (20.9) of the cross variogram $\gamma_{ij}(\mathbf{h})$.

Assuming for the expectation of the cross increments

$$\mathrm{E}[\,Z_i(\mathbf{x}+\mathbf{h}) - Z_j(\mathbf{x})\,] \quad = \quad 0 \tag{20.17}$$

the *pseudo cross variogram* comes as

$$\pi_{ij}(\mathbf{h}) \quad = \quad \frac{1}{2}\,\mathrm{E}[\;(Z_i(\mathbf{x}+\mathbf{h}) - Z_j(\mathbf{x}))^2\;]. \tag{20.18}$$

The function $\pi_{ij}(\mathbf{h})$ has the advantage of not being even. The assumption of stationary cross increments is however unrealistic: it usually does not make sense to take the difference between two variables measured in different physical units (even if they are rescaled) as the variables often also do not have the same support. PAPRITZ et al. [241], PAPRITZ & FLÜHLER [240] have experienced limitations in the usefulness of the pseudo cross variogram and they argue that it applies only to second-order stationary functions. Another drawback of the pseudo cross variogram function, which takes only positive values, is that it is not adequate for modeling negatively correlated variables. For these reasons we shall not consider further this approach and deal only with the more classical cross covariance function to which $\pi_{ij}(\mathbf{h})$ is related by

$$\pi_{ij}(\mathbf{h}) = \frac{C_{ii}(\mathbf{0}) + C_{jj}(\mathbf{0})}{2} - C_{ij}(\mathbf{h}), \tag{20.19}$$

assuming second-order stationarity.

A different viewpoint is found in CRESSIE & WIKLE [66].

Difficult characterization of the cross covariance function

For a pair of second order stationary random functions the following inequality is valid

$$C_{ii}(\mathbf{0})\, C_{jj}(\mathbf{0}) \geq |C_{ij}(\mathbf{h})|^2. \tag{20.20}$$

This inequality tells us that a cross covariance function is bounded by the product of the values of the direct covariance functions at the origin.

It should be remembered that there are no inequalities of the following type

$$C_{ij}(\mathbf{0}) \not\geq |C_{ij}(\mathbf{h})|, \tag{20.21}$$

$$C_{ii}(\mathbf{h})\, C_{jj}(\mathbf{h}) \not\geq |C_{ij}(\mathbf{h})|^2, \tag{20.22}$$

as the quantities on the left hand side of the expressions can in both cases be negative (or zero). In particular we do not usually have the inequality

$$|C_{ij}(\mathbf{0})| \not\geq |C_{ij}(\mathbf{h})|, \tag{20.23}$$

because the maximal absolute value of $C_{ij}(\mathbf{h})$ may not be located at the origin.

The matrix $\mathbf{C}(\mathbf{h})$ of direct and cross covariances is in general neither positive nor negative definite at any specific lag $\mathbf{h}$. It is hard to describe the properties of a covariance function matrix, while it easy to characterize the corresponding matrix of spectral densities or distribution functions.

21 Covariance Function Matrices

It is actually difficult to characterize directly a covariance function matrix. This becomes easy in the spectral domain on the basis of Cramer's generalization of the Bochner theorem, which is presented in this chapter. We consider complex covariance functions.

Covariance function matrix

The matrix $\mathbf{C}(\mathbf{h})$ of direct and cross covariance functions of a vector of complex random functions (with zero means without loss of generality),

$$\mathbf{z}(\mathbf{x}) \quad = \quad (Z_1, \ldots, Z_i, \ldots, Z_N)^\top \qquad \text{with} \quad \mathrm{E}[\,\mathbf{z}(\mathbf{x})\,] = \mathbf{0}, \tag{21.1}$$

is defined as

$$\mathbf{C}(\mathbf{h}) \quad = \quad \mathrm{E}\Big[\,\mathbf{z}(\mathbf{x})\,\overline{\mathbf{z}(\mathbf{x}+\mathbf{h})}^\top\,\Big]. \tag{21.2}$$

The covariance function matrix is a *Hermitian positive semi-definite* function, that is to say, for any set of points we have $\mathbf{x}_\alpha \in \mathcal{D}$ and any set of complex weights w_α^i

$$\sum_{i=1}^{N}\sum_{j=1}^{N}\sum_{\alpha=0}^{n}\sum_{\beta=0}^{n} w_\alpha^i\,\overline{w}_\beta^j\,C_{ij}(\mathbf{x}_\alpha-\mathbf{x}_\beta) \;\geq\; 0. \tag{21.3}$$

A *Hermitian matrix* is the generalization to complex numbers of a real symmetric matrix. The diagonal elements of a Hermitian $N \times N$ matrix $\mathbf{A}$ are real and the off-diagonal elements are equal to the complex conjugates of the corresponding elements with transposed indices: $a_{ij} = \overline{a_{ji}}$.

For the matrix of direct and cross covariance functions of a set of complex variables this means that the direct covariances are real, while the cross covariance functions are generally complex.

Cramer's theorem

Following a generalization of Bochner's theorem due to CRAMER [61] (see also [109], [362]), each element $C_{ij}(\mathbf{h})$ of a matrix $\mathbf{C}(\mathbf{h})$ of continuous direct and cross covariance functions has the spectral representation

$$C_{ij}(\mathbf{h}) \quad = \quad \int_{-\infty}^{+\infty}\cdots\int_{-\infty}^{+\infty} \mathrm{e}^{\,\mathrm{i}\boldsymbol{\omega}^\top\mathbf{h}}\,dF_{ij}(\boldsymbol{\omega}), \tag{21.4}$$

where the $F_{ij}(\boldsymbol{\omega})$ are spectral distribution functions and $\boldsymbol{\omega}$ is a vector of the same dimension as $\mathbf{h}$. The diagonal terms $F_{ii}(\boldsymbol{\omega})$ are real, non decreasing and bounded. The off-diagonal terms $F_{ij}(\boldsymbol{\omega})$ $(i \neq j)$ are in general complex valued and of finite variation. Conversely, any matrix of continuous functions $\mathbf{C}(\mathbf{h})$ is a matrix of covariance functions, if the matrices of increments $\Delta F_{ij}(\boldsymbol{\omega})$ are Hermitian positive semi-definite for any block $\Delta\boldsymbol{\omega}$ (using the terminology of [300]).

Spectral densities

If attention is restricted to absolutely integrable covariance functions, we have the following representation

$$C_{ij}(\mathbf{h}) \;=\; \int_{-\infty}^{+\infty} \cdots \int_{-\infty}^{+\infty} \mathrm{e}^{\,\mathrm{i}\boldsymbol{\omega}^\top \mathbf{h}}\, f_{ij}(\boldsymbol{\omega})\, d\boldsymbol{\omega}. \tag{21.5}$$

This Fourier transform can be inverted and the *spectral density functions* $f_{ij}(\boldsymbol{\omega})$ can be computed from the cross covariance functions

$$f_{ij}(\boldsymbol{\omega}) \;=\; \frac{1}{(2\pi)^n} \int_{-\infty}^{+\infty} \cdots \int_{-\infty}^{+\infty} \mathrm{e}^{\,-\mathrm{i}\boldsymbol{\omega}^\top \mathbf{h}}\, C_{ij}(\mathbf{h})\, d\mathbf{h}. \tag{21.6}$$

The matrix of spectral densities of a set of covariance functions is positive semi-definite for any value of $\boldsymbol{\omega}$. For any pair of random functions this implies the inequality

$$|f_{ij}(\boldsymbol{\omega})|^2 \;\leq\; f_{ii}(\boldsymbol{\omega})\, f_{jj}(\boldsymbol{\omega}). \tag{21.7}$$

With functions that have spectral densities it is thus simple to check whether a given matrix of functions can be considered as a covariance function matrix.

EXERCISE 21.1 *Compute the spectral density of the exponential covariance function (in one spatial dimension)* $C(h) = b\,\mathrm{e}^{-a|h|}$, $b > 0$, $a > 0$ *using the formula*

$$f(\omega) \;=\; \frac{1}{2\pi} \int_{-\infty}^{+\infty} \mathrm{e}^{-\mathrm{i}\omega h}\, C(h)\, dh. \tag{21.8}$$

EXERCISE 21.2 *Show (in one dimensional space) that the function*

$$C_{ij}(h) \;=\; \mathrm{e}^{-\left(\frac{a_i + a_j}{2}\right)|h|} \tag{21.9}$$

can only be the cross covariance function of two random functions with an exponential cross covariance function $C_i(h) = \mathrm{e}^{-a_i|h|}$, $a_i > 0$, *if* $a_i = a_j$.

Phase shift

In one dimensional space, for example along the time axis, phase shifts can easily be interpreted. Considering the inverse Fourier transforms (admitting their existence) of the even and the odd term of the cross covariance function, which are traditionally called the ***cospectrum*** $c_{ij}(\omega)$ and the ***quadrature spectrum*** $q_{ij}(\omega)$, we have the following decomposition of the spectral density

$$f_{ij}(\omega) \quad = \quad c_{ij}(\omega) - \mathrm{i}\, q_{ij}(\omega). \tag{21.10}$$

The cospectrum represents the covariance between the frequency components of the two processes which are in phase, while the quadrature spectrum characterizes the covariance of the out of phase components. Further details are found in [251], [362].

In polar notation the complex function $f_{ij}(\omega)$ can be expressed as

$$f_{ij}(\omega) \quad = \quad |f_{ij}(\omega)|\, \mathrm{e}^{\,\mathrm{i}\,\varphi_{ij}(\omega)}. \tag{21.11}$$

The ***phase spectrum*** $\varphi_{ij}(\omega)$ is the average phase shift of the proportion of $Z_i(x)$ with the frequency ω on the proportion of $Z_j(x)$ at the same frequency and

$$\tan\, \varphi_{ij}(\omega) \quad = \quad \frac{\mathrm{Im}(f_{ij}(\omega))}{\mathrm{Re}(f_{ij}(\omega))} = \frac{-q_{ij}(\omega)}{c_{ij}(\omega)}. \tag{21.12}$$

Absence of phase shift at a frequency ω implies a zero imaginary part $\mathrm{Im}(f_{ij}(\omega))$ for this frequency. When there is no phase shift at any frequency, the cross spectral density $f_{ij}(\omega)$ is real and its Fourier transform is an even function.

22 Intrinsic Multivariate Correlation

Is the multivariate correlation structure of a set of variables independent of the spatial correlation? When the answer is positive, the multivariate correlation is said to be *intrinsic*. It is an interesting question in applications to know whether or not a set of variables can be considered intrinsically correlated, because the answer may imply considerable simplifications for subsequent handling of this data.

Intrinsic correlation model

The simplest multivariate covariance model that can be adopted for a covariance function matrix consists in describing the relations between variables by the variance-covariance matrix $\mathbf{V}$ and the relations between points in space by a spatial correlation function $\rho(\mathbf{h})$ which is the same for all variables

$$\mathbf{C}(\mathbf{h}) \quad = \quad \mathbf{V}\, \rho(\mathbf{h}). \tag{22.1}$$

This model is called the *intrinsic correlation* model because it has the particular property that the correlation ρ_{ij} between two variables does not depend upon spatial scale,

$$\frac{\sigma_{ij}\, \rho(\mathbf{h})}{\sqrt{\sigma_{ii}\, \rho(\mathbf{h})\, \sigma_{jj}\, \rho(\mathbf{h})}} \quad = \quad \frac{\sigma_{ij}}{\sqrt{\sigma_{ii}\, \sigma_{jj}}} = \rho_{ij}. \tag{22.2}$$

In practice the intrinsic correlation model is obtained when direct and cross covariance functions are chosen which are all proportional to a same basic spatial correlation function,

$$C_{ij}(\mathbf{h}) \quad = \quad b_{ij}\, \rho(\mathbf{h}), \tag{22.3}$$

and the coefficients b_{ij} are subsequently interpreted as variances σ_{ii} or covariances σ_{ij}, depending whether i is equal to j or not.

EXERCISE 22.1 *Show that the intrinsic correlation model is a valid model for covariance function matrices.*

The intrinsic correlation model implies even cross covariance functions because $\rho(\mathbf{h})$ is an even function as it is a normalized direct covariance function.

An intrinsic correlation model can be formulated for intrinsically stationary random functions (please note the two different uses of the adjective *intrinsic*). For variograms the intrinsic correlation model is defined as the product of a positive semi-definite matrix $\mathbf{B}$ of coefficients b_{ij}, called the *coregionalization matrix*, multiplied with a direct variogram $\gamma(\mathbf{h})$

$$\mathbf{\Gamma}(\mathbf{h}) = \mathbf{B}\,\gamma(\mathbf{h}). \tag{22.4}$$

EXERCISE 22.2 *Show that the intrinsic correlation model is a valid model for variogram matrices.*

Linear model

The linear random function model, that can be associated with the intrinsically correlated multivariate variogram model, consists of a linear combination of coefficients a_p^i with factors $Y_p(\mathbf{x})$

$$Z_i(\mathbf{x}) = \sum_{p=1}^{N} a_p^i\, Y_p(\mathbf{x}). \tag{22.5}$$

The factors $Y_p(\mathbf{x})$ have pairwise uncorrelated increments,

$$\mathrm{E}[\,(Y_p(\mathbf{x}+\mathbf{h}) - Y_p(\mathbf{x}))\cdot(Y_q(\mathbf{x}+\mathbf{h}) - Y_q(\mathbf{x}))\,] = 0 \qquad \text{for} \quad p \neq q, \tag{22.6}$$

and all have the same variogram

$$\mathrm{E}[\,(Y_p(\mathbf{x}+\mathbf{h}) - Y_p(\mathbf{x}))^2\,] = 2\,\gamma(\mathbf{h}) \qquad \text{for} \quad p = 1, \ldots, N. \tag{22.7}$$

The variogram of a pair of random functions in this model is proportional to one basic model

$$\begin{aligned}
\gamma_{ij}(\mathbf{h}) &= \frac{1}{2}\,\mathrm{E}[\,(Z_i(\mathbf{x}+\mathbf{h}) - Z_i(\mathbf{x}))\cdot(Z_j(\mathbf{x}+\mathbf{h}) - Z_j(\mathbf{x}))\,] \\
&= \frac{1}{2}\,\mathrm{E}\Big[\sum_{p=1}^{N}\sum_{q=1}^{N} a_p^i\, a_q^j\,(Y_p(\mathbf{x}+\mathbf{h}) - Y_p(\mathbf{x}))\cdot(Y_q(\mathbf{x}+\mathbf{h}) - Y_q(\mathbf{x}))\Big] \\
&= \frac{1}{2}\sum_{p=1}^{N} a_p^i\, a_p^j\,\mathrm{E}[\,(Y_p(\mathbf{x}+\mathbf{h}) - Y_p(\mathbf{x}))^2\,] \\
&= b_{ij}\,\gamma(\mathbf{h}),
\end{aligned} \tag{22.8}$$

where each coefficient b_{ij}, element of the coregionalization matrix $\mathbf{B}$, is the result of the summation over the index p of products of a_p^i with a_p^j,

$$b_{ij} = \sum_{p=1}^{N} a_p^i\, a_p^j. \tag{22.9}$$

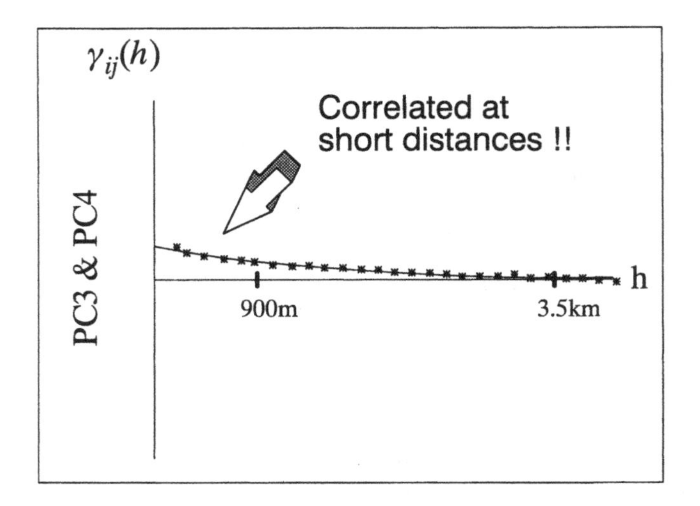

Figure 22.1: Cross variogram of a pair of principal components.

One possibility to decompose a given coregionalization matrix **B** for specifying the linear model is to perform a principal component analysis with **B** playing the role of the variance-covariance matrix,

$$a_p^i = \sqrt{\lambda_p}\, q_p^i, \tag{22.10}$$

where λ_p is an eigenvalue and q_p^i is an element of an eigenvector of **B**.

Codispersion coefficients

A sensitive question to explore for a given data set is to know whether the intrinsic correlation model is adequate. One possibility to approach the problem is to examine plots of the ratios of the cross versus the direct variograms, which are called *codispersion coefficients* (MATHERON, 1965)

$$\mathrm{cc}_{ij}(\mathbf{h}) = \frac{\gamma_{ij}(\mathbf{h})}{\sqrt{\gamma_{ii}(\mathbf{h})\,\gamma_{jj}(\mathbf{h})}}. \tag{22.11}$$

If the codispersion coefficients are constant, the correlation of the variable pair does not depend on spatial scale. This is an obvious implication of the intrinsic correlation model

$$\mathrm{cc}_{ij}(\mathbf{h}) = \frac{b_{ij}\,\gamma(\mathbf{h})}{\sqrt{b_{ii}\,b_{jj}}\,\gamma(\mathbf{h})}, = \frac{b_{ij}}{\sqrt{b_{ii}\,b_{jj}}} = \rho_{ij} \tag{22.12}$$

where ρ_{ij} is the correlation coefficient between two variables, computed from elements of the coregionalization matrix **B**.

Another possible test for intrinsic correlation is based on the principal components of the set of variables. If the cross variograms between the principal components are not zero at all lags $\mathbf{h}$, the principal components are not uncorrelated at all spatial scales of the study and the intrinsic correlation model is not appropriate.

EXAMPLE 22.3 (PRINCIPAL COMPONENTS' CROSS VARIOGRAM) *Ten chemical elements were measured in a region for gold prospection (see [344]). Figure 22.1 shows the cross variogram between the third and the fourth principal component computed on the basis of the variance-covariance matrix of this geochemical data for which a hypothesis of intrinsic correlation is obviously false. Indeed, near the origin of the cross variogram the two components are significantly correlated although the statistical correlation coefficient of the principal components is equal to zero. This is a symptom that the correlation structure of the data is different at small scale, an aspect that a non-regionalized principal component analysis cannot incorporate.*

We defer to Chapter 26 an explanation on how the ordinate on Figure 22.1 has been scaled. The graph has actually been scaled in the same way as Figure 26.2, using the so-called hull of perfect correlation *defined in expression (26.16) on p177 — but not displaying it.*

23 Heterotopic Cokriging

The cokriging procedure is a natural extension of kriging when a multivariate variogram or covariance model and multivariate data are available. A variable of interest is cokriged at a specific location from data about itself and about auxiliary variables in the neighborhood. The data set may not cover all variables at all sample locations. Depending on how the measurements of the different variables are scattered in space we distinguish between isotopic and heterotopic data sets. After defining these situations we examine cokriging in the heterotopic case.

Isotopy and heterotopy

The measurements available for different variables $Z_i(\mathbf{x})$ in a given domain may be located either at the same sample points or at different points for each variable as illustrated on Figure 23.1. The following situations can be distinguished

- *entirely heterotopic data*: the variables have been measured on different sets of sample points and have no sample locations in common;
- *partially heterotopic data*: some variables share some sample locations;
- *isotopy*: data is available for each variable at all sampling points.

Entirely heterotopic data poses a problem for inferring the cross variogram or covariance model. Experimental cross variograms cannot be computed for entirely heterotopic data. Experimental cross covariances, though they can be computed, are still problematic as the corresponding direct covariance values refer to different sets of points (and sometimes subregions). The value at the origin of the cross covariances cannot be computed.

With partially heterotopic data it is advisable, whenever possible, to infer the cross variogram or the covariance function model on the basis of the isotopic subset of the data.

Actually heterotopy for spatial data is as much a problem as missing values in multivariate statistics (remember the Example 3.1 on p22), even if a model is built in between in the case of geostatistics.

A particular case of partial heterotopy important for cokriging is when the set of sample points of the variable of interest is included in the sets of sample points of other variables, which serve as auxiliary variables in the estimation procedure. In this

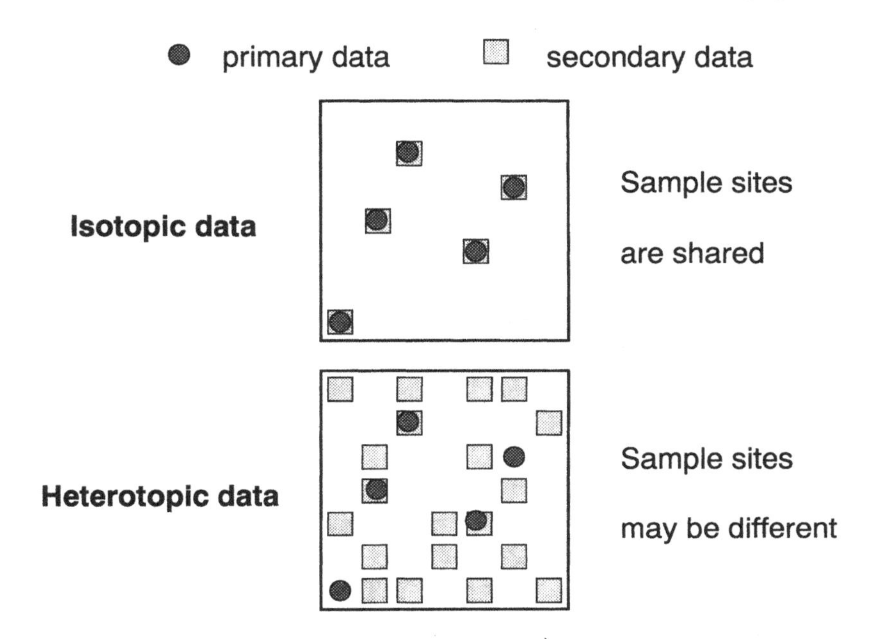

Figure 23.1: Isotopic and partially heterotopic data.

case, when the auxiliary variables are available at more points than the main variable, cokriging is typically of advantage.

In cokriging problems with heterotopic data we can distinguish between sparsely and densely sampled auxiliary variables. In the second case, when an auxiliary variable is available everywhere in the domain, particular techniques like collocated cokriging (presented in this chapter) or the external drift method (see Chapter 37) can be of interest.

The question whether cokriging is still interesting in the case of isotopy, when the auxiliary variables are available at the same locations as the variable of interest, will be examined in the next chapter.

Ordinary cokriging

The ordinary cokriging estimator is a linear combination of weights w_α^i with data from different variables located at sample points in the neighborhood of a point $\mathbf{x}_0$. Each variable is defined on a set of samples of (possibly different) size n_i and the estimator is defined as

$$Z_{i_0}^{\star}(\mathbf{x}_0) \quad = \quad \sum_{i=1}^{N} \sum_{\alpha=1}^{n_i} w_\alpha^i \, Z_i(\mathbf{x}_\alpha), \tag{23.1}$$

where the index i_0 refers to a particular variable of the set of N variables. The number of samples n_i depends upon the index i of the variables, so as to include into the notation the possibility of heterotopic data.

In the framework of a joint intrinsic hypothesis we wish to estimate a particular variable of a set of N variables on the basis of an estimation error which should be nil on average. This condition is satisfied by choosing weights which sum up to one for the variable of interest and which have a zero sum for the auxiliary variables

$$\sum_{\alpha=1}^{n_i} w_\alpha^i \;=\; \delta_{ii_0} = \begin{cases} 1 & \text{if } i = i_0 \\ 0 & \text{otherwise.} \end{cases} \tag{23.2}$$

Expanding the expression for the average estimation error, we get

$$\begin{aligned} \mathrm{E}[\, Z_{i_0}^{\star}(\mathbf{x}_0) - Z_{i_0}(\mathbf{x}_0) \,] \;&=\; \mathrm{E}\Big[\sum_{i=1}^{N} \sum_{\alpha=1}^{n_i} w_\alpha^i \, Z_i(\mathbf{x}_\alpha) \\ &\qquad - \underbrace{\sum_{\alpha=1}^{n_{i_0}} w_\alpha^{i_0}}_{1} \, Z_{i_0}(\mathbf{x}_0) - \sum_{\substack{i=0 \\ i \neq i_0}}^{N} \underbrace{\sum_{\alpha=1}^{n_i} w_\alpha^i}_{0} \, Z_i(\mathbf{x}_0) \Big] \\ &=\; \sum_{i=1}^{N} \sum_{\alpha=1}^{n_i} w_\alpha^i \, \underbrace{\mathrm{E}[\, Z_i(\mathbf{x}_\alpha) - Z_i(\mathbf{x}_0) \,]}_{0} \\ &=\; 0. \end{aligned} \tag{23.3}$$

For the variance of the estimation error we thus have

$$\sigma_{\mathrm{E}}^2 \;=\; \mathrm{E}\Big[\Big(\sum_{i=1}^{N} \sum_{\alpha=1}^{n_i} w_\alpha^i \, Z_i(\mathbf{x}_\alpha) - Z_{i_0}(\mathbf{x}_0) \Big)^2 \Big]. \tag{23.4}$$

Introducing weights w_0^i defined as

$$w_0^i \;=\; -\delta_{ii_0} = \begin{cases} -1 & \text{if } i = i_0, \\ 0 & \text{if } i \neq i_0, \end{cases} \tag{23.5}$$

which are included into the sums, we can shorten the expression of the estimation variance to

$$\sigma_{\mathrm{E}}^2 \;=\; \mathrm{E}\Big[\Big(\sum_{i=1}^{N} \sum_{\alpha=0}^{n_i} w_\alpha^i \, Z_i(\mathbf{x}_\alpha) \Big)^2 \Big]. \tag{23.6}$$

Then, inserting fictitious random variables $Z_i(\mathbf{0})$ positioned arbitrarily at the origin, increments can be formed

$$\begin{aligned} \sigma_{\mathrm{E}}^2 \;&=\; \mathrm{E}\Big[\Big(\sum_{i=1}^{N} \Big(\sum_{\alpha=0}^{n_i} w_\alpha^i \, Z_i(\mathbf{x}_\alpha) - Z_i(\mathbf{0}) \underbrace{\sum_{\alpha=0}^{n_i} w_\alpha^i}_{0} \Big) \Big)^2 \Big] \\ &=\; \mathrm{E}\Big[\Big(\sum_{i=1}^{N} \sum_{\alpha=0}^{n_i} w_\alpha^i \underbrace{(Z_i(\mathbf{x}_\alpha) - Z_i(\mathbf{0}))}_{\text{increments}} \Big)^2 \Big]. \end{aligned} \tag{23.7}$$

Defining cross covariances of increments $C^I_{ij}(\mathbf{x}_\alpha, \mathbf{x}_\beta)$ (which are not translation invariant) we have

$$\sigma^2_{\mathrm{E}} = \sum_{i=1}^{N}\sum_{j=1}^{N}\sum_{\alpha=0}^{n_i}\sum_{\beta=0}^{n_j} w^i_\alpha\, w^j_\beta\, C^I_{ij}(\mathbf{x}_\alpha, \mathbf{x}_\beta). \tag{23.8}$$

In order to convert the increment covariances to variograms, the additional assumption has to be made that the cross covariances of increments are symmetric. With this hypothesis we obtain the translation invariant quantity

$$\begin{aligned}\sigma^2_{\mathrm{E}} &= 2\sum_{i=1}^{N}\sum_{\alpha=1}^{n_i} w^i_\alpha\, \gamma_{ii_0}(\mathbf{x}_\alpha-\mathbf{x}_0) - \gamma_{i_0i_0}(\mathbf{x}_0-\mathbf{x}_0)\\ &\quad -\sum_{i=1}^{N}\sum_{j=1}^{N}\sum_{\alpha=1}^{n_i}\sum_{\beta=1}^{n_j} w^i_\alpha\, w^j_\beta\, \gamma_{ij}(\mathbf{x}_\alpha-\mathbf{x}_\beta).\end{aligned} \tag{23.9}$$

After a minimization in which the constraints on the weights generate N parameters of Lagrange μ_i, we have the ordinary cokriging system

$$\begin{cases} \displaystyle\sum_{j=1}^{N}\sum_{\beta=1}^{n_j} w^j_\beta\, \gamma_{ij}(\mathbf{x}_\alpha-\mathbf{x}_\beta) + \mu_i = \gamma_{ii_0}(\mathbf{x}_\alpha-\mathbf{x}_0) & \text{for } i=1,\dots N;\ \alpha=1,\dots n_i \\ \displaystyle\sum_{\beta=1}^{n_i} w^i_\beta = \delta_{ii_0} & \text{for } i=1,\dots N, \end{cases} \tag{23.10}$$

and the cokriging variance

$$\sigma^2_{\mathrm{CK}} = \sum_{i=1}^{N}\sum_{\alpha=1}^{n_i} w^i_\alpha\, \gamma_{ii_0}(\mathbf{x}_\alpha-\mathbf{x}_0) + \mu_{i_0} - \gamma_{i_0i_0}(\mathbf{x}_0-\mathbf{x}_0). \tag{23.11}$$

Simple cokriging

Ordinary cokriging has no meaning when no data is available for the variable of interest in a given neighborhood. On the other hand, simple kriging leans on the knowledge of the means of the variables, so that an estimation of a variable can be calibrated without having any data value for this variable in the cokriging neighborhood.

The simple cokriging estimator is made up of the mean of the variable of interest plus a linear combination of weights w^i_α with the residuals of the variables

$$Z^\star_{i_0}(\mathbf{x}_0) = m_{i_0} + \sum_{i=1}^{N}\sum_{\alpha=1}^{n_i} w^i_\alpha\, (Z_i(\mathbf{x}_\alpha) - m_i). \tag{23.12}$$

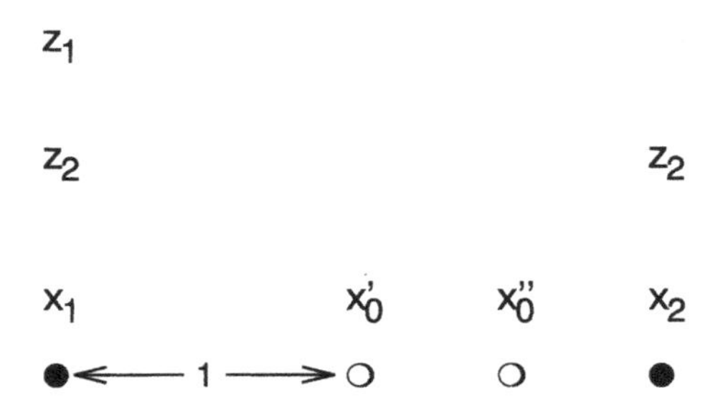

Figure 23.2: Three data values at two locations $\mathbf{x}_1$, $\mathbf{x}_2$.

To this estimator we can associate a simple cokriging system, written in matrix form

$$\begin{pmatrix} \mathbf{C}_{11} & \dots & \mathbf{C}_{1j} & \dots & \mathbf{C}_{1N} \\ \vdots & \ddots & & & \vdots \\ \mathbf{C}_{i1} & & \mathbf{C}_{ii} & & \mathbf{C}_{iN} \\ \vdots & & & \ddots & \vdots \\ \mathbf{C}_{N1} & \dots & \mathbf{C}_{Nj} & \dots & \mathbf{C}_{NN} \end{pmatrix} \begin{pmatrix} \mathbf{w}_1 \\ \vdots \\ \mathbf{w}_i \\ \vdots \\ \mathbf{w}_N \end{pmatrix} = \begin{pmatrix} \mathbf{c}_{1i_0} \\ \vdots \\ \mathbf{c}_{ii_0} \\ \vdots \\ \mathbf{c}_{Ni_0} \end{pmatrix}, \tag{23.13}$$

where the left hand side matrix is built up with square symmetric $n_i \times n_i$ blocs $\mathbf{C}_{ii}$ on the diagonal and with rectangular $n_i \times n_j$ blocs $\mathbf{C}_{ij}$ off the diagonal, with

$$\mathbf{C}_{ij} = \mathbf{C}_{ji}^{\top}. \tag{23.14}$$

The blocks $\mathbf{C}_{ij}$ contain the covariances between sample points for a fixed pair of variables. The vectors $\mathbf{c}_{ii_0}$ list the covariances with the variable of interest, for a specific variable of the set, between sample points and the estimation location. The vectors $\mathbf{w}_i$ represent the weights attached to the samples of a fixed variable.

EXAMPLE 23.1 (HETEROTOPIC BIVARIATE COKRIGING) *Let us examine the following simple cokriging problem. We are interested in cokriging estimates on the basis of 3 data values at two points* $\mathbf{x}_1$ *and* $\mathbf{x}_2$ *as shown on Figure 23.2. In this 1D neighborhood with two sample points we have one data value for* $Z_1(x)$ *and two data values for* $Z_2(x)$. *The direct covariance functions and the cross covariance function (which is assumed an even function) are modeled with a spherical structure with a range parameter* $a = 1$,

$$C_{11}(h) = 2\,\rho_{sph}(h), \qquad C_{22}(h) = \rho_{sph}(h), \tag{23.15}$$

$$C_{12}(h) = -1\,\rho_{sph}(h). \tag{23.16}$$

This is by the way an intrinsic correlation model.

EXERCISE 23.2 *Is the coregionalization matrix positive definite? What is the value of the correlation coefficient?*

We want to estimate $Z_1(x)$ using simple cokriging assuming known means m_1 and m_2, so the estimator is

$$\begin{aligned} Z_1^\star(x_0) &= m_1 + w_1(Z_1(x_1) - m_1) \\ &\quad + w_2(Z_2(x_1) - m_2) + w_3(Z_2(x_2) - m_2). \end{aligned} \tag{23.17}$$

Let us write out the simple cokriging system for the isotopic case (as if we had a measurement for $Z_1(x_2)$)

$$\left(\begin{array}{cc|cc} C_{11}(x_1-x_1) & C_{11}(x_1-x_2) & C_{12}(x_1-x_1) & C_{12}(x_1-x_2) \\ C_{11}(x_2-x_1) & C_{11}(x_2-x_2) & C_{12}(x_2-x_1) & C_{12}(x_2-x_2) \\ \hline C_{21}(x_1-x_1) & C_{21}(x_1-x_2) & C_{22}(x_1-x_1) & C_{22}(x_1-x_2) \\ C_{21}(x_2-x_1) & C_{21}(x_2-x_2) & C_{22}(x_2-x_1) & C_{22}(x_2-x_2) \end{array}\right)$$

$$\times \left(\begin{array}{c} w_1^1 \\ w_2^1 \\ \hline w_1^2 \\ w_2^2 \end{array}\right) = \left(\begin{array}{c} C_{11}(x_1-x_0) \\ C_{11}(x_2-x_0) \\ \hline C_{21}(x_1-x_0) \\ C_{21}(x_2-x_0) \end{array}\right), \tag{23.18}$$

where the blocks are marked by lines.

EXERCISE 23.3 *How does the system shrink due to the fact that actually no data value is available for $Z_1(x)$ at the point x_2?*

Computing the numerical values of the covariances we have the following system with the three right hand sides for cokriging at x_0', x_0'' and x_2, as shown on Figure 23.2,

$$\begin{pmatrix} 2 & -1 & 0 \\ -1 & 1 & 0 \\ 0 & 0 & 1 \end{pmatrix} \begin{pmatrix} w_1 \\ w_2 \\ w_3 \end{pmatrix} = \begin{pmatrix} 0 \\ 0 \\ 0 \end{pmatrix}, \begin{pmatrix} 0 \\ 0 \\ -5/16 \end{pmatrix}, \begin{pmatrix} 0 \\ 0 \\ -1 \end{pmatrix}. \tag{23.19}$$

EXERCISE 23.4 *What are the solutions w_1, w_2, w_3 of the simple cokriging system for each target point x_0', x_0'' and x_2?*

The cokriging estimate $z_1^\star(x_o')$ is equal to the mean m_1 because the point x_o' is out of range with respect to both data points x_1 and x_2.

For the point x_0'' the cokriging estimate is equal to

$$z_1^\star(x_0'') = m_1 - \frac{5}{16}(z_2(x_2) - m_2) \tag{23.20}$$

as this point is now within the range of the point x_2.

When cokriging at the target point x_2 we have

$$z_1^\star(x_2) = m_1 + \frac{\sigma_{12}}{\sigma_{22}}(z_2(x_2) - m_2) = m_1 - (z_2(x_2) - m_2), \tag{23.21}$$

which is a linear regression equation (compare with Eq. 3.21 on p19).

The simple kriging of $Z_1(x)$ at the three target points (without using the auxiliary variable) would have given each time the mean m_1 as a solution because all three *points are out of range of* x_1. *In comparison the cokriging solutions are more interesting soon as the target point is within range of the point* x_2 *where a value of the auxiliary variable is available.*

We notice in passing that in the heterotopic case cokriging with an intrinsic correlation model does generally not boil down to kriging.

24 Collocated Cokriging

A particular heterotopic situation encountered in practice is when we have a variable of interest known at a few points and an auxiliary variable known everywhere in the domain (or at least at all nodes of a given estimation grid and at the data locations of the variable of interest. With plenty of data available for the auxiliary variable the question at hand is how to choose a parsimonious neighborhood.

Cokriging neighborhood

Cokriging with many variables using all data easily generates a very large linear system to solve. This means that the choice of a subset of data around a given estimation location, called a *neighborhood,* is a crucial step in cokriging. It is of particular importance to know when, due to the particular structure of a coregionalization, the full cokriging with all data is actually equivalent to a cokriging using a subset of data, so that the neighborhood can be reduced a priori and the cokriging system simplified accordingly, thus reducing in the end the numerical effort to a considerable extent. For isotopic data the most important aspect is to know whether the coregionalization shows direct and cross variograms that are proportional to one direct variogram which entails the cokriging to be equivalent to a separate kriging, leaving out the secondary variables. This topic will be exposed in Chapter 25. Concerning heterotopic data we will focus on a case that has attracted most attention recently as it is increasingly frequently encountered in applications: the case of a *dense secondary* variable.

Figure 24.1 sketches three different neighborhoods for a given central estimation location (denoted by a star), primary data (denoted by full circles) as well as three alternate subsets of data from a secondary variable (denoted by squares). The neighborhood:

(A) uses all data available for the secondary variable,

(B) restricts the secondary information to the subset of locations where primary data is available as well as to the estimation location,

(C) merely includes a sample value of the secondary variable at the estimation location.

Case (A) can be termed the *full* neighborhood, while case (C) was called a *collocated* neighborhood by Xu et al. [360] as the secondary data is collocated with the estimation location. Whereas case (B) was termed a *multicollocated* neighborhood

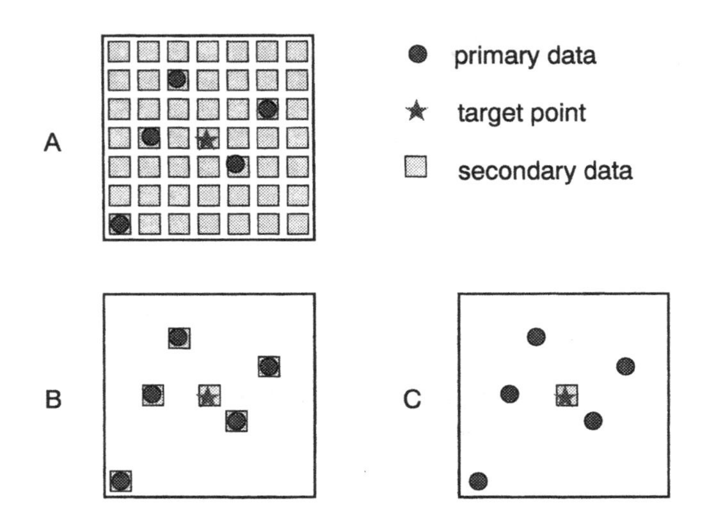

Figure 24.1: Three possible neighborhoods with a dense secondary variable.

by Chilès & Delfiner [51] as additionally the secondary data is also collocated with the primary data.

Using the full neighborhood (A) with secondary data dense in space will easily lead to linear dependencies for neighboring samples in the cokriging system, causing it to be singular. The size of the system can also be numerically challenging. Vargas-Guzman & Yeh [328] suggest a way out of numerical difficulties by starting from a small neighborhood and progressively extending the neighborhood in the framework of what they call a sequential cokriging.

Collocated simple cokriging

The collocated neighborhood (C) is only valid for simple cokriging because using a single sample for the secondary variable in ordinary cokriging leads to a trivial cokriging weight for that sample due to the constraint. For simple cokriging the approach solely requires the inference of a correlation coefficient instead of a cross-covariance function, a simplification that can only be meaningful for coregionalization models with proportionalities. However, as shown below the full simple cokriging with such a model does not reduce to a collocated simple cokriging.

With reference to XU et al. [360] we call *collocated simple cokriging* a neighborhood definition strategy in which the neighborhood of the auxiliary variable is arbitrarily reduced to only one point: the estimation location. The value $S(\mathbf{x}_0)$ is said to be *collocated* with the target point of $Z(\mathbf{x})$. The collocated simple cokriging estimator is

$$Z^{\star}(\mathbf{x}_0) \quad = \quad m_z + w_0\,(S(\mathbf{x}_0) - m_S) + \sum_{\alpha=1}^{n} w_\alpha\,(Z(\mathbf{x}_\alpha) - m_z). \qquad (24.1)$$

The simple cokriging system for such a neighborhood is written

$$\begin{pmatrix} \mathbf{C}_{ZZ} & \mathbf{c}_{ZS} \\ \mathbf{c}_{ZS}^{\top} & \sigma_{SS} \end{pmatrix} \begin{pmatrix} \mathbf{w}_Z \\ w_0 \end{pmatrix} = \begin{pmatrix} \mathbf{c}_{ZZ} \\ \sigma_{ZS} \end{pmatrix}, \tag{24.2}$$

where $\mathbf{C}_{ZZ}$ is the left hand matrix of the simple kriging system of $Z(\mathbf{x})$ and $\mathbf{c}_{ZZ}$ is the corresponding right hand side. The vector $\mathbf{c}_{ZS}$ contains the cross covariances between the n sample points of $Z(\mathbf{x})$ and the target point $\mathbf{x}_0$ with its collocated value $S(\mathbf{x}_0)$.

In the case of intrinsic correlation the collocated simple cokriging can be expressed as

$$\begin{pmatrix} \sigma_{ZZ}\,\mathbf{R} & \sigma_{ZS}\,\mathbf{r}_0 \\ \sigma_{ZS}\,\mathbf{r}_0^{\top} & \sigma_{SS} \end{pmatrix} \begin{pmatrix} \mathbf{w}_Z \\ w_0 \end{pmatrix} = \begin{pmatrix} \sigma_{ZZ}\,\mathbf{r}_0 \\ \sigma_{ZS} \end{pmatrix}, \tag{24.3}$$

where $\mathbf{R}$ is the matrix of spatial correlations $\rho(\mathbf{x}_\alpha - \mathbf{x}_\beta)$, while $\mathbf{r}_0$ is the vector of spatial correlations $\rho(\mathbf{x}_\alpha - \mathbf{x}_0)$.

To perform collocated cokriging with intrinsic correlation of the variable pair we only need to know the covariance function

$$C_{ZZ}(\mathbf{h}) = \sigma_{ZZ}\,\rho(\mathbf{h}) \tag{24.4}$$

and the correlation coefficient ρ_{ZS} between $Z(\mathbf{x})$ and $S(\mathbf{x})$, as well as the variance σ_{SS} of $S(\mathbf{x})$.

An interesting question is to know whether simple cokriging reduces to collocated cokriging with the intrinsic correlation model. Actually the answer is negative (as shown in an exercise below) and we have to conclude that there is no theoretical justification for selecting only one collocated sample for cokriging the auxiliary variable.

EXERCISE 24.1 *Suppose that $Z(\mathbf{x})$ and $S(\mathbf{x})$ are intrinsically correlated with unit variances. We choose to use as a cokriging neighborhood the data $Z(\mathbf{x}_\alpha)$ at n sample locations $\mathbf{x}_\alpha$ and corresponding values $S(\mathbf{x}_\alpha)$ at the same locations. We also include into the cokriging neighborhood the value $S(\mathbf{x}_0)$ available at the grid node $\mathbf{x}_0$. Does this cokriging boil down to collocated cokriging?*

Collocated ordinary cokriging

The collocated neighborhood (C) used for simple cokriging in the previous section would yield a trivial result if applied in ordinary cokriging: because of the constraint that the weights of the auxiliary variable should sum up to zero, the weight w_0 is zero and the auxiliary variable does not come into play.

An ordinary cokriging needs to use more data together with the value $S(\mathbf{x}_0)$. If the values $S(\mathbf{x}_\alpha)$ that are *collocated* with the sample points of the main variable are also included we get a multicollocated neighborhood (B). The ordinary cokriging estimator is

$$Z^{\star}(\mathbf{x}_0) = w_0\,S(\mathbf{x}_0) + \sum_{\alpha=1}^{n} \left(w_Z^{\alpha}\,Z(\mathbf{x}_\alpha) + w_S^{\alpha}\,S(\mathbf{x}_\alpha)\right). \tag{24.5}$$

The corresponding cokriging system can be written

$$\begin{pmatrix} \mathbf{C}_{ZZ} & \mathbf{C}_{ZS} & \mathbf{c}_{ZS} & \mathbf{1} & \mathbf{0} \\ \mathbf{C}_{SZ} & \mathbf{C}_{SS} & \mathbf{c}_{SS} & \mathbf{0} & \mathbf{1} \\ \mathbf{c}_{ZS}^{\top} & \mathbf{c}_{SS}^{\top} & \sigma_{SS} & 0 & 1 \\ \mathbf{1}^{\top} & \mathbf{0}^{\top} & 0 & 0 & 0 \\ \mathbf{0}^{\top} & \mathbf{1}^{\top} & 1 & 0 & 0 \end{pmatrix} \begin{pmatrix} \mathbf{w}_Z \\ \mathbf{w}_S \\ w_0 \\ -\mu_1 \\ -\mu_2 \end{pmatrix} = \begin{pmatrix} \mathbf{c}_{ZZ} \\ \mathbf{c}_{SZ} \\ \sigma_{ZS} \\ 1 \\ 0 \end{pmatrix}. \tag{24.6}$$

As the weights are constrained, the system can also be expressed on the basis of variograms instead of covariance functions (contrarily to simple collocated cokriging).

Simplification with a particular covariance model

Cokriging in a multicollocated neighborhood is examined in detail by Rivoirard [271] who has shown in the bivariate case that cokriging with a full neighborhood is equivalent to cokriging with a multicollocated neighborhood when the covariance structure is of the type:

$$\begin{cases} C_{ZZ}(\mathbf{h}) & = p^2\, C(\mathbf{h}) \;+\; C^R(\mathbf{h}), \\ C_{SS}(\mathbf{h}) & = C(\mathbf{h}), \\ C_{ZS}(\mathbf{h}) & = p\, C(\mathbf{h}), \end{cases} \tag{24.7}$$

where p is a proportionality coefficient, $C^R(\mathbf{h})$ is a covariance structure that is particular to the primary variable and $C(\mathbf{h})$ is common to both variables. Implications of this model are that the covariance function $C_{SS}(\mathbf{h})$ is more regular than $C_{ZZ}(\mathbf{h})$, if $C^R(\mathbf{h})$ is less regular than $C(\mathbf{h})$.

The corresponding cokriging estimator $Z^{\star\star}$ is then shown to reduce to

$$Z^{\star\star}(\mathbf{x}_0) \quad = \quad Z^{\star}(\mathbf{x}_0) + p\,(S(\mathbf{x}_0) - S^{\star}(\mathbf{x}_0))\,, \tag{24.8}$$

where the two krigings $Z^{\star}$ and $S^{\star}$ are both obtained using the weights from an ordinary kriging system set up with the covariance function $C^R(\mathbf{h})$ for the n_z sample locations of the primary variable:

$$Z^{\star}(\mathbf{x}_0) = \sum_{\alpha=1}^{n_z} w_\alpha^R Z(\mathbf{x}_\alpha), \qquad S^{\star}(\mathbf{x}_0) = \sum_{\alpha=1}^{n_z} w_\alpha^R S(\mathbf{x}_\alpha). \tag{24.9}$$

Rivoirard [271] obtained this result for Gaussian random functions with conditional independence of $Z(\mathbf{x}_0)$ and $S(\mathbf{x})$ knowing $S(\mathbf{x}_0)$. It is straightforward to derive it more generally for second-order stationary random functions simplifying the cokriging system on starting directly from the covariance function model (24.7).

This result can be generalized to the multivariate case for uncorrelated secondary variables (e.g. principal components of remote sensing channels). Let the first variable

Z_1 be the variable of interest, let $j = 2, \ldots, N$ be the indices of $N-1$ principal components with covariance functions $C_{jj}(\mathbf{h})$ and assume the coregionalization model,

$$\begin{cases} C_{11}(\mathbf{h}) = \sum_{j=2}^{N} {p_j}^2\, C_{jj}(\mathbf{h}) + C^R(\mathbf{h}), & \\ C_{1j}(\mathbf{h}) = p_j\, C_{jj}(\mathbf{h}) & \text{for } j = 2, \ldots, N, \\ C_{jj'}(\mathbf{h}) = 0 & \text{for } j \neq j', \end{cases} \tag{24.10}$$

where p_j are coefficients. Knowing that remote sensing channels in our experience are not seldom intrinsically correlated [183, 47] it makes sense to compute their principal components and this type of coregionalization model could be applied in that context.

The cokriging estimator is then

$$Z_1^{\star\star}(\mathbf{x}_0) \quad = \quad Z_1^{\star}(\mathbf{x}_0) + \sum_{j=2}^{N} p_j \left(Z_j(\mathbf{x}_0) - Z_j^{\star}(\mathbf{x}_0) \right), \tag{24.11}$$

reducing the cokriging to a linear combination of coefficients p_j with differences between the known values $Z_j(\mathbf{x}_0)$ and the ordinary krigings $Z_j^{\star}(\mathbf{x}_0)$ computed from the n_1 sample locations of Z_1 using $C^R(\mathbf{h})$.

25 Isotopic Cokriging

When a set of variables is intrinsically correlated, cokriging is equivalent to kriging for each variable, if all variables have been measured at all sample locations. For a particular variable in a given data set in the isotopic case, cokriging can be equivalent to kriging, even if the variable set is not intrinsically correlated. This is examined in detail in this chapter.

Cokriging with isotopic data

Let us consider the case of isotopic data. With isotopy the cokriging of a set of variables has the important advantage over a separate kriging of each variable that it preserves the coherence of the estimators. This can be seen when estimating a sum $S(\mathbf{x})$ of variables

$$S(\mathbf{x}) \;=\; \sum_{i=1}^{N} Z_i(\mathbf{x}). \tag{25.1}$$

The cokriging of $S(\mathbf{x})$ is equal to the sum of the cokrigings of the variables $Z_i(\mathbf{x})$ (using for each cokriging N out of the $N+1$ variables)

$$S^{\mathrm{CK}}(\mathbf{x}) \;=\; \sum_{i=1}^{N} Z_i^{\mathrm{CK}}(\mathbf{x}). \tag{25.2}$$

However, if we krige each term of the sum and add up the krigings we generally do not get the same result as when we krige directly the added up data: the two estimations are not coherent.

EXAMPLE 25.1 *The thickness $T(\mathbf{x})$ of a geologic layer is defined as the difference between its upper and lower limits,*

$$\underbrace{T(\mathbf{x})}_{\textit{thickness}} \;=\; \underbrace{Z_U(\mathbf{x})}_{\textit{upper limit}} \;-\; \underbrace{Z_L(\mathbf{x})}_{\textit{lower limit}} \tag{25.3}$$

The cokriging estimators of each term using the information of two out of the three variables (the third being redundant) are coherent,

$$T^{CK}(\mathbf{x}) \;=\; Z_U^{CK}(\mathbf{x}) - Z_L^{CK}(\mathbf{x}). \tag{25.4}$$

But if kriging is used, the left hand result is in general different from the right hand difference,

$$T^K(\mathbf{x}) \neq Z_U^K(\mathbf{x}) - Z_L^K(\mathbf{x}), \tag{25.5}$$

and there is no criterion to help decide as to which estimate of thickness is the better.

In some situations isotopic cokriging is equivalent to kriging. The trivial case is when all cross variograms or cross covariance functions are zero. A more subtle case is when cross variograms or covariances are proportional to a direct variogram or covariance function.

Autokrigeability

A variable is said to be *autokrigeable* with respect to a set of variables, if the direct kriging of this variable is equivalent to the cokriging. The trivial case is when all variables are uncorrelated (at any scale!) and it is easy to see that this implies that all kriging weights are zero except for the variable of interest.

In a situation of isotopy any variable is autokrigeable if it belongs to a set of intrinsically correlated variables. Assuming that the matrix $\mathbf{V}$ of the intrinsic correlation model is positive definite, we can rewrite the simple cokriging equations using Kronecker products $\otimes$:

$$(\mathbf{V} \otimes \mathbf{R})\,\mathbf{w} \;=\; \mathbf{v}_{i_0} \otimes \mathbf{r}_0, \tag{25.6}$$

where

$$\mathbf{R} = [\rho(\mathbf{x}_\alpha - \mathbf{x}_\beta)] \qquad \text{and} \qquad \mathbf{r}_0 = (\rho(\mathbf{x}_\alpha - \mathbf{x}_0)). \tag{25.7}$$

The $nN \times nN$ left hand matrix of the isotopic simple cokriging with intrinsic correlation is expressed as the Kronecker product between the $N \times N$ variance-covariance matrix $\mathbf{V}$ and the $n \times n$ left hand side matrix $\mathbf{R}$ of simple kriging

$$\mathbf{V} \otimes \mathbf{R} \;=\; \begin{pmatrix} \sigma_{11}\,\mathbf{R} & \dots & & \dots & \sigma_{1N}\,\mathbf{R} \\ \vdots & \ddots & & & \vdots \\ & & \sigma_{ij}\,\mathbf{R} & & \\ \vdots & & & \ddots & \vdots \\ \sigma_{N1}\,\mathbf{R} & \dots & & \dots & \sigma_{NN}\,\mathbf{R} \end{pmatrix}. \tag{25.8}$$

The nN right hand vector is the Kronecker product of the vector of covariances

σ_{ii_0} times the right hand vector $\mathbf{r}_0$ of simple kriging

$$\mathbf{v}_{i_0} \otimes \mathbf{r}_0 = \begin{pmatrix} \sigma_{1i_0}\,\mathbf{r}_0 \\ \vdots \\ \sigma_{ii_0}\,\mathbf{r}_0 \\ \vdots \\ \sigma_{Ni_0}\,\mathbf{r}_0 \end{pmatrix}. \tag{25.9}$$

Now, as $\mathbf{v}_{i_0}$ is also present in the left hand side of the cokriging system, a solution exists for which the weights of the vector $\mathbf{w}_{i_o}$ are those of the simple kriging of the variable of interest and all other weights are zero

$$\begin{pmatrix} \ldots & \mathbf{v}_{i_0} \otimes \mathbf{R} & \ldots \end{pmatrix} \begin{pmatrix} 0 \\ \vdots \\ \mathbf{w}_{i_0} \\ \vdots \\ 0 \end{pmatrix} = \mathbf{v}_{i_0} \otimes \mathbf{r}_0. \tag{25.10}$$

This is the only solution of the simple cokriging system as $\mathbf{V}$ and $\mathbf{R}$ are positive definite.

It should be noted that in this demonstration the autokrigeability of the variable of interest only depends on the block structure $\mathbf{v}_{i_0} \otimes \mathbf{R}$ and not on the structure $\mathbf{V} \otimes \mathbf{R}$ of the whole left hand matrix. Thus for a variable to be autokrigeable we only need that the cross variograms or covariances with this variable are proportional to the direct variogram or covariance.

To test on an isotopic data set whether a variable Z_{i_0} is autokrigeable, we can compute *autokrigeability coefficients* defined as the ratios of the cross variograms against the direct variogram

$$\mathrm{ac}_{i_0 j}(\mathbf{h}) = \frac{\gamma_{i_0 j}(\mathbf{h})}{\gamma_{i_0 i_0}(\mathbf{h})}. \tag{25.11}$$

If the autokrigeability coefficients are constant at any working scale for each variable $j = 1, \ldots, N$ of a variable set, then the variable of interest is autokrigeable with respect to this set of variables.

A set of N variables for which each variable is autokrigeable with respect to the $N-1$ other variables is intrinsically correlated.

COMMENT 25.2 *The property of autokrigeability is used explicitly in nonlinear geostatistics in the formulation of models with orthogonal indicator residuals (see [267, 270]).*

Bivariate ordinary cokriging

We now examine in detail the ordinary cokriging of a variable of interest $Z(\mathbf{x})$ with only one auxiliary variable $Y(\mathbf{x})$. The ordinary cokriging estimator for $Z(\mathbf{x})$ is written

$$Z^\star(\mathbf{x}_0) = \sum_{\alpha=1}^{n_1} w_\alpha^1 \, Z(\mathbf{x}_\alpha) + \sum_{\beta=1}^{n_2} w_\beta^2 \, Y(\mathbf{x}_\beta) \tag{25.12}$$

$$= \mathbf{w}_1^\top \mathbf{z} + \mathbf{w}_2^\top \mathbf{y}. \tag{25.13}$$

The unbiasedness conditions are $\mathbf{1}^\top \mathbf{w}_1 = 1$ and $\mathbf{1}^\top \mathbf{w}_2 = 0$.
The ordinary cokriging system in matrix notation is

$$\begin{pmatrix} \mathbf{C}_{ZZ} & \mathbf{C}_{ZY} & \mathbf{1} & \mathbf{0} \\ \mathbf{C}_{YZ} & \mathbf{C}_{YY} & \mathbf{0} & \mathbf{1} \\ \mathbf{1}^\top & \mathbf{0}^\top & 0 & 0 \\ \mathbf{0}^\top & \mathbf{1}^\top & 0 & 0 \end{pmatrix} \begin{pmatrix} \mathbf{w}_1 \\ \mathbf{w}_2 \\ -\mu_1 \\ -\mu_2 \end{pmatrix} = \begin{pmatrix} \mathbf{c}_{ZZ} \\ \mathbf{c}_{YZ} \\ 1 \\ 0 \end{pmatrix}. \tag{25.14}$$

Suppose the two variables are uncorrelated, both in the spatial and in the multivariate sense, i.e. $\mathbf{C}_{ZY} = \mathbf{C}_{YZ} = 0$ and $\mathbf{c}_{YZ} = 0$. Then the cokriging system has the form

$$\begin{cases} \mathbf{C}_{ZZ} \, \mathbf{w}_1 \quad -\mu_1 = \mathbf{c}_{ZZ} \\ \mathbf{1}^\top \mathbf{w}_1 = 1 \\ \mathbf{C}_{YY} \, \mathbf{w}_2 \quad -\mu_2 = 0 \\ \mathbf{1}^\top \mathbf{w}_2 = 0. \end{cases} \tag{25.15}$$

Obviously $\mathbf{w}_2 = 0$ and $\mu_2 = 0$, so that $\mathbf{w}_1$, μ_1 are the solution of the ordinary kriging of Z alone, even in the heterotopic case.

For the isotopic case, let us look now at ordinary cokriging with the intrinsic correlation model. The matrices in the system have the following structure

$$\mathbf{C}_{ZZ} = \sigma_{ZZ} \, \mathbf{R} \qquad \mathbf{C}_{ZY} = \sigma_{ZY} \, \mathbf{R} \qquad \mathbf{c}_{ZZ} = \sigma_{ZZ} \, \mathbf{r}_0 \tag{25.16}$$

$$\mathbf{C}_{YZ} = \sigma_{YZ} \, \mathbf{R} \qquad \mathbf{C}_{YY} = \sigma_{YY} \, \mathbf{R} \qquad \mathbf{c}_{YZ} = \sigma_{YZ} \, \mathbf{r}_0. \tag{25.17}$$

Thus the system is

$$\begin{cases} \sigma_{ZZ} \, \mathbf{R} \, \mathbf{w}_1 + \sigma_{ZY} \, \mathbf{R} \, \mathbf{w}_2 - \mu_1 = \sigma_{ZZ} \, \mathbf{r}_0 \\ \mathbf{1}^\top \mathbf{w}_1 = 1 \\ \sigma_{YZ} \, \mathbf{R} \, \mathbf{w}_1 + \sigma_{YY} \, \mathbf{R} \, \mathbf{w}_2 - \mu_2 = \sigma_{YZ} \, \mathbf{r}_0 \\ \mathbf{1}^\top \mathbf{w}_2 = 0. \end{cases} \tag{25.18}$$

Setting $\mathbf{w}_2 = 0$ and $\mu_1 = \sigma_{ZZ} \, \mu$, $\mu_2 = \sigma_{YZ} \, \mu$ we obtain

$$\begin{cases} \sigma_{ZZ} \, \mathbf{R} \, \mathbf{w}_1 - \sigma_{ZZ} \, \mu = \sigma_{ZZ} \, \mathbf{r}_0 \\ \mathbf{1}^\top \mathbf{w}_1 = 1 \\ \sigma_{YZ} \, \mathbf{R} \, \mathbf{w}_1 - \sigma_{YZ} \, \mu = \sigma_{YZ} \, \mathbf{r}_0. \end{cases} \tag{25.19}$$

Dividing out the σ_{ZZ} and σ_{YZ} in the first and the third set of equations, we see that $\mathbf{w}_1$ and μ are solution of the simple direct kriging Z: the variable of interest is autokrigeable.

26 Multivariate Nested Variogram

The multivariate regionalization of a set of random functions can be represented with a spatial multivariate linear model. The associated multivariate nested variogram model is easily fitted to the multivariate data. Several coregionalization matrices describing the multivariate correlation structure at different scales of a phenomenon result from the variogram fit. The relation between the coregionalization matrices and the classical variance-covariance matrix is examined.

Linear model of coregionalization

A set of real second-order stationary random functions $\{Z_i(\mathbf{x}); i = 1, \ldots, N\}$ can be decomposed into sets $\{Z_u^i(\mathbf{x}); u = 0, \ldots, S\}$ of spatially uncorrelated components

$$Z_i(\mathbf{x}) \quad = \quad \sum_{u=0}^{S} Z_u^i(\mathbf{x}) + m_i, \tag{26.1}$$

where for all values of the indices i, j, u and v,

$$\mathrm{E}[\, Z_i(\mathbf{x})\,] \quad = \quad m_i, \tag{26.2}$$

$$\mathrm{E}[\, Z_u^i(\mathbf{x})\,] \quad = \quad 0, \tag{26.3}$$

and

$$\operatorname{cov}(Z_u^i(\mathbf{x}), Z_u^j(\mathbf{x}+\mathbf{h})) \quad = \quad \mathrm{E}[\, Z_u^i(\mathbf{x})\, Z_u^j(\mathbf{x}+\mathbf{h})\,] = C_{ij}^u(\mathbf{h}), \tag{26.4}$$

$$\operatorname{cov}(Z_u^i(\mathbf{x}), Z_v^j(\mathbf{x}+\mathbf{h})) \quad = \quad 0 \qquad \text{when} \qquad u \neq v. \tag{26.5}$$

The cross covariance functions $C_{ij}^u(\mathbf{h})$ associated with the spatial components are composed of real coefficients b_{ij}^u and are proportional to real correlation functions $\rho_u(\mathbf{h})$

$$C_{ij}(\mathbf{h}) \quad = \quad \sum_{u=0}^{S} C_{ij}^u(\mathbf{h}) = \sum_{u=0}^{S} b_{ij}^u\, \rho_u(\mathbf{h}), \tag{26.6}$$

which implies that the cross covariance functions are even in this model.

Coregionalization matrices $\mathbf{B}_u$ of order $N \times N$ can be set up and we have a multivariate nested covariance function model

$$\mathbf{C}(\mathbf{h}) \quad = \quad \sum_{u=0}^{S} \mathbf{B}_u\, \rho_u(\mathbf{h}) \tag{26.7}$$

with positive semi-definite coregionalization matrices $\mathbf{B}_u$.

EXERCISE 26.1 *When is the above covariance function model equivalent to the intrinsic correlation model?*

EXERCISE 26.2 *Show that a correlation function $\rho_u(\mathbf{h})$ having a non zero sill b_{ij}^u on a given cross covariance function has necessarily non zero sills b_{ii}^u and b_{jj}^u on the corresponding direct covariance functions.*

Conversely, if a sill b_{ii}^u is zero for a given structure of a variable, all sills of the structure on all cross covariance functions with this variable are zero.

Each spatial component $Z_u^i(\mathbf{x})$ can itself be represented as a set of uncorrelated factors $Y_u^p(\mathbf{x})$ with transformation coefficients a_{up}^i,

$$Z_u^i(\mathbf{x}) \;=\; \sum_{p=1}^{N} a_{pu}^i \, Y_u^p(\mathbf{x}), \tag{26.8}$$

where for all values of the indices i, j, u, v, p and q,

$$\mathrm{E}[\, Y_u^p(\mathbf{x}) \,] \;=\; 0, \tag{26.9}$$

and,

$$\operatorname{cov}(Y_u^p(\mathbf{x}), Y_u^p(\mathbf{x}+\mathbf{h})) \;=\; \rho_u(\mathbf{h}), \tag{26.10}$$

$$\operatorname{cov}(Y_u^p(\mathbf{x}), Y_v^q(\mathbf{x}+\mathbf{h})) \;=\; 0 \quad \text{when} \quad u \neq v \quad \text{or} \quad p \neq q. \tag{26.11}$$

Combining the spatial with the multivariate decomposition, we obtain the *linear model of coregionalization*

$$Z_i(\mathbf{x}) \;=\; \sum_{u=0}^{S} \sum_{p=1}^{N} a_{pu}^i \, Y_u^p(\mathbf{x}). \tag{26.12}$$

In practice first a set of correlation functions $\rho_u(\mathbf{h})$ (i.e. normalized variograms $g_u(\mathbf{h})$) is selected, taking care to keep S reasonably small. Then the coregionalization matrices are fitted using a weighted least squares algorithm (described below). The weighting coefficients are chosen by the practitioner so as to provide a graphically satisfactory fit which downweighs arbitrarily distance classes which do not comply with the shape suggested by the experimental variograms. Finally the coregionalization matrices are decomposed, yielding the transformation coefficients a_{ip}^u which specify the linear coregionalization model

$$\mathbf{B}_u = \mathbf{A}_u \, \mathbf{A}_u^\top \qquad \text{where } \mathbf{A}_u = [a_{pu}^i]. \tag{26.13}$$

The decomposition of the $\mathbf{B}_u$ into the product of $\mathbf{A}_u$ with its transpose is usually based on the eigenvalue decomposition of each coregionalization matrix. Several decompositions for the purpose of a regionalized multivariate data analysis are discussed in the next chapter.

Bivariate fit of the experimental variograms

The multivariate nested variogram model associated with a linear model of intrinsically stationary random functions is

$$\boldsymbol{\Gamma}(\mathbf{h}) = \sum_{u=0}^{S} \mathbf{B}_u \, g_u(\mathbf{h}), \tag{26.14}$$

where the $g_u(\mathbf{h})$ are normalized variograms and the $\mathbf{B}_u$ are positive semi-definite matrices.

In the case of two variables it is simple to design a procedure for fitting the variogram model to the experimental variograms. We start by fitting the two direct variograms using a nested model. At least one structure $g_u(\mathbf{h})$ should be common to both variograms to obtain a non trivial coregionalization model. Then we are able to fit the sills b_{ij}^u of the cross variogram, using the sills of the direct variograms to set bounds within which the coregionalization model is authorized,

$$|b_{ij}^u| \leq \sqrt{b_{ii}^u \, b_{jj}^u}, \tag{26.15}$$

because the second order principal minors of $\mathbf{B}_u$ are positive.

Constrained weighted least squares routines exist, which allow integrating these constraints into an automated fit for a set of predefined structures.

The extension of this bivariate procedure to more than two variables does not guarantee a priori an authorized model, because higher order principal minors of the coregionalization matrices are not constrained to be positive.

EXAMPLE 26.3 *The inequality relation between sills can be used to provide a graphical criterion to judge the goodness of the fit of a variogram model to an experimental variogram.* A hull of perfect correlation *is defined by replacing the sills b_{ij}^u of the cross variogram by the square root of the product of the sills of the corresponding direct variograms (which have to be fitted first) and setting the sign of the total to + or −:*

$$\text{hull}(\gamma_{ij}(\mathbf{h})) = \pm \sum_{u=0}^{S} \sqrt{b_{ii}^u \, b_{jj}^u} \, g_u(\mathbf{h}). \tag{26.16}$$

Figure 26.1 shows the cross variogram between nickel and arsenic for the Loire geochemical data set. The fit of a nugget-effect plus two spherical models with ranges of 3.5km and 6.5km seems inappropriate on this graph.

The same fit viewed on Figure 26.2 within the hull of perfect (positive or negative) correlation now looks satisfactory. This shows that a cross variogram fit should be judged in the context of the spatial correlation between two variables: in this case the statistical correlation between the two variables is very poor (the correlation coefficient is equal to $r = .12$). The b_{ij}^u coefficients of the two spherical structures are very small in absolute value and the model is actually close to a pure nugget-effect, indicating that correlation between arsenic and nickel only exists at the microscale.

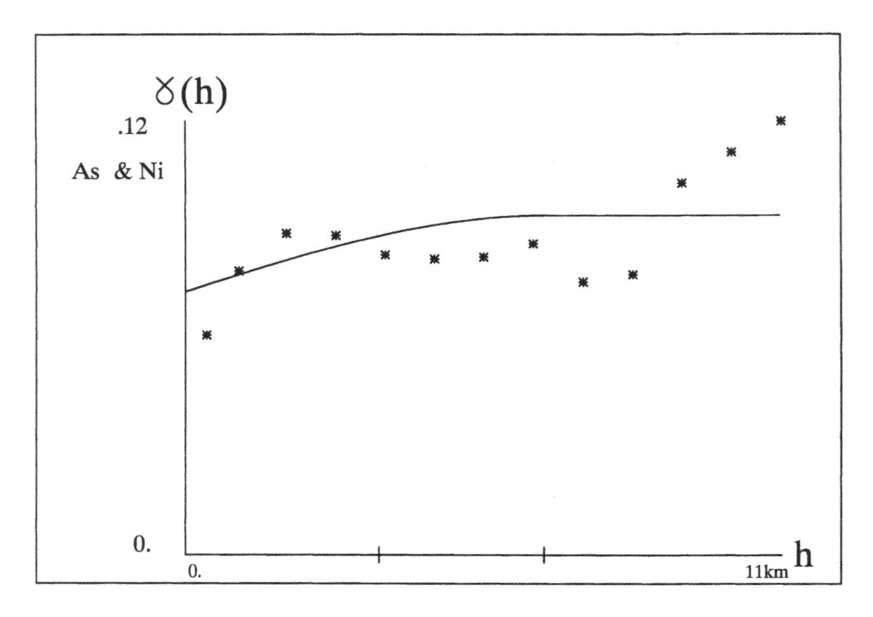

Figure 26.1: Experimental cross variogram of arsenic and nickel with a seemingly badly fitted model.

Multivariate fit

A multivariate fit is sought for a model $\mathbf{\Gamma}(\mathbf{h})$ to the matrices $\mathbf{\Gamma}^{\star}(\mathfrak{H}_k)$ of the experimental variograms calculated for n_c classes $\mathfrak{H}_k$ of separation vectors $\mathbf{h}$. An iterative algorithm due to GOULARD [125] is based on the least squares criterion

$$\sum_{k=1}^{n_c} \text{tr}[\,(\mathbf{\Gamma}^{\star}(\mathfrak{H}_k) - \mathbf{\Gamma}(\mathfrak{H}_k)\,)^2\,], \tag{26.17}$$

which will be minimized under the constraint that the coregionalization matrices $\mathbf{B}_u$ of the model $\mathbf{\Gamma}(\mathbf{h})$ have to be positive semi-definite.

Starting with S arbitrarily chosen positive semi-definite matrices $\mathbf{B}_u$, we look for an S+1-th matrix $\mathbf{B}_v$ which best fills up the gap $\mathrm{d}\mathbf{\Gamma}^{\star}_{vk}$ between a model with S matrices and the experimental matrices $\mathbf{\Gamma}^{\star}(\mathfrak{H}_k)$

$$\mathrm{d}\mathbf{\Gamma}^{\star}_{vk} \;=\; \mathbf{\Gamma}^{\star}(\mathfrak{H}_k) - \sum_{\substack{u=0\\u\neq v}}^{S} \mathbf{B}_u\, g_u(\mathfrak{H}_k). \tag{26.18}$$

The sum of the differences weighted by $g_v(\mathfrak{H}_k)$ is a symmetric matrix $\mathrm{d}\mathbf{\Gamma}^{\star}_v$ which is in general not positive semi-definite. Decomposing it into eigenvalues and eigenvectors,

$$\sum_{k=1}^{n_c} \mathrm{d}\mathbf{\Gamma}^{\star}_{vk}\, g_v(\mathfrak{H}_k) \;=\; \mathrm{d}\mathbf{\Gamma}^{\star}_v = \mathbf{Q}_v\, \mathbf{\Lambda}_v\, \mathbf{Q}_v^{\top} \qquad \text{with } \mathbf{Q}_v^{\top}\mathbf{Q}_v = \mathbf{I}, \tag{26.19}$$

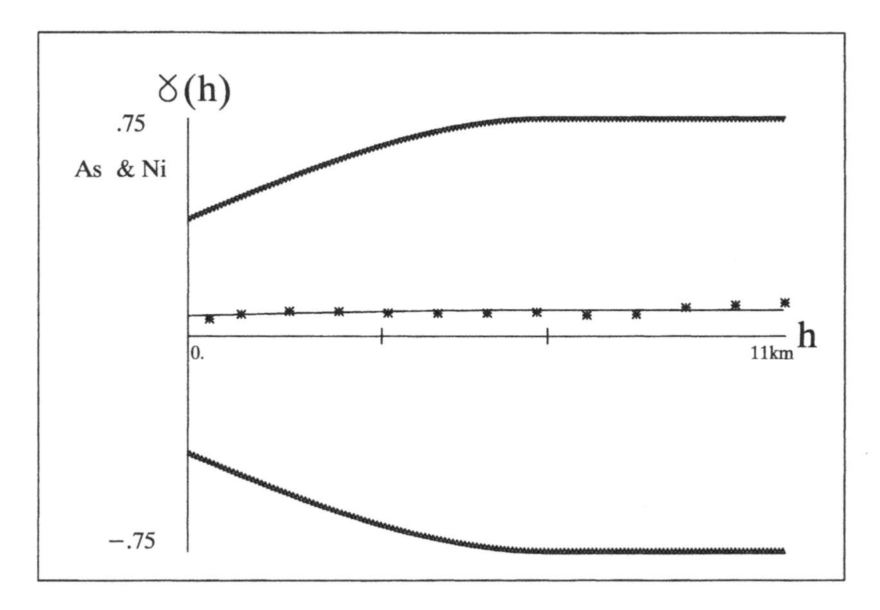

Figure 26.2: The same cross variogram fit is displayed together with the hull of perfect positive or negative correlation provided by the previous fitting of the corresponding direct variograms. The cross variogram model is defined within the bounds of the hull of perfect correlation.

we can build with the positive eigenvalues a new matrix $\overline{\mathrm{d}\boldsymbol{\Gamma}_v^\star}$ which is the positive semi-definite matrix nearest to $\mathrm{d}\boldsymbol{\Gamma}_v^\star$ in the least squares sense

$$\overline{\mathrm{d}\boldsymbol{\Gamma}_v^\star} \;=\; \mathbf{Q}_v\,\boldsymbol{\Lambda}_v^o\,\mathbf{Q}_v^\top, \tag{26.20}$$

where $\boldsymbol{\Lambda}_v^o$ is equal to the diagonal eigenvalues matrix $\boldsymbol{\Lambda}_v$ except for the negative eigenvalues, which have been set to zero.

Dividing by the square of the variogram $g_v(\mathbf{h})$ of the $S+1$-th structure, we obtain a positive semi-definite matrix $\mathbf{B}_v^\star$ which minimizes the fitting criterion

$$\mathbf{B}_v^\star \;=\; \frac{\overline{\mathrm{d}\boldsymbol{\Gamma}_v^\star}}{\sum\limits_{k=1}^{n_c}\left(g_v(\mathfrak{H}_k)\right)^2}. \tag{26.21}$$

This procedure is applied in turn to each matrix $\mathbf{B}_u$ and iterated. The algorithm converges well in practice, although convergence is theoretically not ensured.

In practice weights $w(\mathfrak{H}_k)$ are included into the criterion to allow the user to put low weight on distance classes which the model should ignore

$$\sum_{k=1}^{n_c} w(\mathfrak{H}_k)\,\mathrm{tr}\big[\,(\boldsymbol{\Gamma}^\star(\mathfrak{H}_k)-\boldsymbol{\Gamma}(\mathfrak{H}_k)\,)^2\,\big]. \tag{26.22}$$

The steps of this weighted least squares algorithm are analogous to the algorithm without weights.

COMMENT 26.4 *The heart of Goulard's algorithm is built on the following result about eigenvalues (see* RAO [254]*, p63).*

Let $\mathbf{A}$ *be a symmetric* $N \times N$ *matrix. The Euclidean norm of* $\mathbf{A}$ *is*

$$||\mathbf{A}|| = \sqrt{\sum_{i=1}^{N}\sum_{j=1}^{N}|a_{ij}|^2} \tag{26.23}$$

and its eigenvalue decomposition

$$\mathbf{A} = \sum_{p=1}^{N}\lambda_p\,\mathbf{q}_p\,\mathbf{q}_p^{\top} \qquad \textit{with} \quad \mathbf{Q}\,\mathbf{Q}^{\top} = \mathbf{I}. \tag{26.24}$$

We shall assume that the eigenvalues λ_p *are in decreasing order.*

We look for a symmetric $N \times N$ *matrix* $\mathbf{B}$ *of rank* $k < N$ *which best approximates a given matrix* $\mathbf{A}$ *i.e.*

$$\inf_{\mathbf{B}} ||\mathbf{A} - \mathbf{B}||. \tag{26.25}$$

Multiplying a matrix by an orthogonal matrix does not modify its length (measured with the Euclidean norm), so we have

$$\begin{aligned} ||\mathbf{A}-\mathbf{B}||^2 &= ||\,\mathbf{Q}^{\top}\mathbf{A}\,\mathbf{Q} - \underbrace{\mathbf{Q}^{\top}\mathbf{B}\,\mathbf{Q}}_{\mathbf{C}}\,||^2 \\ &= ||\mathbf{\Lambda} - \mathbf{C}||^2 \\ &= \sum_{p=1}^{N}\sum_{q=1}^{N}(\lambda_p\,\delta_{pq} - c_{pq})^2, \end{aligned} \tag{26.26}$$

where δ_{pq} *is one for* $p{=}\,q$ *and zero otherwise. The best choice for* $\mathbf{C}$ *is obviously a diagonal matrix as*

$$\sum_{p=1}^{N}(\lambda_p - c_{pp})^2 + \underbrace{\sum_{p=1}^{N}\sum_{q=1}^{N}}_{p\neq q}(c_{pq})^2 \;\geq\; \sum_{p=1}^{N}(\lambda_p - c_{pp})^2. \tag{26.27}$$

The smallest diagonal matrix $\mathbf{C}$ *of rank* k *is the one for which the diagonal elements* c_{pp} *are equal to the eigenvalues* λ_p *for* $p \leq k$ *and zero for* $p > k$

$$\sum_{p=1}^{N}(\lambda_p - c_{pp})^2 \;\geq\; \sum_{p=k+1}^{N}(\lambda_p)^2. \tag{26.28}$$

This means that

$$\mathbf{C} = \mathbf{Q}^{\top}\mathbf{B}\,\mathbf{Q} = \mathbf{\Lambda}^{o}, \tag{26.29}$$

where $\mathbf{\Lambda}^{o}$ *is the diagonal matrix of eigenvalues of* $\mathbf{A}$ *in which the last* $k{-}N$ *(lowest) eigenvalues have been set to zero.*

The best rank-k *approximation of* $\mathbf{A}$ *is* $\mathbf{B} = \mathbf{Q}\,\mathbf{\Lambda}^{o}\,\mathbf{Q}^{\top}$.

	Statistical correlation	Micro-scale correlation (0–10 m)	Small scale correlation (10–130 m)	Large scale correlation (130–2300 m)
Cu–Pb	-.08	0.0	-.04	-.36
Cu–Zn	.42	.57	.31	.42
Pb–Zn	.35	.23	.46	.11

Table 26.1: Correlations of Brilon geochemical variables.

The need for an analysis of the coregionalization

We can use classical principal component analysis to define the values of factors at sample locations and then krige the factors over the whole region to make maps. What would be the benefit of a spatial multivariate analysis based on the linear model of coregionalization and the corresponding multivariate nested variogram?

To answer this question we restrict the discussion to a second order stationary context, to make sure that the variance-covariance matrix $\mathbf{V}$, on which classical principal component analysis is based, exists from the point of view of the model.

If we let the lag $\mathbf{h}$ go to infinity in a second order stationary context with structures $g_u(\mathbf{h})$ having unit sills, we notice that the multivariate variogram model is equal to the variance-covariance matrix for large $\mathbf{h}$

$$\boldsymbol{\Gamma}(\mathbf{h}) \quad \rightarrow \quad \mathbf{V} \qquad \text{for } \mathbf{h} \rightarrow \infty. \tag{26.30}$$

In this setting the variance-covariance matrix is simply a sum of coregionalization matrices

$$\boxed{\mathbf{V} = \sum_{u=0}^{S} \mathbf{B}_u}\;. \tag{26.31}$$

Relation (26.31) teaches us that when the variables are not intrinsically correlated, it is necessary to analyze separately each coregionalization matrix $\mathbf{B}_u$. The variance-covariance matrix $\mathbf{V}$ is a mixture of different correlation structures stemming from all scales covered by the sampling grid and this amalgamate is likely to be meaningless from the point of view of the linear model of coregionalization.

Furthermore, it should be noted that coregionalization matrices $\mathbf{B}_u$ can be obtained under any type of stationarity hypothesis, while the variance-covariance matrix $\mathbf{V}$ is only meaningful with data fitting into a framework of second-order stationarity.

EXAMPLE 26.5 *A multivariate nested variogram has been fitted to the direct and cross variograms of the three elements copper, lead and zinc sampled in a forest near the town of Brilon, Germany (as described in [340, 338]).*

Regionalized correlation coefficients

$$\rho_{ij}^{u} = \frac{b_{ij}^{u}}{\sqrt{b_{ii}^{u}\, b_{jj}^{u}}} \tag{26.32}$$

have been computed for each variable pair at three characteristic spatial scales and are listed on Table 26.1 together with the classical correlation coefficient

$$r_{ij} = \frac{s_{ij}}{\sqrt{s_{ii}\, s_{jj}}}. \tag{26.33}$$

The three characteristic scales were defined on the basis of the nested variogram model in the following way

- micro-scale*: the variation below the minimal sample spacing of 10m is summarized by a nugget-effect model;*
- small scale*: the variation at a scale of 10m to 130m is represented by a spherical model of 130m range;*
- large scale*: the variation above 130m is captured by a spherical model with a range of 2.3km.*

At the micro-scale, the best correlated pair is copper and zinc with a regionalized correlation of .57: the geochemist to whom these figures were presented thought this was normal, because when copper has a high value at a sample point, zinc also tends to have a high value at the same location.

At the small scale, lead and zinc have a coefficient of .46 which is explained by the fact that when there are high values of lead at some sample points, there are high values of zinc at points nearby within the same zones.

Finally, at the large scale, we notice the negative coefficient of -.36 between copper and lead, while at the two other characteristic scales the correlation is (nearly) zero. This is due to the fact that when copper is present in a zone, this excludes lead and vice-versa.

Naturally this tentative interpretation does not answer all questions. Interpretation of correlation coefficients is difficult because they do not obey an equivalence relation: if A is correlated with B and B with C, this does not imply that A is correlated with C.

At least we can draw one conclusion from the analysis of this table of coefficients: the set of three variables is clearly not intrinsically correlated and correlation strongly depends on spatial scale. The variance-covariance matrix does not describe well the relations between the variables. Multiple linear regression would be a bad idea and cokriging based on the multivariate nested variogram the alternative to go for in a regression problem.

27 Case Study: Ebro Estuary

The present chapter illustrates the application of cokriging combining in unique neighborhood heterotopic data from two different sources in the Ebro river estuary (Spain): on the one hand a sounder measuring conductivity, on the other water samples analyzed for chlorophyll and salinity. We also discuss the problem of choosing the variogram model by showing conditional simulations obtained with models having a different behavior at the origin.

Kriging conductivity

The Ebro is the second largest river of Spain. We shall use data collected on the 5th October 1999 by the Polytechnic Universities of Barcelona and Valencia. The measurements were performed between 11am and 6pm, navigating upstream estuary. Details on the campaign and on the interpretation of the data are given in [304, 117, 303]. We shall use these data only for demonstrative purpose, to discuss some problems in the application of kriging, cokriging and conditional simulations.

Conductivity was measured employing a multiparametric sounding Hydrolab Surveyor III with the aim of locating the freshwater-seawater interface. The measurements were performed at five locations along the river, sampling vertically with a 10 centimeter spacing. This resulted in a total of 185 conductivity values. A plot of the five profiles is shown on Figure 27.1 using symbols proportional to the value of conductivity. Conductivity expresses the salinity of the water. The transition zone between freshwater and seawater is easily identified between 3 and 4 meter depth. The abscissa indicates the distance from the mouth of the Ebro river in kilometers.

The river bed displayed on Figure 27.1 is actually based on bathymetric measurements stemming from a previous campaign in the month of July 1999. We can assume that the bottom did not experiment great changes, but obviously, if there are different river discharges, the water levels (and as a consequence the depths) will be different. We made a fast computation of these differences using the following approach: the average discharge during the July campaign was 129 m^3/s while it amounted to 184 m^3/s in the October campaign (measured at a station upstream in the city of Tortosa). For this difference of river discharge, the water level is evaluated as about 20 cm higher on average and the bathymetry was corrected accordingly.

We will consider the problem of interpolating by kriging the conductivity profiles. Experimental variograms were computed between and within the profiles, using 60 lags of 10cm in the vertical and 100m in the horizontal. They are shown on Fig-

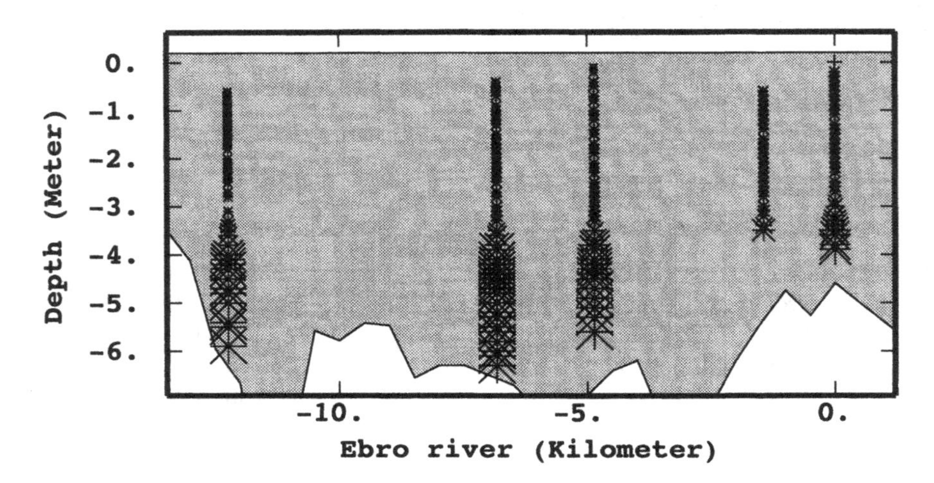

Figure 27.1: The 185 Hydrolab Surveyor III sample locations are plotted with symbols proportional to measured conductivity.

ure 27.2. In the horizontal direction (with a tolerance of ± 45 degrees) the variogram is denoted D1, while in the vertical it is denoted D2. Note the difference in scale in horizontal (kilometers) and vertical (meters) directions.

Considering that we know only well the variogram structure in the vertical direction we can do nothing better than to adopt the same model in the horizontal using a geometrical anisotropy. The main axes of the anisotropy ellipse are taken parallel to the horizontal and the vertical. The cubic model (M2) fitted in the vertical with a range of 7.5m was adjusted to the horizontal with a range of 17km. A nugget-effect of 2 $(mS/cm)^2$ was added to reflect measurement uncertainty and the sill of the cubic model is 1150 $(mS/cm)^2$, about three times the variance which is represented as a horizontal dotted line on Figure 27.2. The smooth parabolic behavior at the origin of the cubic model may be interpreted as reflecting the averaging over a non-point support by the physical measurement device.

An interpolation grid of 137 × 75 nodes with 100m × 10cm cells was defined, starting from an origin at (-12.9km, -6.8m). This interpolation grid and a neighborhood including all data (unique neighborhood) will be used in all examples.

The ordinary kriging of conductivity using the cubic model and filtering the nugget effect is shown on Figure 27.3. The map represents well the two phases, freshwater and seawater, suggested by the data on Figure 27.1. This picture of the spatial distribution of chlorophyll relies heavily on the geometric anisotropy built into the geostatistical model, which emphasizes the horizontal dependence between the profiles.

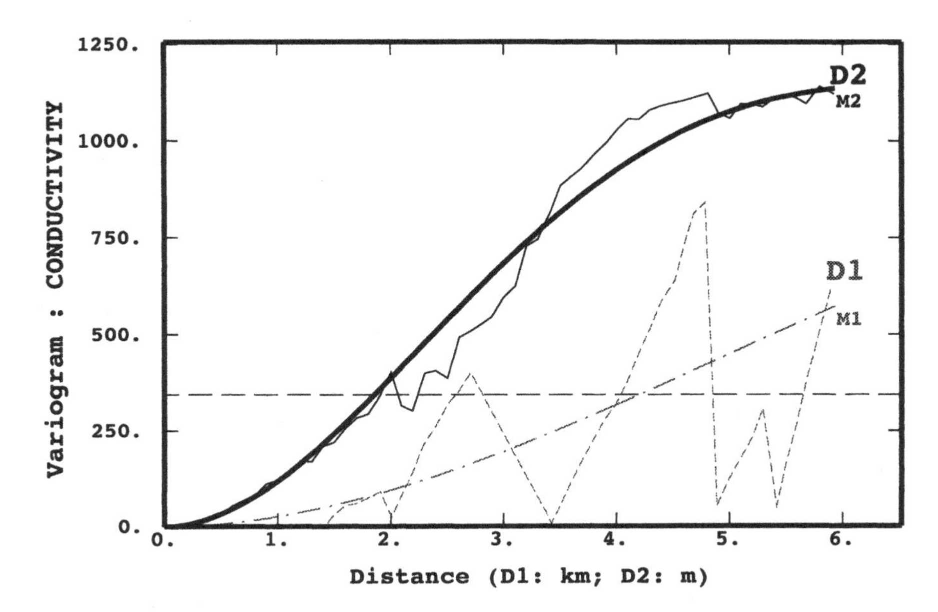

Figure 27.2: Experimental variogram of conductivity in two directions (D1: horizontal, D2: vertical). Model variogram in both directions (M1, M2). The abscissa should be read as kilometers for the horizontal and as meters for the vertical.

Cokriging of chlorophyll

The Hydrolab device has been used to obtain quickly an indication about the depth of the freshwater/seawater interface. Water samples were obtained using a new device called SWIS (Salt Wedge Interface System) developed jointly by the Polytecnic Universities of Barcelona and Valencia. It consists of six tubes connected to a vacuum system that can be operated from the surface. Spacing the tubes at 10cm from each other, the roughly half a meter wide interface can be sampled with up to six samples in one go. Additional samples (at zero, 1.5, 3 and 4.5 meter depth) were taken, leading to a total of 47 samples for the five measurement points along the Ebro river. The number of locations where both water samples and Hydrolab samples are available is 31.

On Figure 27.4 the water sample locations are plotted using symbols proportional to the value of chlorophyll. At the same locations salinity measurements are available. The scatter plot of salinity against chlorophyll is shown on Figure 27.5. Nine water samples located in the freshwater are plotted as stars, while the samples in the salt wedge are represented with crosses. While the relationship within both media can be assumed linear, this is not the case when considering all data. As cokriging requires a linear relationship between the variable of interest and the auxiliary variables, the logarithm (basis 10) was taken for both salinity and conductivity.

Direct and cross variograms were computed for the three variables and were fitted

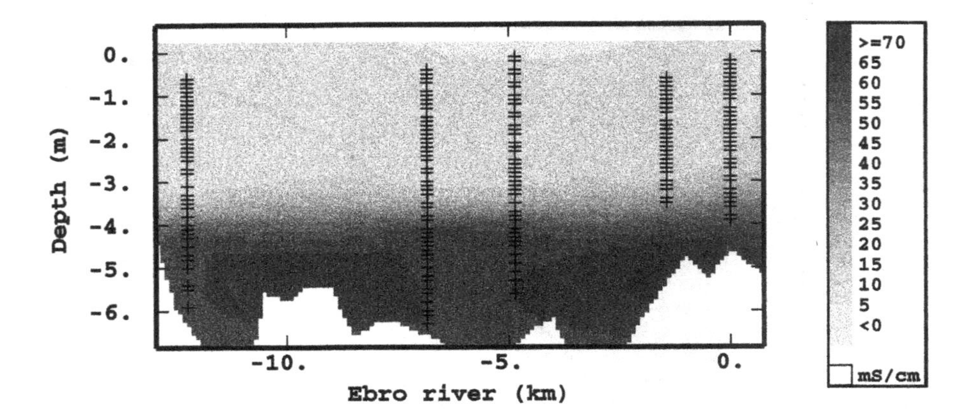

Figure 27.3: Map of conductivity obtained by ordinary kriging, filtering the nugget effect component.

using a nugget effect and a cubic model with a geometric anisotropy, taking a maximal of 17km along the horizontal and a minimal range of 7.5m in the vertical. The fitting was done using an improved version of the algorithm of Goulard & Voltz [127], running 200 iterations, restraining the fitting to distances less than 3km in the horizontal and 3m in the vertical, and taking weights proportional to the number of pairs in each direction, divided also by the average distances. The set of fitted direct and cross variograms is shown on Figure 27.6. The variogram of conductivity is fitted with a more generous nugget effect and a lower sill than it was by hand on Figure 27.2. The cross variograms are displayed together with the correlation hulls computed from the corresponding direct variograms.

The cokriging of chlorophyll taking logarithms of salinity and conductivity as auxiliary variables is displayed on Figure 27.7. The nugget effect, viewed as measurement error, was filtered. The use of conductivity as an auxiliary variable permits to extrapolate chlorophyll, quite successfully, at depths greater than where it was measured.

Extrapolation of chlorophyll in greater depth than where water samples are available depends much on how the model is formulated. Three different cokrigings using a nugget effect plus cubic model with ranges as defined above were experimented:

A the ordinary cokriging of chlorophyll with (untransformed) salinity and conductivity;

B the ordinary cokriging of chlorophyll with logarithms of salinity and conductivity (reference case described in detail above);

C the universal cokriging of chlorophyll with logarithms of salinity and conductivity, adding a linear drift in the vertical direction, to take account explicitly of the vertical non stationarity.

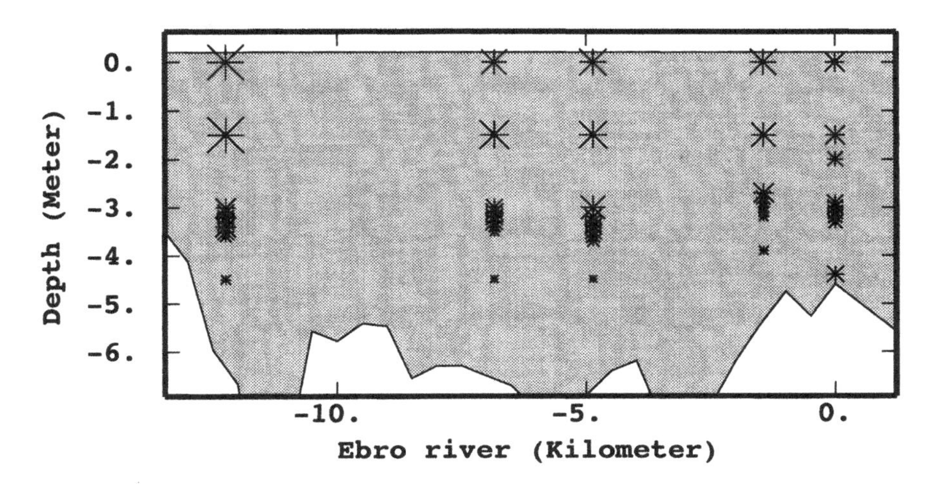

Figure 27.4: The 47 water sample locations are plotted with symbols proportional to chlorophyll.

The three cokrigings essentially differ in the extrapolation behavior below a depth of 5m. The scatter diagrams of cokrigings A and C against B are plotted on Figure 27.8. The cokriging A (i.e. without taking logarithms, and thus neglecting the non linear relation with the auxiliary variables) grossly extrapolates at depths below 5m. This generates the cloud of values for which cokriging A differs much from cokriging B. The cokriging C, adding a linear drift in the vertical, also yields higher values than case B in areas at greater depth, far from the water sample profiles.

Conditional simulation of chlorophyll

The variogram models of chlorophyll and salinity shown on Figure 27.6 cannot decently be qualified as "fitting" the corresponding experimental variograms, because the latter exhibit little structure due to the small amount of data. So let us discuss this problem using the 47 chlorophyll data.

The experimental variogram of chlorophyll is shown on Figure 27.9 together with two models differing in their behavior at the origin:

- a geometrically anisotropic *cubic* model with a sill of 60 $(\text{mg/m}^3)^2$, with ranges 17km in the horizontal and 7.5m in the vertical.

- a geometrically anisotropic *exponential* model with a sill of 30 $(\text{mg/m}^3)^2$, with ranges 17km in the horizontal and 7.5m in the vertical.

A grid set at an origin (-12.9km,-5.2m) with 137 $\times$ 59 nodes using 100m $\times$ 10cm cells was defined for interpolation and stochastic simulation.

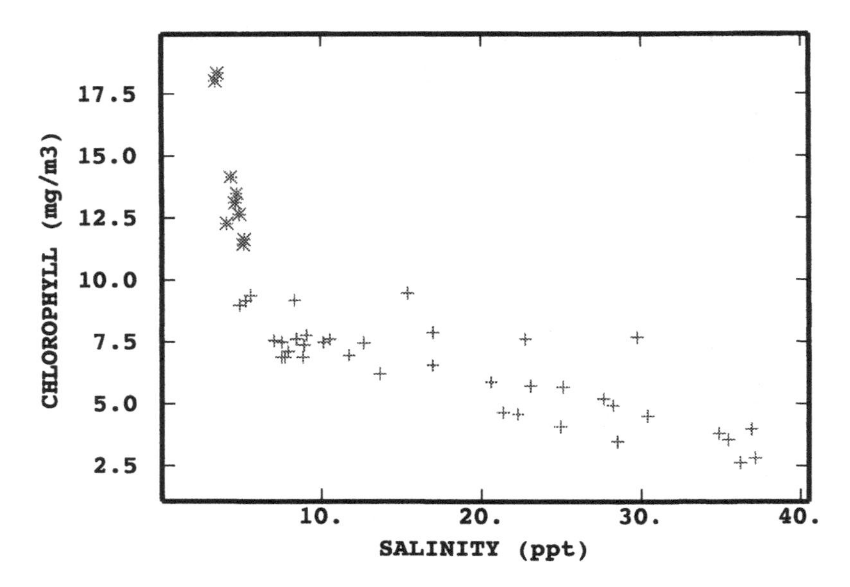

Figure 27.5: Scatter diagram between salinity and chlorophyll for the 47 water samples. Nine samples located in freshwater are plotted as stars, the others as crosses.

Ordinary kriging was performed with variogram models and is shown on Figure 27.10 (the greytone scale used is the same as on Figure 27.7). The upper map represents the kriging with the cubic model while the lower map was obtained with the exponential model. We note that the cubic model with such a large range as compared to the dimensions of the domain has an extrapolative behavior, generating a maximum at -3km which is not supported by data.

The exponential model is more conservative: the maxima and minima in the map obviously refer to the highest and lowest values in the chlorophyll data. However this has also another implication: the kriged value will be more alike the kriged mean of the domain the more we move away from data locations.

It is well known that kriging does not represent an attempt to reconstruct the regionalized variable (had we the ability to measure it everywhere in the domain). The regionalized variable in the geostatistical model is but one realization of the random function. Kriging can be thought of as the average of many realizations that coincide with the data at sample locations. Thus if we are interested in how the regionalized variable at hand might look like, we have to employ *conditional simulation* instead of kriging.

We use the turning bands method [203, 51, 178] for simulating realizations of the random function. We shall assume that the 47 chlorophyll values can be considered as a few samples of a realization of a Gaussian random function as the histogram does not indicate an asymmetric distribution. One thousand bands were used for simulation – with only 100 bands some bands could be seen with naked eye on the simulated

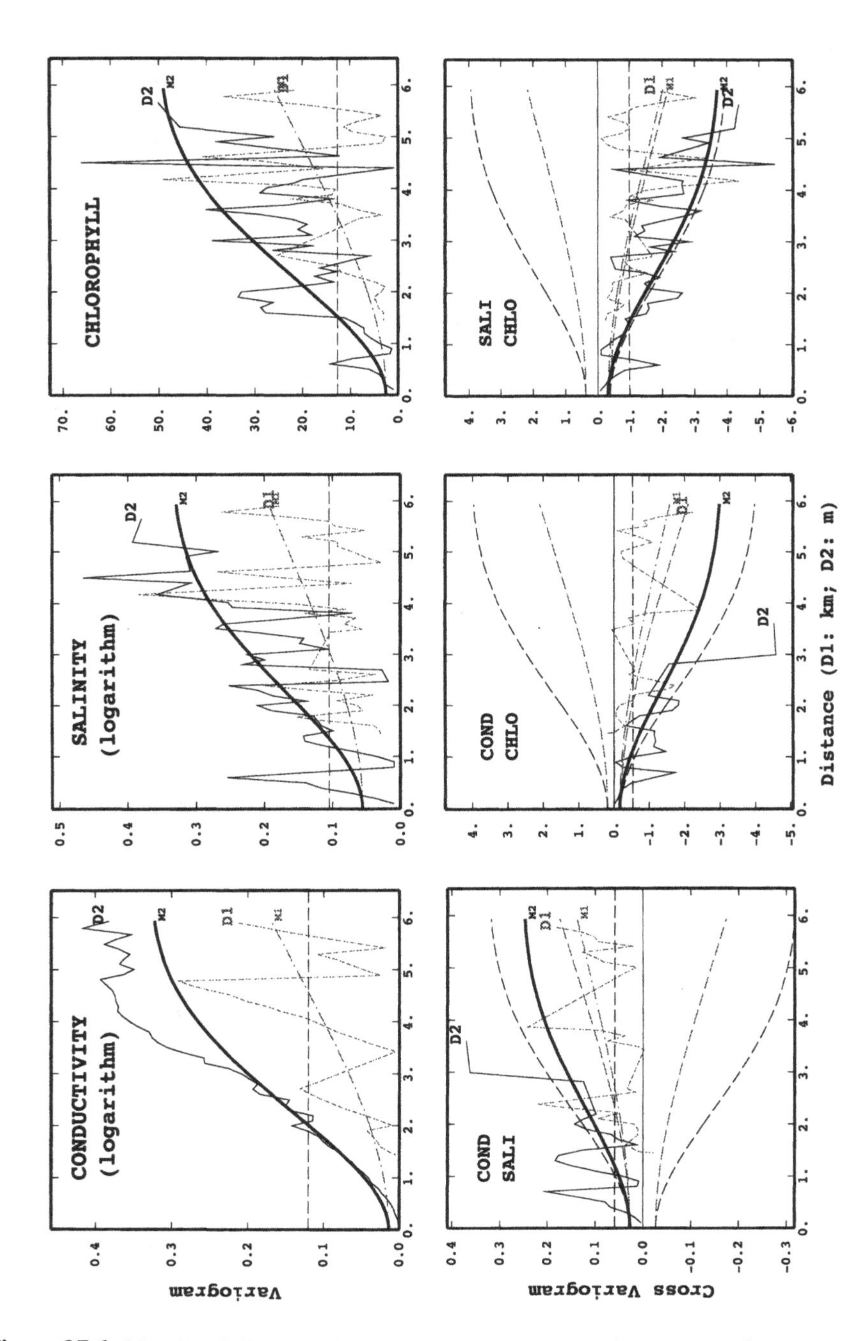

Figure 27.6: Matrix of direct and cross variograms in two directions. The cross variograms additionally represent the hulls of correlation in both directions.

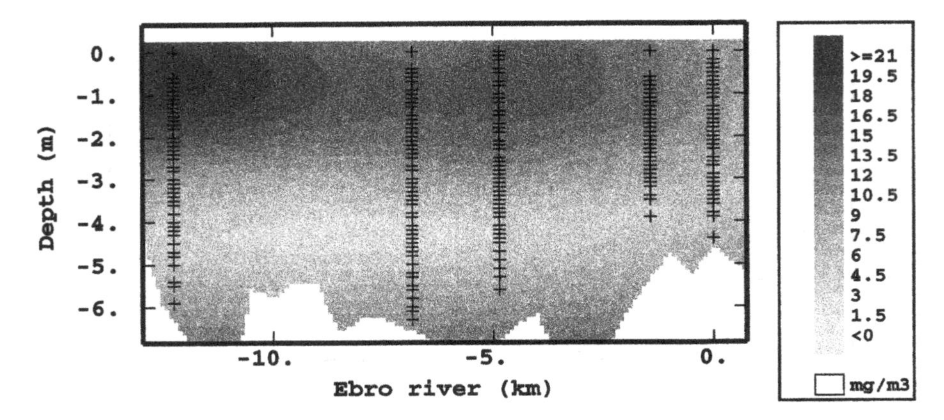

Figure 27.7: Map of chlorophyll obtained by ordinary cokriging, filtering the nugget effect component, and including the logarithms of salinity and conductivity as covariates.

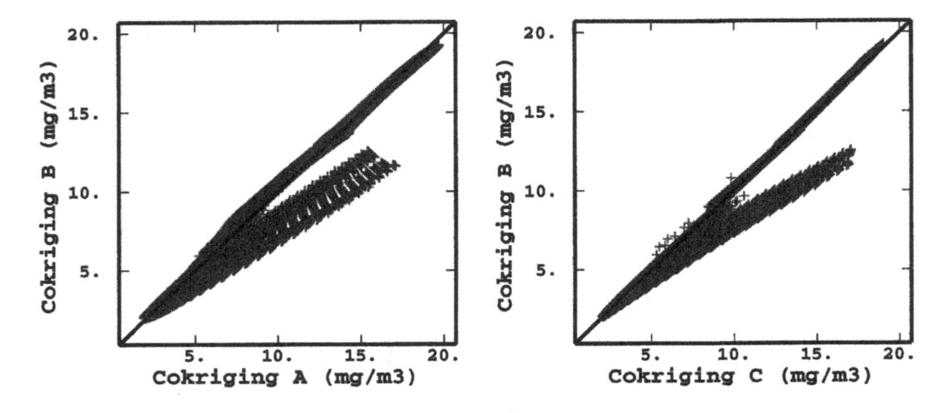

Figure 27.8: Scatter diagrams of chlorophyll cokriging experiments A and C against B. The axes are in the same scale and the first bisector is drawn.

map of the exponential model [112]. The latter is no surprise when the range of the variogram is larger than the size of the domain. The simulation with a thousand bands is no big deal on modern computers.

Two conditional simulations are shown on Figure 27.11 (the greytone scale used is the same as on Figure 27.7). The upper map illustrates the fact that a surface corresponding to the realization of a random function with a cubic variogram is smooth, which is related to the parabolic shape of this model at the origin. The lower map shows that the realization of a random function with an exponential variogram has a rough aspect, due to the non differentiability of that model at the origin.

In applications the smoothness/roughness of the regionalized variable, when it is

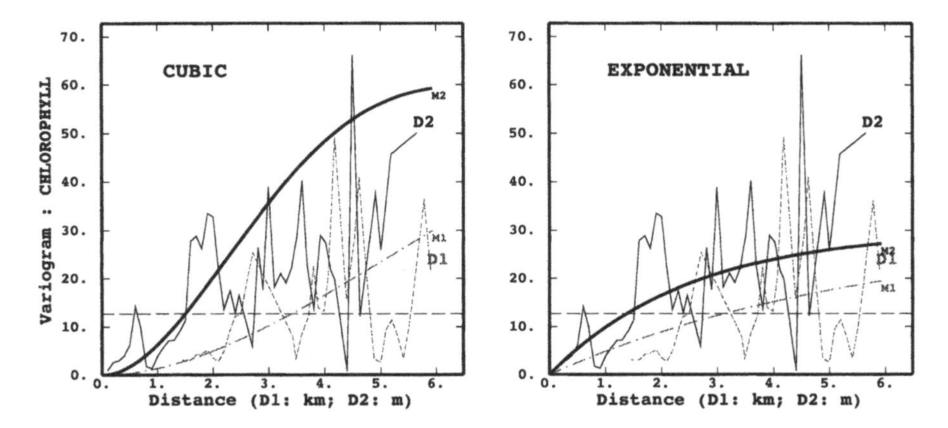

Figure 27.9: The variogram of chlorophyll has been "fitted" with two models having a different behavior at the origin.

known, can be an important criterion for selecting the type of variogram model to be employed when the data, like in the present case, are not sufficient to characterize properly the behavior at the origin of the variogram.

From a practical point of view it has to be decided whether reality looks rather like the upper or the lower picture of Figure 27.11. If the biologist is not able to tell, we can still rely on the shape of the vertical experimental variogram as seen on Figure 27.9: a linear behavior near the origin, as it is expressed by the exponential variogram model on the right graph, seems to be the more adequate interpretation. Choosing this option would imply that the bottom simulation on Figure 27.11 would be the one retained for further use.

The Ebro estuary case study is continued in Section 37 on p297 where the output from a numerical model is used as external drift for further improvement of kriging and conditional simulations.

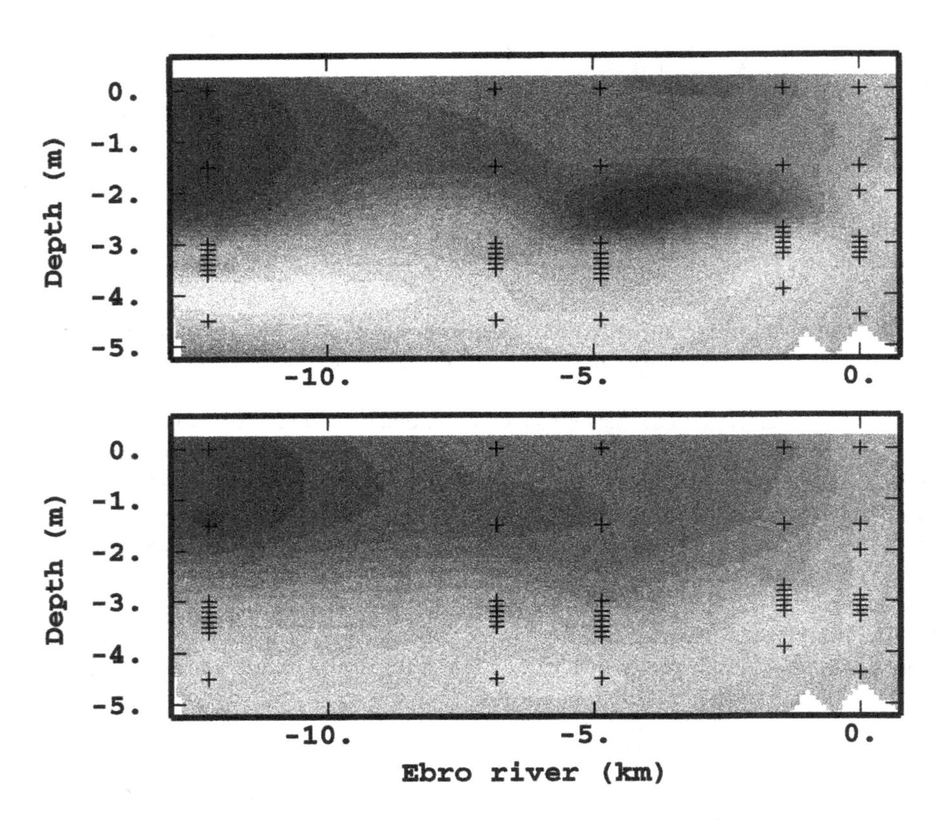

Figure 27.10: Ordinary kriging of chlorophyll using the cubic (upper map) and the exponential (lower map) variogram models.

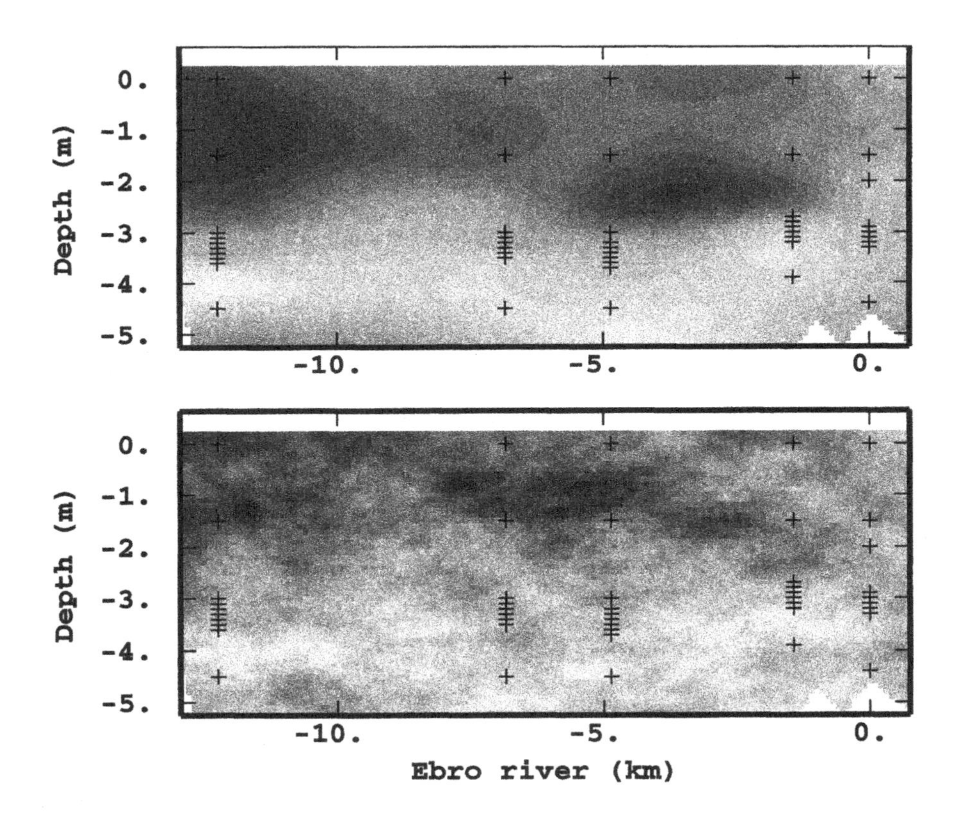

Figure 27.11: Conditional simulations of chlororophyll with the turning bands method using the cubic (upper map) and the exponential (lower map) variogram models.

28 Coregionalization Analysis

The geostatistical analysis of multivariate spatial data can be subdivided into two steps

- the analysis of the coregionalization of a set of variables leading to the definition of a linear model of coregionalization;
- the cokriging of specific factors at characteristic scales.

These techniques have originally been called *factorial kriging analysis* (from the French *analyse krigeante* [214]). They allow to isolate and to display sources of variation acting at different spatial scales with a different correlation structure.

Regionalized principal component analysis

Principal component analysis can be applied to coregionalization matrices, which are the variance-covariance matrices describing the correlation structure of a set of variables at characteristic spatial scales.

Regionalized principal component analysis consists in decomposing each matrix $\mathbf{B}_u$ into eigenvalues and eigenvectors

$$\mathbf{B}_u = \mathbf{A}_u \mathbf{A}_u^\top \qquad \text{with } \mathbf{A}_u = \mathbf{Q}_u \sqrt{\mathbf{\Lambda}_u} \text{ and } \mathbf{Q}_u \mathbf{Q}_u^\top = \mathbf{I}. \tag{28.1}$$

The matrices $\mathbf{A}_u$ specify the coefficients of the linear model of coregionalization. The transformation coefficients

$$a_{pu}^i = \sqrt{\lambda_u^p}\, q_p^i \tag{28.2}$$

are the covariances between the original variables $Z_i(\mathbf{x})$ and the factors $Y_u^p(\mathbf{x})$. They can be used to plot the position of the variables on correlation circles for each characteristic spatial scale of interest. These plots are helpful to compare the correlation structure of the variables at the different spatial scales.

The correlation circle plots can be used to identify intrinsic correlation: if the plots show the same patterns of correlation, this means that the eigenvectors of the coregionalization matrices are similar and the matrices only differ by their eigenvalues. Thus the matrices $\mathbf{B}_u$ are all proportional to a matrix $\mathbf{B}$, the coregionalization matrix of the intrinsic correlation model.

Applications of regionalized principal component analysis (or factor analysis) have been performed in the fields of geochemical exploration [337, 129, 307, 29, 344, 20], petroleum exploration [365], soil science [346, 126, 119, 127, 256], hydrogeology [278, 123], plant ecology [227], volcanic tremor intensity time series [151], to mention just a few.

Generalizing the analysis

The analysis can be generalized by choosing eigenvectors which are orthogonal with respect to a symmetric matrix $\mathbf{M}_u$ representing a metric

$$\mathbf{A}_u = \mathbf{Q}_u \mathbf{M}_u \sqrt{\boldsymbol{\Lambda}_u} \qquad \text{with } \mathbf{Q}_u \mathbf{M}_u \mathbf{Q}_u^{\top} = \mathbf{I}. \tag{28.3}$$

A possibility is to use the metric

$$\mathbf{M}_u = \sum_{v=0}^{S} \mathbf{B}_v, \tag{28.4}$$

which is equivalent to the variance-covariance matrix $\mathbf{V}$ in a second order stationary model. This metric generates a contrast between the global variation and the variation at a specific scale described by a coregionalization matrix $\mathbf{B}_u$.

COMMENT 28.1 *This type of metric has also been used frequently to decompose the matrix $\boldsymbol{\Gamma}^*(\mathfrak{H}_1)$ of experimental direct and cross variograms for the first distance class $\mathfrak{H}_1$ by numerous authors [350, 279, 320, 25, 20, 88], looking for the eigenvectors of*

$$\frac{\boldsymbol{\Gamma}^*(\mathfrak{H}_1)}{\mathbf{V}}. \tag{28.5}$$

This noise removal procedure has been compared to coregionalization analysis in [346] and [343]. The basic principle is illustrated with a simple model which shows why the approach can be successful in removing micro-scale variation.

Suppose we have a multivariate nested covariance function model

$$\mathbf{C}(\mathbf{h}) = \mathbf{B}_0\, \rho_{nug}(\mathbf{h}) + \mathbf{B}_1\, \rho_{sph}(\mathbf{h}), \tag{28.6}$$

where $\rho_{nug}(\mathbf{h})$ is a nugget-effect and $\rho_{sph}(\mathbf{h})$ a spherical correlation function. The variance-covariance matrix in this model is

$$\mathbf{V} = \mathbf{B}_0 + \mathbf{B}_1. \tag{28.7}$$

Let $\mathfrak{H}_1$ be the distance class grouping vectors $\mathbf{h}$ greater than zero and shorter than the range of the spherical correlation function. Then

$$\mathbf{C}(\mathbf{h}) = \mathbf{B}_1\, \rho_{sph}(\mathbf{h}) \qquad \text{for} \quad \mathbf{h} \in \mathfrak{H}_1 \tag{28.8}$$

and the eigenvectors of

$$\frac{\mathbf{C}(\mathfrak{H}_1)}{\mathbf{V}} \tag{28.9}$$

are equivalent to those of

$$\frac{\mathbf{B}_1}{\mathbf{B}_0 + \mathbf{B}_1}, \tag{28.10}$$

what reminds the setting of discriminant analysis. Corresponding factors reduce to a minimum the influence of noise (micro-scale variation), as reflected in the matrix $\mathbf{B}_0$.

Regionalized canonical and redundancy analysis

When the set of variables is split into two groups any coregionalization matrix can be partitioned into

$$\mathbf{B}_u = \begin{pmatrix} \mathbf{B}_u^{11} & \mathbf{B}_u^{12} \\ \mathbf{B}_u^{21} & \mathbf{B}_u^{22} \end{pmatrix}, \tag{28.11}$$

where $\mathbf{B}_u^{11}$, $\mathbf{B}_u^{22}$ are the within group coregionalization matrices while $\mathbf{B}_u^{12}$ represents the between groups coregionalization.

The formalism of canonical analysis can be applied to that specific spatial scale of index u

$$\mathbf{B}_u^{11} = \mathbf{A}_u^{11} \, (\mathbf{A}_u^{11})^\top, \tag{28.12}$$

where

$$\mathbf{A}_u^{11} = \mathbf{Q}_u^{11} \, \mathbf{M}_u^{11} \, \sqrt{\mathbf{\Lambda}_u^{11}} \qquad \text{with } \mathbf{Q}_u^{11} \, \mathbf{M}_u^{11} \, (\mathbf{Q}_u^{11})^\top = \mathbf{I}, \tag{28.13}$$

and

$$\mathbf{M}_u^{11} = \mathbf{B}_u^{12} \, (\mathbf{B}_u^{22})^{-1} \, \mathbf{B}_u^{21}, \tag{28.14}$$

with non-singular matrices $\mathbf{B}_u^{22}$.

The results of canonical analysis are often deceptive. Principal component analysis with instrumental variables (RAO [253]), which has been reinvented in psychometrics under the name of *redundancy analysis* (an account is given in [325]), can be a more appropriate alternative. It is based on the metric

$$\mathbf{M}_u^{11} = \mathbf{B}_u^{12} \, \mathbf{B}_u^{21}. \tag{28.15}$$

GOOVAERTS [120] has first applied a regionalized redundancy analysis to examine the links between soil properties and chemical variables from banana leaves. See also [183] for an application to remote sensing and geochemical ground data, which turned out to be intrinsically correlated, so that redundancy analysis was based on $\mathbf{V}$ instead of $\mathbf{B}_u$.

Cokriging regionalized factors

The linear model of coregionalization defines factors at particular spatial scales. We wish to estimate a regionalized factor from data in a local neighborhood around each estimation location $\mathbf{x}_0$.

The estimator of a specific factor $Y_{u_0}^{p_0}(\mathbf{x})$ at a location $\mathbf{x}_0$ is a weighted average of data from variables in the neighborhood with unknown weights w_α^i

$$Y_{p_0 u_0}^{\star}(\mathbf{x}_0) = \sum_{i=1}^{N} \sum_{\alpha=1}^{n_i} w_\alpha^i \, Z_i(\mathbf{x}_\alpha). \tag{28.16}$$

In the framework of local second-order stationarity, in which local means m_l^i for the neighborhood around $\mathbf{x}_0$ are meaningful, an unbiased estimator is built for the factor (of zero mean, by construction) by using weights summing up to zero for each variable

$$\mathrm{E}[\, Y^\star_{p_0 u_0}(\mathbf{x}) - Y^{p_0}_{u_0}(\mathbf{x})\,] \;=\; \sum_{i=1}^{N} m_l^i \underbrace{\sum_{\alpha=1}^{n} w_\alpha^i}_{0} = 0. \tag{28.17}$$

The effect of the constraints on the weights is to filter out the local means of the variables $Z_i(\mathbf{x})$.

The estimation variance σ_{E}^2 is

$$\begin{aligned}\sigma_{\mathrm{E}}^2 \;&=\; \mathrm{E}[\,(Y^\star_{p_0 u_0}(\mathbf{x}) - Y^{p_0}_{u_0}(\mathbf{x}))^2\,] \qquad (28.18)\\ &=\; 1 + \sum_{i=1}^{N}\sum_{j=1}^{N}\sum_{\alpha=1}^{n}\sum_{\beta=1}^{n} w_\alpha^i\, w_\beta^j\, C_{ij}(\mathbf{x}_\alpha - \mathbf{x}_\beta) - 2\sum_{i=1}^{N}\sum_{\alpha=1}^{n} w_\alpha^i\, a^i_{u_0 p_0}\, \rho_{u_0}(\mathbf{x}_\alpha - \mathbf{x}_0).\end{aligned}$$

The minimal estimation variance is realized by the cokriging system

$$\begin{cases} \displaystyle\sum_{j=1}^{N}\sum_{\beta=1}^{n_j} w_\beta^j\, C_{ij}(\mathbf{x}_\alpha - \mathbf{x}_\beta) - \mu_i = \boxed{a^i_{p_0 u_0}\, \rho_{u_0}(\mathbf{x}_\alpha - \mathbf{x}_0)} & \text{for } i = 1, \ldots N; \\ & \alpha = 1, \ldots n \\ \displaystyle\sum_{\beta=1}^{n_i} w_\beta^i = 0 & \text{for } i = 1, \ldots N. \end{cases} \tag{28.19}$$

We find in the right hand side of this system the transformation coefficients $a^i_{p_0 u_0}$ of the factor of interest. These coefficients are multiplied by values of the spatial correlation function $\rho_{u_0}(\mathbf{h})$ which describes the correlation at the scale of interest.

The factor cokriging is used to estimate a regionalized factor at the nodes of a regular grid which serves to draw a map.

Regionalized multivariate analysis

Cokriging a factor is more cumbersome and computationally more intensive than kriging it. Coregionalization analysis is more lengthy than a traditional analysis which ignores spatial scale. When is all this effort necessary and worthwhile? When can it be avoided? The answer is based on the notion of intrinsic correlation.

The steps of both a classical or a regionalized multivariate analysis (MVA) for spatial data are summarized on Figure 28.1. The key question to investigate is whether the correlation between variables is dependent on spatial scale. Three ways to test for scale-dependent correlation have been described

1. codispersion coefficients $cc_{ij}(\mathbf{h})$ can be computed and plotted: if they are not constant for each variable pair, the correlation structure of the variable set is affected by spatial scale;

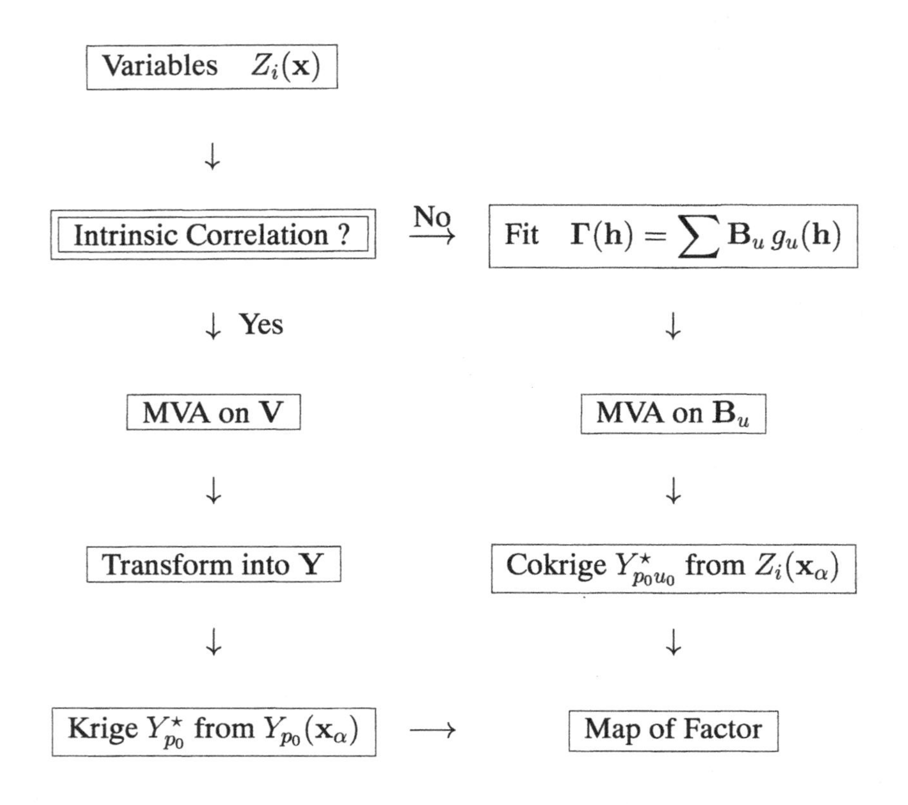

Figure 28.1: Classical multivariate analysis followed by a kriging of the factors (on the left) versus regionalized multivariate analysis (on the right).

2. cross variograms between principal components of the variables can be computed: if they are not zero for each principal component pair at any lag $\mathbf{h}$ (like on Figure 22.1 on page 156), the classical principal components are meaningless because the variance-covariance matrix of the variable set is merely a mixture of different variance-covariance structures at various spatial scales;

3. plots of correlation circles in a regionalized principal component analysis can be examined: if the patterns of association between the variables are not identical for the coregionalization matrices, the intrinsic correlation model is not appropriate for this data set. With only few variables it is possible to look directly at a table of regionalized correlation coefficients instead of the regionalized principal components (like in Example 26.5 on page 181).

If the data appears to be intrinsically correlated, we can apply any classical method of multivariate analysis, calculate the direct variograms of the factors, krige them on a grid and represent them as maps. But if correlation is affected by spatial scale, we need to fit a linear model of coregionalization and to cokrige the factors, as suggested by the right hand path of Figure 28.1.

It is interesting and important to note that the samples need not to be without autocorrelation (as implied by the usual hypothesis of sample independence) to allow for a classical multivariate analysis to be performed independently of the geostatistical treatment. For this purpose the multivariate regionalized data need only to comply with the intrinsic correlation model, which factorizes the multivariate correlation and the autocorrelation.

29 Kriging a Complex Variable

The question of how to model and krige a complex variable has been analyzed by LAJAUNIE & BÉJAOUI [170] and GRZEBYK [130]. The covariance structure can be approached in two ways: either by modeling the real and imaginary parts of the complex covariance or by modeling the coregionalization of the real and complex parts of the random function. This opens several possibilities for (co-)kriging the complex random function.

Coding directional data as a complex variable

Complex random functions can be useful to model directional data in two spatial dimensions. For analyzing and mapping a wind field we code the direction $\Theta(\mathbf{x})$ and the intensity $R(\mathbf{x})$ of wind at each location of the field as a complex variable

$$Z(\mathbf{x}) \quad = \quad R(\mathbf{x})\, \mathrm{e}^{\,\mathrm{i}\,\Theta(\mathbf{x})}. \tag{29.1}$$

An alternate representation of $Z(\mathbf{x})$ can be set up using the coordinates of the complex plane,

$$Z(\mathbf{x}) \quad = \quad U(\mathbf{x}) + \mathrm{i}\, V(\mathbf{x}), \tag{29.2}$$

as shown on Figure 29.1.

We shall use this second representation of the random function.

Complex covariance function

We assume that $Z(\mathbf{x})$, $U(\mathbf{x})$ and $V(\mathbf{x})$ are second order stationary and centered. The covariance function of $Z(\mathbf{x})$ is defined as

$$C(\mathbf{h}) \quad = \quad \mathrm{E}[\, Z(\mathbf{x}{+}\mathbf{h})\, \overline{Z(\mathbf{x})}\,] = C^{\mathrm{Re}}(\mathbf{h}) + \mathrm{i}\, C^{\mathrm{Im}}(\mathbf{h}), \tag{29.3}$$

where $\overline{Z(\mathbf{x})} = U(\mathbf{x}) - \mathrm{i}\, V(\mathbf{x})$ is the complex conjugate.

Let $C_{UU}(\mathbf{h})$, $C_{VV}(\mathbf{h})$ and $C_{UV}(\mathbf{h})$ be the (real) direct and cross covariance functions of $U(\mathbf{x})$ and $V(\mathbf{x})$. Then the complex covariance $C(\mathbf{h})$ can be expressed as

$$\begin{aligned} C(\mathbf{h}) &= C_{UU}(\mathbf{h}) + C_{VV}(\mathbf{h}) - \mathrm{i}\, C_{UV}(\mathbf{h}) + \mathrm{i}\, C_{VU}(\mathbf{h}) \\ &= C_{UU}(\mathbf{h}) + C_{VV}(\mathbf{h}) + \mathrm{i}\,(C_{UV}(-\mathbf{h}) - C_{UV}(\mathbf{h})). \end{aligned} \tag{29.4}$$

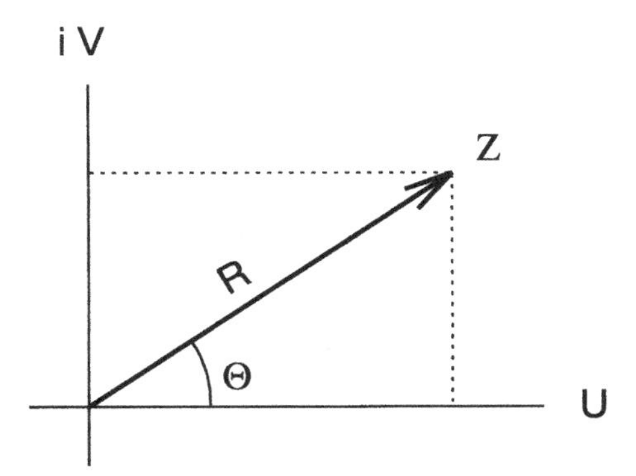

Figure 29.1: Coding a wind vector in the complex plane.

The real part of the complex covariance,

$$C^{\mathrm{Re}}(\mathbf{h}) \quad = \quad C_{UU}(\mathbf{h}) + C_{VV}(\mathbf{h}), \tag{29.5}$$

is even and a covariance function, while the imaginary part,

$$C^{\mathrm{Im}}(\mathbf{h}) \quad = \quad C_{UV}(-\mathbf{h}) - C_{UV}(\mathbf{h}), \tag{29.6}$$

is odd and not a covariance function.

Complex kriging

The linear combination for estimating, on the basis of a complex covariance (CC), the centered variable $Z(\mathbf{x})$ at a location $\mathbf{x}_0$ from data at locations $\mathbf{x}_\alpha$ with weights $w_\alpha \in \mathbb{C}$ is

$$Z^\star_{\mathrm{CC}}(\mathbf{x}_0) \quad = \quad \sum_{\alpha=1}^{n} w_\alpha \, Z(\mathbf{x}_\alpha). \tag{29.7}$$

In terms of the real and imaginary parts of $Z(\mathbf{x})$ it can be written as

$$\begin{aligned} Z^\star_{\mathrm{CC}}(\mathbf{x}_0) \quad = \quad & \sum_{\alpha=1}^{n} \big[\, (w^{\mathrm{Re}}_\alpha \, U(\mathbf{x}_\alpha) - w^{\mathrm{Im}}_\alpha \, V(\mathbf{x}_\alpha)) \\ & \qquad + \mathrm{i}(w^{\mathrm{Re}}_\alpha \, V(\mathbf{x}_\alpha) + w^{\mathrm{Im}}_\alpha \, U(\mathbf{x}_\alpha) \,) \, \big]. \end{aligned} \tag{29.8}$$

EXERCISE 29.1 *From the variance of the linear combination*

$$\mathrm{var}\Big(\sum_{\alpha=1}^{n} w_\alpha \, Z(\mathbf{x}_\alpha) \Big) \quad = \quad \sum_{\alpha=1}^{n} \sum_{\beta=1}^{n} w_\alpha \, \overline{w}_\beta \, C(\mathbf{x}_\alpha - \mathbf{x}_\beta) \tag{29.9}$$

the following inequality can be deduced

$$\sum_{\alpha=1}^{n}\sum_{\beta=1}^{n}\left(w_{\alpha}^{\mathrm{Re}}\, w_{\beta}^{\mathrm{Re}} + w_{\alpha}^{\mathrm{Im}}\, w_{\beta}^{\mathrm{Im}}\right) C^{\mathrm{Re}}(\mathbf{x}_{\alpha}-\mathbf{x}_{\beta}) \\ + \sum_{\alpha=1}^{n}\sum_{\beta=1}^{n}\left(w_{\alpha}^{\mathrm{Re}}\, w_{\beta}^{\mathrm{Im}} - w_{\beta}^{\mathrm{Re}}\, w_{\alpha}^{\mathrm{Im}}\right) C^{\mathrm{Im}}(\mathbf{x}_{\alpha}-\mathbf{x}_{\beta}) \;\geq\; 0. \qquad (29.10)$$

Show that this inequality implies that $C^{\mathrm{Re}}(\mathbf{h})$ *is a positive definite function.*

EXERCISE 29.2 *On the basis of the linear combination* $(1+\mathrm{i})\, Z(\mathbf{0}) + (1-\mathrm{i})\, Z(\mathbf{h})$ *show that*

$$|C^{\mathrm{Im}}(\mathbf{h})| \;\leq\; \sigma^2. \qquad (29.11)$$

Complex kriging consists in computing the complex weights which are solution of the simple kriging

$$\sum_{\beta=1}^{n} w_{\beta}\, C(\mathbf{x}_{\alpha}-\mathbf{x}_{\beta}) \;=\; C(\mathbf{x}_0-\mathbf{x}_{\alpha}) \qquad \text{for } \alpha = 1,\ldots,n. \qquad (29.12)$$

This is equivalent to the following system, in which the real and imaginary parts of the weights have been separated

$$\begin{cases} \sum\limits_{\beta=1}^{n} \left(w_{\beta}^{\mathrm{Re}}\, C^{\mathrm{Re}}(\mathbf{x}_{\alpha}-\mathbf{x}_{\beta}) - w_{\beta}^{\mathrm{Im}}\, C^{\mathrm{Im}}(\mathbf{x}_{\alpha}-\mathbf{x}_{\beta})\right) \;=\; C^{\mathrm{Re}}(\mathbf{x}_0-\mathbf{x}_{\alpha}) & \text{for } \forall\alpha \\ \sum\limits_{\beta=1}^{n} \left(w_{\beta}^{\mathrm{Re}}\, C^{\mathrm{Im}}(\mathbf{x}_{\alpha}-\mathbf{x}_{\beta}) + w_{\beta}^{\mathrm{Im}}\, C^{\mathrm{Re}}(\mathbf{x}_{\alpha}-\mathbf{x}_{\beta})\right) \;=\; C^{\mathrm{Im}}(\mathbf{x}_0-\mathbf{x}_{\alpha}) & \text{for } \forall\alpha. \end{cases} \qquad (29.13)$$

In matrix notation we have

$$\begin{pmatrix} \mathbf{C}^{\mathrm{Re}} & \mathbf{C}^{\mathrm{Im}} \\ \mathbf{C}^{\mathrm{Im}^{\top}} & \mathbf{C}^{\mathrm{Re}} \end{pmatrix} \begin{pmatrix} \mathbf{w}^{\mathrm{Re}} \\ \mathbf{w}^{\mathrm{Im}} \end{pmatrix} \;=\; \begin{pmatrix} \mathbf{c}^{\mathrm{Re}} \\ \mathbf{c}^{\mathrm{Im}} \end{pmatrix}. \qquad (29.14)$$

To implement complex kriging it is necessary to model $C^{\mathrm{Re}}(\mathbf{h})$ and $C^{\mathrm{Im}}(\mathbf{h})$ in a consistent way such that $C(\mathbf{h})$ is a complex covariance function. This question is postponed to the end of this chapter.

Cokriging of the real and imaginary parts

An alternative approach consists in relying on the coregionalization of the real and imaginary parts to make the simple cokriging of $Z(\mathbf{x})$ with the estimator

$$Z^{\star}_{\mathrm{CK}}(\mathbf{x}_0) \;=\; U^{\star}_{\mathrm{CK}}(\mathbf{x}_0) + \mathrm{i}\, V^{\star}_{\mathrm{CK}}(\mathbf{x}_0), \qquad (29.15)$$

where

$$U^{\star}_{\text{CK}}(\mathbf{x}_0) = \sum_{\alpha=1}^{n} \left(\mu^1_\alpha\, U(\mathbf{x}_\alpha) + \nu^1_\alpha\, V(\mathbf{x}_\alpha) \right), \tag{29.16}$$

$$V^{\star}_{\text{CK}}(\mathbf{x}_0) = \sum_{\alpha=1}^{n} \left(\mu^2_\alpha\, U(\mathbf{x}_\alpha) + \nu^2_\alpha\, V(\mathbf{x}_\alpha) \right). \tag{29.17}$$

EXERCISE 29.3 *Show that the estimation variance of $Z^{\star}_{\text{CK}}(\mathbf{x}_0)$ is equal to the sum of the estimation variances of the real and imaginary parts.*

As the estimators for $U(\mathbf{x})$ and $V(\mathbf{x})$ both have the same structure, we shall only treat the cokriging of the former.

EXERCISE 29.4 *Compute the estimation variance* $\text{var}(U(\mathbf{x}_0) - U^{\star}_{\text{CK}}(\mathbf{x}_0))$.

For cokriging $U(\mathbf{x})$ from the real and imaginary parts of the data $Z(\mathbf{x}_\alpha)$ we have the system

$$\begin{pmatrix} \mathbf{C}_{UU} & \mathbf{C}_{UV} \\ \mathbf{C}_{VU} & \mathbf{C}_{VV} \end{pmatrix} \begin{pmatrix} \boldsymbol{\mu}^1 \\ \boldsymbol{\nu}^1 \end{pmatrix} = \begin{pmatrix} \mathbf{c}_{UU} \\ \mathbf{c}_{UV} \end{pmatrix}, \tag{29.18}$$

where the blocks $\mathbf{C}_{UU}$ and $\mathbf{C}_{VV}$ are the left hand sides of the simple kriging systems of $U(\mathbf{x})$ and $V(\mathbf{x})$, respectively. The block $\mathbf{C}_{UV} = \mathbf{C}_{VU}^{\top}$ contains the cross covariance values between all data locations. The vector $\mathbf{c}_{UU}$ is the right hand side of the simple kriging of $U(\mathbf{x})$ while the vector $\mathbf{c}_{UV}$ groups the cross covariances between all data points and the estimation location.

Complex kriging and cokriging versus a separate kriging

Considering either the data on the complex variable or the corresponding real and imaginary parts, we have three ways of building an estimate of the complex quantity

- the estimator of complex kriging:

$$\begin{aligned} Z^{\star}_{\text{CC}}(\mathbf{x}_0) &= \mathbf{w}^{\top}\mathbf{z} && (29.19)\\ &= \mathbf{u}^{\top}\mathbf{w}^{\text{Re}} - \mathbf{v}^{\top}\mathbf{w}^{\text{Im}} + \mathrm{i}\,(\mathbf{u}^{\top}\mathbf{w}^{\text{Im}} + \mathbf{v}^{\top}\mathbf{w}^{\text{Re}}), && (29.20)\end{aligned}$$

- the estimator based on the coregionalization of $U(\mathbf{x})$ and $V(\mathbf{x})$:

$$Z^{\star}_{\text{CK}}(\mathbf{x}_0) = \mathbf{u}^{\top}\boldsymbol{\mu}^1 + \mathbf{v}^{\top}\boldsymbol{\nu}^1 + \mathrm{i}\,(\mathbf{u}^{\top}\boldsymbol{\mu}^2 + \mathbf{v}^{\top}\boldsymbol{\nu}^2), \tag{29.21}$$

- an estimator obtained by kriging separately $U(\mathbf{x})$ and $V(\mathbf{x})$:

$$Z^{\star}_{\text{KS}}(\mathbf{x}_0) = \mathbf{u}^{\top}\mathbf{w}^1_{\text{K}} + \mathrm{i}\,\mathbf{v}^{\top}\mathbf{w}^2_{\text{K}}. \tag{29.22}$$

Let us examine the case of intrinsic correlation of $U(\mathbf{x})$ and $V(\mathbf{x})$. The imaginary part of the complex covariance is zero and we have the complex kriging system

$$\begin{pmatrix} \mathbf{C}^{\mathrm{Re}} & \mathbf{0} \\ \mathbf{0} & \mathbf{C}^{\mathrm{Re}} \end{pmatrix} \begin{pmatrix} \mathbf{w}^{\mathrm{Re}} \\ \mathbf{w}^{\mathrm{Im}} \end{pmatrix} = \begin{pmatrix} \mathbf{c}^{\mathrm{Re}} \\ \mathbf{0} \end{pmatrix}, \tag{29.23}$$

and more specifically,

$$\begin{pmatrix} (\sigma_{UU}+\sigma_{VV})\,\mathbf{R} & \mathbf{0} \\ \mathbf{0} & (\sigma_{UU}+\sigma_{VV})\,\mathbf{R} \end{pmatrix} \begin{pmatrix} \mathbf{w}^{\mathrm{Re}} \\ \mathbf{w}^{\mathrm{Im}} \end{pmatrix} = \begin{pmatrix} (\sigma_{UU}+\sigma_{VV})\,\mathbf{r}_0 \\ \mathbf{0} \end{pmatrix}. \tag{29.24}$$

For the cokriging system of $U(\mathbf{x})$ with intrinsic correlation we write

$$\begin{pmatrix} \sigma_{UU}\,\mathbf{R} & \sigma_{UV}\,\mathbf{R} \\ \sigma_{VU}\,\mathbf{R} & \sigma_{VV}\,\mathbf{R} \end{pmatrix} \begin{pmatrix} \boldsymbol{\mu}^1 \\ \boldsymbol{\nu}^1 \end{pmatrix} = \begin{pmatrix} \sigma_{UU}\,\mathbf{r}_0 \\ \sigma_{UV}\,\mathbf{r}_0 \end{pmatrix}. \tag{29.25}$$

For both estimators the solution is equivalent to the separate simple kriging of the real and imaginary parts and we find

$$\mathbf{w}^{\mathrm{Re}} = \boldsymbol{\mu}^1 = \boldsymbol{\nu}^2 = \mathbf{w}_{\mathrm{K}}^1 = \mathbf{w}_{\mathrm{K}}^2, \tag{29.26}$$

while all other weights are zero.

The case of an even cross covariance function is remarkable. With an even cross covariance $C_{UV}(\mathbf{h}) = C_{VU}(\mathbf{h})$ the imaginary part of the complex covariance is zero and the imaginary weights of complex kriging vanish. The estimator reduces to

$$Z^{\star}_{\mathrm{CC}}(\mathbf{x}_0) = \mathbf{z}^{\top}\,\mathbf{w}^{\mathrm{Re}}. \tag{29.27}$$

The difference between the kriging variance σ^2_{CC} of complex kriging and the kriging variance σ^2_{KS} of the separate simple kriging is non negative, so that complex kriging provides a poorer solution than the one obtained by a separate kriging of $U(\mathbf{x})$ and $V(\mathbf{x})$ when the cross covariance function is even.

EXERCISE 29.5 *Show that with an even cross covariance function the difference of the kriging variances is equal to*

$$\begin{aligned} \sigma^2_{\mathrm{CC}} - \sigma^2_{\mathrm{KS}} &= (\mathbf{w}_{\mathrm{K}}^1 - \mathbf{w}^{\mathrm{Re}})^{\top}\,\mathbf{C}_{UU}\,(\mathbf{w}_{\mathrm{K}}^1 - \mathbf{w}^{\mathrm{Re}}) \\ &\quad + (\mathbf{w}_{\mathrm{K}}^2 - \mathbf{w}^{\mathrm{Re}})^{\top}\,\mathbf{C}_{VV}\,(\mathbf{w}_{\mathrm{K}}^2 - \mathbf{w}^{\mathrm{Re}}), \end{aligned} \tag{29.28}$$

and thus non negative.

When the cross covariance function $C_{UV}(\mathbf{h})$ is not even, complex kriging represents an intermediate solution, from the point of view of precision, between the separate kriging and the cokriging of the real and imaginary parts.

Complex covariance function modeling

Knowing the real part of a continuous complex covariance function, its imaginary part can be defined (as a consequence of the Radon-Nikodym theorem) by the relation

$$\mathrm{i}\, C^{\mathrm{Im}}(\mathbf{h}) \;=\; (\Phi * C^{\mathrm{Re}})(\mathbf{h}), \tag{29.29}$$

where Φ is a complex distribution whose Fourier transform is a real odd function φ with values in the interval $[-1,1]$. The class of $C^{\mathrm{Im}}(\mathbf{h})$ corresponding to a given real continuous covariance function $C^{\mathrm{Re}}(\mathbf{h})$, such that

$$C(\mathbf{h}) \;=\; C^{\mathrm{Re}}(\mathbf{h}) + \mathrm{i}\, C^{\mathrm{Im}}(\mathbf{h}) = C^{\mathrm{Re}}(\mathbf{h}) + (\Phi * C^{\mathrm{Re}})(\mathbf{h}) \tag{29.30}$$

is a complex covariance function, are called *compatible* imaginary parts.

A simple class of compatible imaginary parts can be obtained using

$$\Phi \;=\; \frac{\mathrm{i}}{2}\,(\nu - \check{\nu}), \tag{29.31}$$

where ν is a real bounded measure such that $|\varphi(\mathbf{u})| < 1$ with

$$\varphi(\mathbf{u}) \;=\; -\int \sin(\mathbf{u}^{\top}\boldsymbol{\tau})\,\nu(d\boldsymbol{\tau}). \tag{29.32}$$

The compatible imaginary parts are given by

$$C^{\mathrm{Im}}(\mathbf{h}) \;=\; \frac{1}{2}\int \left[C^{\mathrm{Re}}(\mathbf{h}-\boldsymbol{\tau}) - C^{\mathrm{Re}}(\mathbf{h}+\boldsymbol{\tau})\right]\nu(d\boldsymbol{\tau}). \tag{29.33}$$

COMMENT 29.6 *The imaginary parts from this class can be viewed as a randomization*

$$\mathbf{C}^{\mathrm{Im}}(h) \;=\; \frac{1}{2}\,\mathrm{E}[\,C^{\mathrm{Re}}(h-T) - C^{\mathrm{Re}}(h+T)\,], \tag{29.34}$$

where the expectation is taken over the random variable T.

The class can be extended (see [130]) by considering real covariance functions $C^{\mathrm{c}}(\mathbf{h})$ compatible with a given $C^{\mathrm{Re}}(\mathbf{h})$ in the sense that the positive measures $\mu^{\mathrm{c}} \leq \mu^{\mathrm{Re}}$. Then

$$C^{\mathrm{Im}}(\mathbf{h}) \;=\; \frac{1}{2}\int \left[C^{\mathrm{c}}(\mathbf{h}-\boldsymbol{\tau}) - C^{\mathrm{c}}(\mathbf{h}+\boldsymbol{\tau})\right]\nu(d\boldsymbol{\tau}) \tag{29.35}$$

is a compatible imaginary part, provided that the real bounded measure ν satisfies $|\int \sin(\mathbf{u}^{\top}\boldsymbol{\tau})\,\nu(d\boldsymbol{\tau})| < 1$.

Some properties of the functions $C^{\mathrm{c}}(\mathbf{h})$ are

- the difference $C^{\mathrm{Re}}(\mathbf{h}) - C^{\mathrm{c}}(\mathbf{h})$ is a positive definite function;
- $C^{\mathrm{c}}(\mathbf{0}) \leq C^{\mathrm{Re}}(\mathbf{0})$;

– when the *integral range* (see [220], [177]) of $C^{\text{Re}}(\mathbf{h})$ exists

$$\int C^{\text{c}}(\mathbf{h})\, d\mathbf{h} \;\leq\; \int C^{\text{Re}}(\mathbf{h})\, d\mathbf{h}. \tag{29.36}$$

– $C^{\text{c}}(\mathbf{h})$ is more regular at the origin than $C^{\text{Re}}(\mathbf{h})$;

– if $C^{\text{c}}(\mathbf{h})$ and $C^{\text{Re}}(\mathbf{h})$ have spectral densities then

$$f^{\text{c}}(\boldsymbol{\omega}) \;\leq\; f^{\text{Re}}(\boldsymbol{\omega}) \qquad \text{for all } \boldsymbol{\omega}. \tag{29.37}$$

In practice, a finite sum based on translations $\boldsymbol{\tau}_k$ is used

$$C^{\text{Im}}(\mathbf{h}) \;=\; \frac{1}{2}\sum_{k=1}^{K} p_k \left(C^{\text{c}}(\mathbf{h}-\boldsymbol{\tau}_k) - C^{\text{c}}(\mathbf{h}+\boldsymbol{\tau}_k)\right) \tag{29.38}$$

with weights $p_k \geq 0$ and $\sum p_k \leq 1$. Instead of only one function $C^{\text{c}}(\mathbf{h})$, K functions $C^{\text{c}}_k(\mathbf{h})$ can be introduced which represent a family of covariance functions compatible with a given $C^{\text{Re}}(\mathbf{h})$. The translations $\boldsymbol{\tau}_k$ and the models $C^{\text{c}}_k(\mathbf{h})$ are chosen after inspection of the graphs of the imaginary part of the experimental covariance function. The coefficients p_k are fitted by least squares.

LAJAUNIE & BÉJAOUI [170] provide a few fitting and complex kriging examples using directional data from simulated ocean waves.

30 Bilinear Coregionalization Model

The linear model of coregionalization of real variables implies even cross covariance functions and it was thus formulated with variograms in the framework of intrinsic stationarity. The use of cross variograms however excludes deferred correlations (due to delay effects or phase shifts). A more general model was set up by GRZEBYK [130] [131] which allows for non even cross covariance functions.

Complex linear model of coregionalization

The complex analogue to the intrinsic correlation model is

$$\mathbf{C}(\mathbf{h}) \;=\; \mathbf{B}\,\rho(\mathbf{h}) = \mathbf{E}\,\chi(\mathbf{h}) - \mathbf{F}\,\kappa(\mathbf{h}) + \mathrm{i}\,(\mathbf{E}\,\kappa(\mathbf{h}) + \mathbf{F}\,\chi(\mathbf{h})), \qquad (30.1)$$

where $\rho(\mathbf{h}) = \chi(\mathbf{h}) + \mathrm{i}\,\kappa(\mathbf{h})$ is a scalar complex covariance function and $\mathbf{B}$ is a Hermitian positive semi-definite matrix with

$$\mathbf{B} \;=\; \mathbf{E} + \mathrm{i}\,\mathbf{F}. \qquad (30.2)$$

The matrix $\mathbf{E}$ is a symmetric positive semi-definite matrix while $\mathbf{F}$ is antisymmetric.

An underlying linear model with complex coefficients a_p^i and second-order stationary uncorrelated complex random functions $Y_p(\mathbf{x})$ can be written as

$$Z_i(\mathbf{x}) \;=\; \sum_{p=1}^{N} a_p^i\, Y_p(\mathbf{x}), \qquad (30.3)$$

where

$$\begin{aligned} \mathrm{cov}(Y_p(\mathbf{x}+\mathbf{h}), Y_p(\mathbf{x})) &= \mathrm{E}[\, Y_p(\mathbf{x}+\mathbf{h})\, \overline{Y_p(\mathbf{x})}\,] = \rho(\mathbf{h}), \\ \mathrm{cov}(Y_p(\mathbf{x}+\mathbf{h}), Y_q(\mathbf{x})) &= 0 \qquad \text{for } p \neq q. \end{aligned} \qquad (30.4)$$

The alternate representation is based on jointly stationary components $U_p(\mathbf{x})$ and $V_p(\mathbf{x})$

$$Z_i(\mathbf{x}) \;=\; \sum_{p=1}^{N} (c_p^i\, U_p(\mathbf{x}) - d_p^i\, V_p(\mathbf{x})) + \mathrm{i} \sum_{p=1}^{N} (c_p^i\, V_p(\mathbf{x}) + d_p^i\, U_p(\mathbf{x})), \qquad (30.5)$$

where the coregionalization of $U_p(\mathbf{x})$ and $V_p(\mathbf{x})$ is identical for all values of the index p and has the form

$$\begin{aligned} \operatorname{cov}(U_p(\mathbf{x}+\mathbf{h}), U_p(\mathbf{x})) &= C_{UU}(\mathbf{h}) = C_{VV}(\mathbf{h}), \\ \operatorname{cov}(U_p(\mathbf{x}+\mathbf{h}), V_p(\mathbf{x})) &= C_{UV}(\mathbf{h}), \end{aligned} \tag{30.6}$$

while all other covariances are zero.

The coefficients of the factor decomposition and the elements of the complex coregionalization matrix $\mathbf{B}$ are related by

$$b_{ij} = \sum_{p=1}^{N} a_p^i \, \overline{a_p^j}, \tag{30.7}$$

whereas the elements of $\mathbf{E}$ and $\mathbf{F}$ are linked to the linear model by

$$e_{ij} = \sum_{p=1}^{N} (c_p^i \, c_p^j + d_p^i \, d_p^j), \qquad f_{ij} = \sum_{p=1}^{N} (c_p^j \, d_p^i - c_p^i \, d_p^j). \tag{30.8}$$

Naturally we can consider a nested complex multivariate covariance function model of the type

$$\mathbf{C}(\mathbf{h}) = \sum_{u=0}^{S} \mathbf{B}_u \, \rho_u(\mathbf{h}) \tag{30.9}$$

with the underlying complex linear model of coregionalization

$$Z_i(\mathbf{x}) = \sum_{u=0}^{S} \sum_{p=1}^{N} a_{pu}^i \, Y_u^p(\mathbf{x}) \tag{30.10}$$

and the orthogonality relations

$$\operatorname{cov}(Y_u^p(\mathbf{x}+\mathbf{h}), Y_v^q(\mathbf{x})) = 0 \qquad \text{for } p \neq q \quad \text{or } u \neq v, \tag{30.11}$$

where u and v are the indices of different characteristic spatial or time scales.

Bilinear model of coregionalization

The real linear model of coregionalization is in particular not adequate for multivariate time series analysis, where delay effects or phase shifts are common and cannot be included in a model with even cross covariances. A model for real random functions with non even real cross covariance functions can be derived from the complex linear model of coregionalization (as defined in the previous section) by taking its real part. In the case of only one spatial scale (the nested case is analog) we can drop the index u and have

$$Z_i(\mathbf{x}) = \sum_{p=1}^{N} c_p^i \, U_p(\mathbf{x}) - \sum_{p=1}^{N} d_p^i \, V_p(\mathbf{x}). \tag{30.12}$$

This model, as it is the sum of two linear models, has received the name of *bilinear model of coregionalization.* We get a handy model by imposing the following relations between covariance function terms

$$C_{UU}(\mathbf{h}) \;=\; C_{VV}(\mathbf{h}) = \frac{1}{2}\,\chi(\mathbf{h}), \tag{30.13}$$

$$C_{UV}(-\mathbf{h}) \;=\; -C_{UV}(\mathbf{h}) = \frac{1}{2}\,\kappa(\mathbf{h}). \tag{30.14}$$

This implies that the cross covariance function $C_{UV}(\mathbf{h})$ is odd. The cross covariance function between two real variables is

$$\begin{aligned} C_{ij}(\mathbf{h}) &= \operatorname{cov}(Z_i(\mathbf{x}+\mathbf{h}), Z_j(\mathbf{x})) \\ &= \sum_{p=1}^{N} c_p^i\, c_p^j\, C_{UU}(\mathbf{h}) + \sum_{p=1}^{N} d_p^i\, d_p^j\, C_{VV}(\mathbf{h}) \\ &\quad - \sum_{p=1}^{N} c_p^j\, d_p^i\, C_{VU}(\mathbf{h}) - \sum_{p=1}^{N} c_p^i\, d_p^j\, C_{UV}(\mathbf{h}) \\ &= \sum_{p=1}^{N} (c_p^i\, c_p^j + d_p^i\, d_p^j)\, \frac{\chi(\mathbf{h})}{2} - \sum_{p=1}^{N} (c_p^j\, d_p^i - c_p^i\, d_p^j)\, \frac{\kappa(\mathbf{h})}{2}. \end{aligned} \tag{30.15}$$

The multivariate covariance function model is real,

$$\mathbf{C}(\mathbf{h}) \;=\; \frac{1}{2}\,(\,\mathbf{E}\,\chi(\mathbf{h}) - \mathbf{F}\,\kappa(\mathbf{h})\,), \tag{30.16}$$

with matrices $\mathbf{E}$ and $\mathbf{F}$ stemming from the Hermitian positive semi-definite matrix $\mathbf{B}$,

$$\mathbf{B} \;=\; \mathbf{E} + \mathrm{i}\,\mathbf{F}, \tag{30.17}$$

and $\rho(\mathbf{h}) = \chi(\mathbf{h}) + \mathrm{i}\,\kappa(\mathbf{h})$ being a complex covariance function.

GRZEBYK [130] has studied different algorithms which combine the approach used for fitting the multivariate nested variogram with a fit of $\kappa(\mathbf{h})$ when $\chi(\mathbf{h})$ is given. He provides an example of the fit of a covariance model with non even cross covariance functions to the coregionalization of data from three remote sensing channels of a Landsat satellite.

The bilinear coregionalization model is especially well adapted for analyzing multiple or multivariate time series, where delay effects at various time scales are common and often are easily interpretable when causal relations between the variables are known.

Part E

Selective Geostatistics

31 Thresholds and Selectivity Curves

We introduce the notion of *threshold* which leads to selecting subsets of values of a random function that can be summarized using different types of selectivity curves. Incidentally this illustrates the geometrical meaning of two measures of dispersion: the variance and the selectivity. This presentation is based on MATHERON [213][218] and LANTUÉJOUL [175][176]. The context is mining economics, but a parallel to analog concerns in the evaluation of time series from environmental monitoring is drawn at the end of the chapter.

Threshold and proportion

There are a number of problems in which we aim at estimating in a spatial or temporal domain what overall proportion of a given quantity is above a fixed threshold.

The traditional example in geostatistics is mining. For what proportion of a deposit is the grade of a given ore above an economic threshold below which mining is not profitable? This will determine the decision whether or not to open a mine.

A similar example is found in soil pollution: what proportion of a site is above a tolerable pollution level? That will have an impact on its further use and on the cost for eventually cleaning it.

There is often not a unique threshold: the price of the metal to be extracted from a mine fluctuates daily and as extraction has to be planned much before the first ounce is on the market, several thresholds have to be considered. Similarly a polluted soil has to be judged on the basis of different thresholds depending upon its future use. A kindergarten will require a lower pollution level and more care in the rehabilitation of the terrain than what is tolerated when the objective is building a new industrial plant.

Tonnage, recovered quantity, investment and profit

For a mining deposit the concentration of a precious metal is denoted Z, which is a positive variable. We call *tonnage* $T(z)$ the proportion of the deposit (considered of unit volume) for which the grade is above a given threshold z. The value z represents in this chapter a fixed value, the so-called *cut-off*, i.e. the grade above which it is economically interesting to remove a volume from the deposit and to let it undergo chemical and mechanical processing for extracting the metal.

Let $F(z)$ be the distribution function of the grades, then the tonnage $T(z)$ is expressed by the integral

$$T(z) \quad = \quad \int_{z}^{+\infty} dF(u) = \mathrm{E}[\, \mathbf{1}_{Z \geq z}\,]. \tag{31.1}$$

where $\mathbf{1}_{Z \geq z}$ is the indicator function testing whether Z is above a given threshold z.

As $F(z)$ is an increasing function, $T(z)$ is a decreasing function. The tonnage is actually complementary to the distribution function

$$T(z) \quad = \quad P(Z \geq z) = 1 - F(z). \tag{31.2}$$

The quantity $Q(z)$ of metal recovered above a cut-off grade z is given by the integral

$$Q(z) \quad = \quad \int_{z}^{+\infty} u\, dF(u) = \mathrm{E}[\, Z\, \mathbf{1}_{Z \geq z}\,]. \tag{31.3}$$

The function $Q(z)$ is decreasing. The value of $Q(z)$ for z= 0 is the mean because the cut-off grade is a positive variable

$$Q(0) \quad = \quad \int_{0}^{+\infty} u\, dF(u) = \mathrm{E}[\, Z\,] = m. \tag{31.4}$$

The recovered quantity of metal can be written

$$\begin{aligned} Q(z) \quad &= \quad \Big[-u\, T(u)\Big]_{z}^{+\infty} + \int_{z}^{+\infty} T(u)\, du \\ &= \quad z\, T(z) + \int_{z}^{+\infty} T(u)\, du \\ &= \quad C(z) + B(z). \end{aligned} \tag{31.5}$$

The recovered quantity is thus split into two terms, which can be interpreted in the following way in mining economics (assuming that the cut-off grade is well adjusted to the economical context)

- the share $C(z)$ of metal which reflects the *investment*, i.e. the part of the recovered metal that will serve to refund the investment necessary to mine and process the ore,
- the amount $B(z)$ of metal, which represents conventional *profit*, i.e. the leftover of metal once the quantity necessary for refunding the investment has been subtracted.

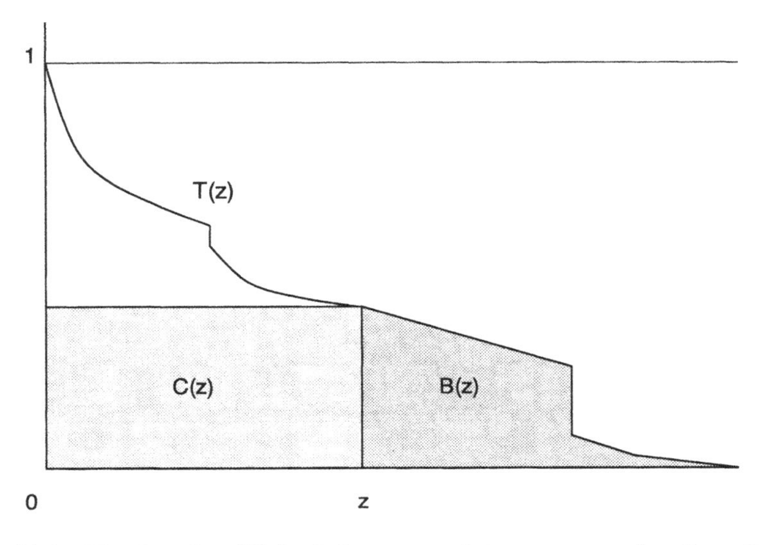

Figure 31.1: The function $T(z)$ of the extracted tonnage as a function of cut-off grade. The shaded surface under $T(z)$ represents the quantity of metal $Q(z)$ which is composed of the conventional investment $C(z)$ and the conventional profit $B(z)$.

These different quantities are all present on the plot of the function $T(z)$ as shown on Figure 31.1. The conventional investment (assuming that the cut-off adequately reflects the economic situation) is a rectangular surface

$$C(z) \quad = \quad z\,T(z), \tag{31.6}$$

while the conventional profit

$$B(z) \quad = \quad \int_{z}^{+\infty} T(u)\,du \tag{31.7}$$

is the surface under $T(z)$ on the right of the cut-off z.

The function $B(z)$ is convex and decreasing. The integral of $B(z)$ for $z=0$ is

$$\int_{0}^{+\infty} B(u)\,du = \frac{1}{2}\,(m^2+\sigma^2), \tag{31.8}$$

where m is the mean and σ^2 is the variance of Z. If we subtract from the surface under $B(z)$ the area corresponding to half the square m^2 we obtain a surface element whose value is half of the variance, as shown on Figure 31.2.

Note that $Q(z)$ divided by $T(z)$ yields the average of the values above cut-off

$$m(z) \quad = \quad \mathrm{E}[\,Z \mid Z \geq z\,] = Q(z)/T(z). \tag{31.9}$$

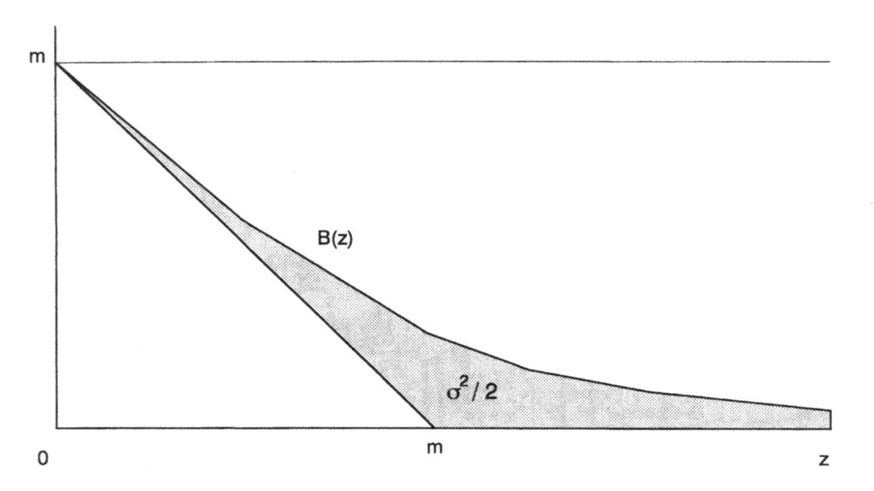

Figure 31.2: Function $B(z)$ of the conventional profit. The shaded surface under $B(z)$ represents half of the variance σ^2.

Selectivity

The selectivity S of a distribution F is defined as

$$S = \frac{1}{2}\,\mathrm{E}[\,|Z - Z'|\,], \tag{31.10}$$

where Z and Z' are two independent random variables with the same distribution F. There is an analogy with the following formula for computing the variance under the same conditions

$$\sigma^2 = \frac{1}{2}\,\mathrm{E}[\,(Z - Z')^2\,]. \tag{31.11}$$

The selectivity and the variance are both measures of dispersion, but the selectivity is more robust than the variance.

The selectivity divided by the mean is known in econometrics as the *Gini coefficient* (see e.g. [157], [4])

$$\mathrm{Gini} = \frac{S}{m} \tag{31.12}$$

and represents an alternative to the *coefficient of variation* σ/m. As $S \leq m$ the Gini coefficient is lower than one

$$0 \leq \mathrm{Gini} \leq 1. \tag{31.13}$$

In geostatistics the Gini coefficient is usually called the *selectivity, index.*

The following inequality between the selectivity and the standard deviation can be shown

$$S \leq \frac{\sigma}{\sqrt{3}}. \tag{31.14}$$

Equality is obtained for the uniform distribution. This provides us with a *uniformity coefficient*

$$U = \frac{\sqrt{3}\,S}{\sigma}, \tag{31.15}$$

where $0 \leq U \leq 1$.

For a Gaussian distribution with a standard deviation σ the selectivity is

$$S_{\text{Gauss}} = \frac{\sigma}{\sqrt{\pi}}, \tag{31.16}$$

which yields a value near to the limit case of the uniform distribution.

For a lognormally distributed variable $Y = m\, \exp(\sigma\, Z - \sigma^2/2)$ we have the selectivity

$$S_{\text{lognorm}} = m\left(2\,G\left(\sigma/\sqrt{2}\right) - 1\right), \tag{31.17}$$

where $G(z)$ is the standardized Gaussian distribution function.

Recovered quantity as a function of tonnage

The function $Q(z)$ of metal with respect to cut-off can be written

$$Q(z) \;=\; z\,T(z) + \int_{z}^{+\infty} T(u)\,du = \int_{0}^{z} T(z)\,du + \int_{z}^{+\infty} T(u)\,du. \tag{31.18}$$

As $Q(z)$ is decreasing, it can be expressed as the integral of the minimum of the two values $T(z)$ and $T(u)$ evaluated over all grades u

$$Q(z) \;=\; \int_{0}^{+\infty} \min\Big(T(z), T(u)\Big)\,du. \tag{31.19}$$

The function $Q(T)$ of metal depending on the tonnage T is defined as

$$Q(T) \;=\; \int_{0}^{+\infty} \min\Big(T, T(u)\Big)\,du, \tag{31.20}$$

for $0 \leq T \leq 1$.

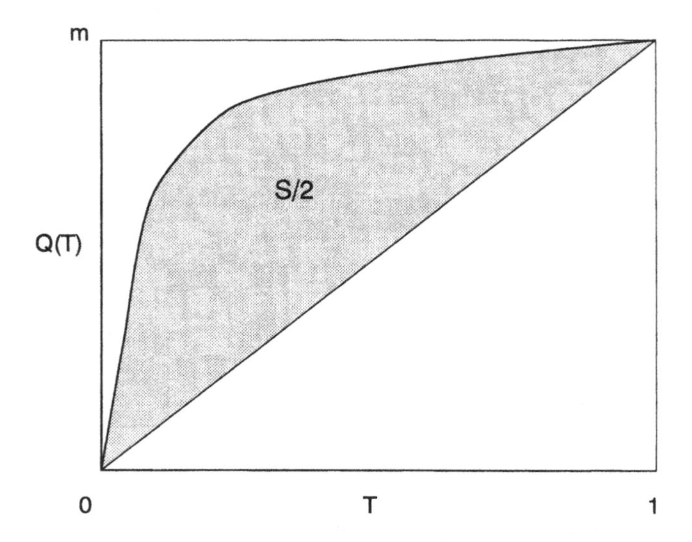

Figure 31.3: The function $Q(T)$ of the quantity of metal with respect to tonnage. The shaded surface between $Q(T)$ and the diagonal line corresponds to half the value of the selectivity S.

The $Q(T)$ curve is equivalent to the *Lorenz diagram*, which is used in economics to represent the concentration of incomes (see e.g. [4]). In the economical context the proportion T of income receivers with an income greater than z would be plotted against the proportion Q of total income held by these income receivers. In economics, however, Q is not plotted against T but against the distribution function F, so that the Lorenz diagram is convex, while the $Q(T)$ curve is concave.

A typical $Q(T)$ curve is shown on Figure 31.3. The area between the curve $Q(T)$ and the diagonal $m\,T$ (between the origin and the point $Q(1) = m$) is worth half of the selectivity S.

It can indeed be shown that

$$\int_0^1 \Big(Q(T) - m\,T\Big)\, dT = \frac{1}{2} \int_0^{+\infty} T(z)\Big(1 - T(z)\Big)\, dz$$

$$= \frac{1}{2} \int_0^{+\infty} B(z)\, dF(z). \qquad (31.21)$$

It can be seen that the selectivity is zero when the distribution F is concentrated at one point (a Dirac measure). Finally we have

$$\int_0^1 \Big(Q(T) - m\,T\Big)\, dT = \frac{1}{2} \int_0^{+\infty}\int_z^{+\infty} (u - z)\, dF(u)\, dF(z) \qquad (31.22)$$

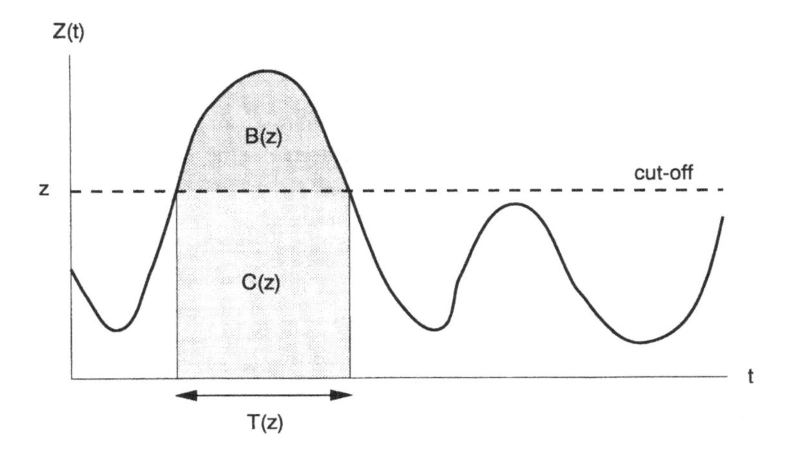

Figure 31.4: Environmental time series $Z(t)$ with a cut-off value z: the shaded part corresponds to the quantity of nuisance $Q(z) = B(z) + C(z)$ during the time $T(z)$ when the tolerance level is trespassed.

$$= \frac{1}{4} \int_0^{+\infty} \int_0^{+\infty} |u - z| \, dF(u) \, dF(z) = \frac{1}{2} S \qquad (31.23)$$

by definition (31.10).

Time series in environmental monitoring

When studying time series in environmental monitoring, interest is generally focused on the trespassing periods of a threshold above which a chemical, a dust concentration or a noise level is thought to be dangerous to health. Such thresholds fixed by environmental regulations are the analog of cut-off grades in mining.

In mining economics the level of profitability is subject to permanent change and it is appropriate to generate graphs with the values of the economic parameters for a whole range of values. Similarly in the environmental sciences the tolerance levels also evolve as the consciousness of dangers to health rises and it is suitable not to generate computations only for a particular set of thresholds.

When time series are evaluated we are interested in knowing during how long in total a tolerance level has been trespassed as shown on Figure 31.4. The tonnage $T(z)$ of mining becomes the fraction of time during which the level was trespassed. On a corresponding $T(z)$ curve like on Figure 31.1 the fraction of the total time can be read during which $Z(t)$ is above the cut-off for any tolerance level z of interest.

A curve of $Q(z)$ (not shown) represents the total quantity of nuisance for the fraction of time when $Z(t)$ is above the tolerance level z. The curve $Q(T)$ as on Figure 31.3 represents the quantity of nuisance as a function of the fraction of time T, which itself depends on the tolerance level.

The selectivity S is an important measure of dispersion to compare distributions with a same mean

- either as an indicator of change in the distribution of a variable for different supports on the same spatial or temporal domain (the mean does not change when changing support);

- or as a coefficient to rank the distributions of different sets of measurements with a similar mean.

It is not clear what meaning the curve $B(z)$ can have in environmental problems, except for the fact that its graph contains a geometric representation of (half of) the variance σ^2 when plotted together with the line joining the value of the mean m on both the abscissa and the ordinate, as suggested on Figure 31.2.

32 Lognormal Estimation

"Considérons la *variable lognormale*,
qui est tout simplement l'exponentielle d'une variable gaussienne.
Dans le long terme, ses propriétés sont dominées par le fait que ses moments
(moyenne, dispersion, etc.)
sont tous finis et s'obtiennent par des formules faciles;
donc la variable lognormale paraît bénigne.
Dans le court ou le moyen terme,
tout se gâte, et son comportement paraît sauvage.
On en traite comme si elle ne sentait pas le soufre,
mais c'est un merveilleux (et dangereux) caméléon."[1]

B MANDELBROT

To motivate the quest for non-linear estimators this chapter starts with a description of the effect of having a limited number of samples when estimating the value of a variable: the intensity of this (lack of) information effect will be different depending on the type of estimator used. The lognormal model is then presented and a few non-linear estimators like lognormal simple and ordinary kriging for point or block support are discussed. The presentation is mainly based on MATHERON [204] and RIVOIRARD [268]. Applications to acoustic data illustrate the performance of lognormal estimation without hiding its weaknesses.

Information effect and quality of estimators

When operating a selection with a threshold, e.g. a cut-off in mining or an environmental regulation limit, we are faced with an *information effect* if this selection is performed on estimated values. When increasing the information, i.e. the number of samples, the dispersion of the estimated values is generally reduced and the effect of not having exhaustive information will have a smaller impact. Another aspect of this question is that for a fixed number of samples the strength of the information effect will depend on the type of estimator used. Let us analyze the information effect in

[1]"Let us consider the *lognormal variable*, which is simply the exponential of a Gaussian variable. In the long run its properties are dominated by the fact that its moments (mean, dispersion, etc.) are all finite et can be obtained by easy formulas; thus the lognormal variable seems harmless. On the short or the medium run however everything is spoilt and its behavior looks wild. One deals with it as if it were not sulfurous, yet it is a marvelous (but dangerous) chameleon."

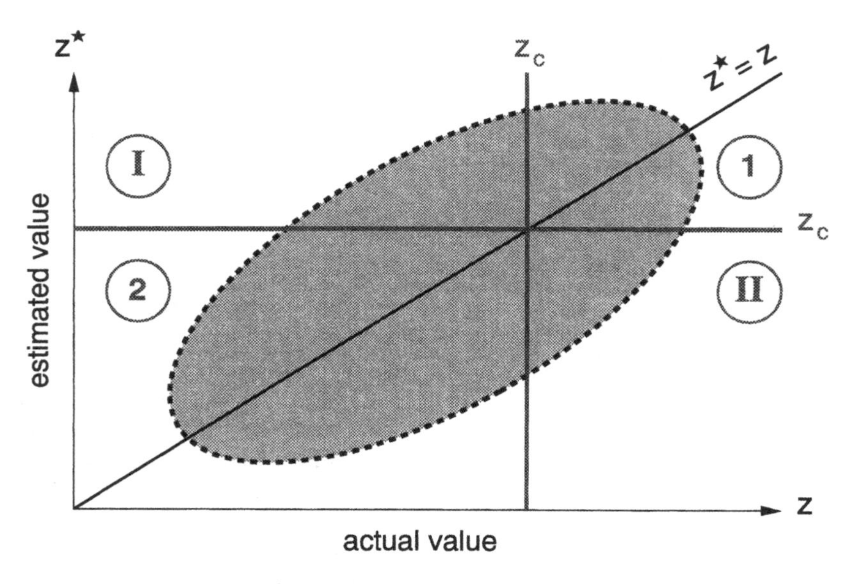

Figure 32.1: Information effect.

detail for the case of a Gaussian variable Z, assuming that the bivariate distribution of Z and an estimator $Z^{\star}$ is bi-Gaussian.

On Figure 32.1 we have drawn schematically the cloud of actual and estimated values using a shaded ellipse. This plot is based on the idea that after estimation and selection we have access to the actual target value and are able to check how the estimator has performed (a situation not always given in practice).

The diagonal line $Z^{\star}= Z$ on Figure 32.1 is the first bisector and represents the points for which the estimated value is identical with the actual value. If the major axis of the ellipse is superposed to the first bisector, the estimator is without bias. Otherwise three cases can occur:

- *over-estimation*: the major axis of the ellipse is parallel to the first bisector and above it. The estimator yields on average a larger value than the actual one and thus tends to over-estimate.

- *under-estimation*: the major axis of the ellipse is parallel to the first bisector and below it.The estimated values are on average lower than the actual ones and we have systematic under-estimation.

- *over- and under-estimation*: if the major axis of the ellipse crosses the first bisector, the estimator is either over-estimating the large actual values and under-estimating the low ones – or the reverse is happening.

Now suppose we have an estimator without bias so that the major axis of the ellipse coincides with the first bisector as drawn on Figure 32.1. When we classify estimated values according to a reference value we are splitting the cloud following the horizontal line labeled "z_c". Let us convene that we select (keep) points that are above the reference value and reject (discard) points that are below it. Whether the selected points were actually worth keeping is described by the vertical line labeled "z_c".

In this manner the cloud has been divided into the following four areas:

- **area 1:** the points of the cloud are estimated above z_c and correspond to actual values above the reference value: the points, selected on the basis of their estimated value, are actually worth keeping.

- **area 2:** the points are estimated below z_c and turn out to be below the reference value: rejected points are actually worth discarding.

- **area I:** in this region we commit the first type of misclassification: points are selected which should have been rejected as they are actually below the threshold.

- **area II:** in that region occurs the second type of error which consists in rejecting points which should have been kept.

The shape and the dispersion of the cloud around the first bisector will depend on the type of estimator used. The information effect can be reduced when using an estimator that is well adapted to the specificities of the regionalized variable from which data is collected. This is a motivation for the study of non-linear estimators in general and in this chapter we start this review by examining a few lognormal estimators.

Logarithmic Gaussian model

The *lognormal model* is based on the assumption that the natural logarithm of a positive random variable Z follows a Gaussian distribution. More specifically, if $L(\mathbf{x})$ is a stationary Gaussian random function,

$$Z(\mathbf{x}) \quad = \quad \exp(L(\mathbf{x})) \tag{32.1}$$

is said to be *lognormal* – a contraction of logarithmic and normal.

The lognormal model has been fashionable in the early times of geostatistics [162, 79, 301, 195]. At that time, in the fifties of the twentieth century, computers were not available and data sets were rather small in size. So, for example in the case of mining samples, the distribution generally appeared left-skew and the logarithmic transformation seemed an adequate way of bringing the data back to a form that could be assumed a realization of a Gaussian random function. MATHERON [195] lists several other examples of data taking positive values and with an asymmetric distribution having a tail stretched towards the high values, like the distribution of incomes in a country, the surface of emerged land as a function of altitude, the granulometry of materials (sand, gravels), to name but a few.

Let $L(\mathbf{x})$ be of mean ν, variance σ^2 and covariance function $\sigma(\mathbf{h})$. The standardized version of $L(\mathbf{x})$ is $Y(\mathbf{x})$ with autocorrelation function $\rho(\mathbf{h})$, so that

$$L(\mathbf{x}) = \nu + \sigma\, Y(\mathbf{x}) \tag{32.2}$$

$$\sigma(\mathbf{h}) = \sigma^2\, \rho(\mathbf{h}). \tag{32.3}$$

The random function $Y(\mathbf{x})$ is related to the lognormal $Z(\mathbf{x})$ by

$$Z(\mathbf{x}) = \exp(\nu + \sigma\, Y(\mathbf{x})) = \mathrm{e}^{\,\nu}\, \mathrm{e}^{\,\sigma Y(\mathbf{x})}. \tag{32.4}$$

Moments of the lognormal variable

For the computation of the moments the following formula is useful:

$$\mathrm{E}[\exp(a\, Y)\,] = \exp\left(\frac{a^2}{2}\right) \qquad \text{for} \quad Y \sim \mathcal{N}(0,1), \tag{32.5}$$

where a is a constant and $\mathcal{N}(0,1)$ is the Gaussian distribution of mean zero and unit variance.

Applying this formula we can express the mean m of $Z(\mathbf{x})$ in terms of the moments of $L(\mathbf{x})$,

$$m = \mathrm{E}[\,Z(\mathbf{x})\,] = \mathrm{E}[\exp(\nu + \sigma\, Y(\mathbf{x}))\,] \tag{32.6}$$

$$= \mathrm{e}^{\,\nu}\, \mathrm{E}[\exp(\sigma\, Y(\mathbf{x}))\,] = \mathrm{e}^{\,\nu} \exp\left(\frac{\sigma^2}{2}\right) \tag{32.7}$$

$$= \mathrm{e}^{\,\nu + \sigma^2/2}. \tag{32.8}$$

We can rewrite $Z(\mathbf{x})$ as

$$Z(\mathbf{x}) = \mathrm{e}^{\,\nu}\, \mathrm{e}^{\,\sigma Y(\mathbf{x})} = \mathrm{e}^{\,\nu+\sigma^2/2}\, \mathrm{e}^{\,\sigma Y(\mathbf{x}) - \sigma^2/2} \tag{32.9}$$

obtaining the new relation

$$Z(\mathbf{x}) = m \exp\left(\sigma\, Y(\mathbf{x}) - \frac{\sigma^2}{2}\right), \tag{32.10}$$

where, interestingly, the transformation includes a corrective term $\sigma^2/2$ depending on the variance of $L(\mathbf{x})$. The lognormal model can now be expressed in a way showing nicely the multiplicative structure with respect to the mean,

$$Z(\mathbf{x}) = m \exp\left(R(\mathbf{x})\right) \tag{32.11}$$

with the logarithmic residual $R(\mathbf{x}) = \log(Z(\mathbf{x})) - \log(m)$.

With the relation (32.10) we compute the second moment of $Z(\mathbf{x})$,

$$\begin{aligned}
\mathrm{E}[\,(Z(\mathbf{x}))^2\,] &= m^2\,\mathrm{E}\Big[\exp\Big(2\,\Big(\sigma\,Y(\mathbf{x})-\frac{\sigma^2}{2}\Big)\Big)\Big] && (32.12)\\
&= m^2\,\mathrm{E}[\exp(2\sigma\,Y(\mathbf{x}))\,]\,\exp(-\sigma^2) && (32.13)\\
&= m^2\,\exp\Big(\frac{4\,\sigma^2}{2}\Big)\,\exp(-\sigma^2) && (32.14)\\
&= m^2\,\exp(\sigma^2). && (32.15)
\end{aligned}$$

The variance of $Z(\mathbf{x})$ is thus

$$\mathrm{var}(Z(\mathbf{x})) = m^2\left(\mathrm{e}^{\sigma^2}-1\right). \tag{32.16}$$

In analogous way the covariance function of $Z(\mathbf{x})$ comes as

$$C(\mathbf{h}) = m^2\,\left(\mathrm{e}^{\,\sigma(\mathbf{h})}-1\right). \tag{32.17}$$

EXERCISE 32.1 *[Distribution of iron grade] A mining company is operating on several iron ore deposits in which the grade is modeled as a stationary random function $Z(\mathbf{x})$. The model for the grade is expressed in terms of a standard normal variable Y with an anamorphosis $Z(\mathbf{x})=\varphi(Y(\mathbf{x}))$ where φ is termed the anamorphosis function. The following analytic model is used to describe the anamorphosis:*

$$Z(\mathbf{x}) = a-b\,\mathrm{e}^{-cY(\mathbf{x})} \qquad \textit{with} \quad Y(\mathbf{x})\sim N(0,1). \tag{32.18}$$

The constant a is known to be 70% Fe because the iron ore cannot be above this grade for geochemical reasons. The two constants b and c are positive ($b,c>0$) and their value is adjusted to each deposit.

All questions in this exercise refer to one specific deposit of the mining company which has been explored using several drill holes. The histogram obtained from the samples taken in the drill hole cores has permitted to fit the parameters specifically for that deposit, i.e. $b=7.8$ and $c=0.5$ (while $a=70$).

1. *Show that the mean of $Z(\mathbf{x})$ can be written as*

$$\mathrm{E}[\,Z(\mathbf{x})\,] = a-b\,\mathrm{e}^{c^2/2} \tag{32.19}$$

and that the variance of $Z(\mathbf{x})$ is:

$$\mathrm{var}(Z(\mathbf{x})) = b^2\,(\mathrm{e}^{\,2c^2}-\mathrm{e}^{\,c^2}). \tag{32.20}$$

Hint: use formula (32.5).

2. *Compute the value of the mean and the variance of the iron ore grade in this deposit.*

3. *Plot the anamorphosis function for a point support for values of $Y(\mathbf{x})$ between -3 and 3, e.g. for $Y(\mathbf{x})$ = -3,-2,-1,0,1,2,3.*

4. *Using a table of the standard normal distribution (or a statistical software), compute and plot the cumulative histogram of the iron grades in the deposit.*

EXERCISE 32.2 *[Reserves: point support] For the iron ore deposit of Exercise 32.1 we calculate reserves on point support. According to the economic situation there is interest in computing global recoverable reserves for a cut-off grade of* $z_c = 58.0\%$. *Let us for now consider an estimation on point support. In that case the tonnage at cut-off* z_c *is defined as* $T(z_c) = \mathrm{E}[\,\mathbf{1}_{Z(\mathbf{x})\geq z_c}\,]$ *and the quantity of metal as* $Q(z_c) = \mathrm{E}[\,Z(\mathbf{x})\,\mathbf{1}_{Z(\mathbf{x})\geq z_c}\,]$.

1. *Show that the tonnage can be computed as*

$$T(z_c) \quad = \quad G\left(\frac{1}{c}\log\left(\frac{a-z_c}{b}\right)\right), \tag{32.21}$$

where $G(y) = P(Y(\mathbf{x}) < y)$ *is the standard normal distribution.*

2. *What proportion of the deposit is above the cut-off grade* $z_c = 58.0\%$?

3. *Show that the quantity of metal is computed as*

$$Q(z_c) \quad = \quad a\,T(z_c) - b\,\mathrm{e}^{c^2/2}\,(1 - G(y_c + c))\,, \tag{32.22}$$

where $z_c = \varphi(y_c)$.
Hint: use the formula $\mathrm{E}[\,\mathrm{e}^{\lambda Y(\mathbf{x})}\,\mathbf{1}_{Y(\mathbf{x})\geq y_c}\,] = \mathrm{e}^{\lambda^2/2}\,(1 - G(y_c - \lambda))$.

4. *What is the quantity of metal corresponding to the cut-off grade of* $z_c = 58\%$?

5. *What is the average grade* $m(z_c)$ *of the selected ore for* $z_c = 58\%$?

Lognormal simple kriging

A set of $n+1$ samples $Y(\mathbf{x}_\alpha)$ ($\alpha = 0,\ldots,n$) of a Gaussian random function $Y(\mathbf{x})$ follows a multivariate normal distribution. The conditional expectation

$$\mathrm{E}[\,Y(\mathbf{x}_0)\ |\ Y(\mathbf{x}_\alpha);\, \alpha = 1,\ldots,n\,] \tag{32.23}$$

is equivalent in this context to the simple kriging,

$$Y^\star(\mathbf{x}_0) \quad = \quad \sum_{\alpha=1}^{n} w_\alpha^{\mathrm{SK}}\, Y(\mathbf{x}_\alpha), \tag{32.24}$$

where the w_α^{SK} are solution of

$$\sum_{\beta=1}^{n} w_\beta^{\mathrm{SK}}\, \rho(\mathbf{x}_\alpha - \mathbf{x}_\beta) \quad = \quad \rho(\mathbf{x}_\alpha - \mathbf{x}_0) \qquad \text{for} \quad \alpha = 1,\ldots,n, \tag{32.25}$$

with the kriging variance $\sigma^2_{\mathrm{SK}} = 1 - \sum_{\alpha=1}^{n} w_\alpha^{\mathrm{SK}}\, \rho(\mathbf{x}_\alpha - \mathbf{x}_0)$.

The conditional variable follows a normal distribution with parameters given by the simple kriging,

$$\Big[Y(\mathbf{x}_0) \mid Y(\mathbf{x}_\alpha) = y(\mathbf{x}_\alpha);\ \alpha = 1,\ldots,n\Big] \quad \sim \quad \mathcal{N}\left(y^\star(\mathbf{x}_0), \sigma^2_{\mathrm{SK}}\right), \tag{32.26}$$

so that the conditional expectation can be rewritten,

$$\mathrm{E}[\,Y(\mathbf{x}_0) \mid Y(\mathbf{x}_\alpha) = y(\mathbf{x}_\alpha),\ \alpha = 1,\ldots,n\,] \quad = \quad \mathrm{E}[\,y^\star(\mathbf{x}_0) + \sigma_{\mathrm{SK}}\, U\,] \tag{32.27}$$

with $U \sim \mathcal{N}(0,1)$.

The lognormal simple kriging consists in computing the conditional expectation of $Z(\mathbf{x})$,

$$\begin{aligned} z^\star(\mathbf{x}_0) &= \mathrm{E}[\,Z(\mathbf{x}_0) \mid Z(\mathbf{x}_\alpha) = z(\mathbf{x}_\alpha),\ \alpha = 1,\ldots,n\,] & (32.28)\\ &= \mathrm{E}[\,Z(\mathbf{x}_0) \mid Y(\mathbf{x}_\alpha) = y(\mathbf{x}_\alpha),\ \alpha = 1,\ldots,n\,] & (32.29) \end{aligned}$$

because conditioning $z(\mathbf{x}_\alpha)$ is the same as conditioning on $y(\mathbf{x}_\alpha)$ as the latter is simply a nonlinear transform of the former.

Applying relation (32.10) and formula (32.5) we get,

$$\begin{aligned} z^\star(\mathbf{x}_0) &= \mathrm{E}[\,m\,\mathrm{e}^{\sigma Y(\mathbf{x}_0) - \sigma^2/2} \mid Y(\mathbf{x}_\alpha) = y(\mathbf{x}_\alpha),\ \alpha = 1,\ldots,n\,] & (32.30)\\ &= m\,\mathrm{E}\Big[\exp\Big(\sigma\left(y^\star(\mathbf{x}_0) + \sigma_{\mathrm{SK}}\, U\right) - \frac{\sigma^2}{2}\Big)\Big] & (32.31)\\ &= m\,\exp\left(\sigma\, y^\star(\mathbf{x}_0) - \frac{\sigma^2}{2}\right)\,\exp\left(\frac{\sigma^2\,\sigma^2_{\mathrm{SK}}}{2}\right). & (32.32) \end{aligned}$$

To express the simple kriging estimator in terms of $L(\mathbf{x})$ we just need to rescale,

$$Y^\star(\mathbf{x}_0) \;=\; \frac{L^\star(\mathbf{x}_0) - \nu}{\sigma}, \qquad \sigma^2_{\mathrm{SK}} \;=\; \frac{\sigma^2_{\mathrm{LSK}}}{\sigma^2}, \tag{32.33}$$

where σ^2_{LSK} is the simple kriging variance associated to $L^\star(\mathbf{x}_0)$.

Finally the lognormal simple kriging estimator is written in a more compact form,

$$Z^\star(\mathbf{x}_0) \;=\; \exp\left(L^\star(\mathbf{x}_0) + \frac{\sigma^2_{\mathrm{LSK}}}{2}\right) \tag{32.34}$$

or, in terms of the logarithmic residual,

$$Z^\star(\mathbf{x}_0) \;=\; m\,\exp\left(R^\star(\mathbf{x}_0) + \frac{\sigma^2_{\mathrm{RSK}}}{2}\right) \tag{32.35}$$

with $\sigma^2_{\mathrm{RSK}} = \sigma^2\,\sigma^2_{\mathrm{SK}}$ as can be seen from Eq. (32.32).

We take note that the lognormal estimate $Z^\star(\mathbf{x}_0)$ is not simply an exponential transform of the simple kriging estimate $L^\star(\mathbf{x}_0)$, but that it also includes a corrective term based on the kriging variance. This makes the lognormal kriging very sensitive to a parameter which depends on the choice of the type of variogram model and on its fit to the experimental variogram.

Proportional effect

In their classical book about the lognormal distribution AITCHISON & BROWN [4] define the law of *proportional effect* as characteristic for a variable subject to a process of change in which the change at each step is proportional to the value of the process. In other words the variation of a positive quantity z obeying the law of proportional effect is more adequately described by dz/z than by the only differential element dz. The addition of a large number of roughly independent effects dz/z will generate a variable y with a normal distribution:

$$y = \int \frac{dz}{z} = \log z \qquad (32.36)$$

and thus z is lognormal.

By analogy the term "proportional effect" is used when the variance or the variogram of z are proportional to the square of its mean. This phenomenon is often an argument to postulate the lognormality of z.

The dispersion of grades of small volumes making up a large volume were described in two apparently contradictory ways in the early fifties of the twentieth century in the mining literature. On one hand DE WIJS [79] proposed a model in which the dispersion variance a random function defined on a support v within a larger volume V is proportional to a function f of the ratio of the two volumes,

$$\sigma^2(v \mid V) = \alpha f\left(\frac{V}{v}\right). \qquad (32.37)$$

On the other hand KRIGE [162] demonstrated in a general manner that a dispersion variance can be expressed as the difference of the values of a function F applied to two supports,

$$\sigma^2(v \mid V) = F(V) - F(v). \qquad (32.38)$$

MATHERON [195] showed that the de Wijsian model implies a similarity principle which is independent of the form of the statistical distribution. He argued that solely a logarithmic function could satisfy both the equations (32.37) and (32.38), such that $F = \alpha f$, yielding

$$\sigma^2(v \mid V) = \alpha \log \frac{V}{v}. \qquad (32.39)$$

Permanence of lognormality

Coming back to the question of lognormality, a *permanence of lognormality* has been reported in various applications over a limited range of spatial or time supports: e.g. KRIGE [162] for gold samples, LARSEN [179] for air pollution concentrations, MALCHAIRE & PIETTE [186] for noise exposure measurements, WILD et al. [357] for airborne concentrations in exposure data from industrial hygiene, to mention but a few [345]. A self-similarity principle has been set up on the basis of this empirical evidence that led various authors to propose formulas of the type of (32.39). It should be noted that the de Wijsian formula (32.39) does not imply a lognormal distribution, but only a de Wijsian variogram,

$$\gamma(\mathbf{h}) \;=\; \alpha \log|\mathbf{h}| \qquad \text{with} \quad 0 < \alpha \leq 1, |\mathbf{h}| > 0. \tag{32.40}$$

MATHERON [195, 196] stated that the permanence of lognormality observed in applications is explained by the lack of independence of the data. Ironically he termed the recurrent rediscovery of this phenomenon in different scientific and technical disciplines as a "serpent de mer" (sort of Loch Ness monster). In [204] (p38) an example is given that shows that for small integration times (with respect to the scaling coefficient of the autocorrelation model) the permanence is verified, while this deteriorates for larger supports.

The model of permanence of the lognormal distribution assumes that the random function on bloc support has the same form as the point variable in Eq. (32.4), that is

$$Z_v(\mathbf{x}) \;=\; \exp(L_v(\mathbf{x})) = \exp(\nu_v + \sigma_v\, Y(\mathbf{x})). \tag{32.41}$$

It is based on an affine change of support model for $L(\mathbf{x})$,

$$\frac{L_v(\mathbf{x}) - \nu_v}{\sigma_v} = \frac{L(\mathbf{x}) - \nu}{\sigma} \sim \mathcal{N}(0,1) \tag{32.42}$$

in which ν and ν_v take different values.

The block lognormal model is viewed as a transform of the standard normal distribution which differs from the point distribution by the coefficients ν_v and σ_v. Actually there is merely one new coefficient to be inferred because in the relation,

$$Z_v \;=\; m \exp\left(\sigma_v\, Y(\mathbf{x}) - \frac{\sigma_v^2}{2}\right), \tag{32.43}$$

the mean m is the same as in relation (32.10) due to (10.17).

In analogy with the affine change of support model (presented in Chapter 10 on p71) we may view the standard deviation of the block variable as the product of the point variance with a change of support coefficient r, with the notable difference however that the two Gaussian variables L and L_v do not have the same mean. In this way the relation (32.43) can be rewritten

$$Z_v \;=\; m \exp\left(\sigma\, r\, Y(\mathbf{x}) - \frac{\sigma^2\, r^2}{2}\right) \qquad \text{with} \quad r = \sigma_v/\sigma. \tag{32.44}$$

In practice the change of support coefficient r can be computed as

$$r \;=\; \frac{\log(1 + \operatorname{var}(Z_v(\mathbf{x}))/m^2)}{\log(1 + \operatorname{var}(Z(\mathbf{x}))/m^2)} \tag{32.45}$$

with $\operatorname{var}(Z(\mathbf{x}))$ computed from formula (32.16) while $\operatorname{var}(Z_v(\mathbf{x}))$ is the average value $\overline{C}(v,v)$ of the covariance function (32.17) over the support v.

In the permanence of lognormality model the expectation of the block lognormal variable conditional upon the data is equivalent to lognormal block simple kriging,

$$Z_v^\star(\mathbf{x}_0) = \mathrm{E}[\, Z_v(\mathbf{x}_0) \mid Z(\mathbf{x}_\alpha) = z(\mathbf{x}_\alpha), \alpha = 1, \ldots, n\,] \tag{32.46}$$

$$= \exp\left(L_v^\star(\mathbf{x}_0) + \frac{\sigma_{\mathrm{LBSK}}^2}{2} \right) \tag{32.47}$$

EXERCISE 32.3 *[Reserves: block support] The calculations in Exercise 32.2 have neglected the fact that the deposit is going to be mined using units v of size 10×20 m^2. It is thus necessary to take into account this change of support. For constructing the change of support model we assume that the grades of the blocks v follow the same type of distribution as the samples, i.e. we assume a permanence of the distribution and have thus, like in relation (32.18),*

$$Z_v = a_v - b_v \, \mathrm{e}^{-c_v Y}, \tag{32.48}$$

where Y is a standard normal random variable. The value of the constant a_v is known to be $a_v = a = 70\%$ because the grade of a block cannot be larger than 70% like for point support.

To define the distribution of the block grades is is still necessary to compute the value of the two unknowns b_v and c_v. For this we use the fact that $\mathrm{E}[\, Z_v \,] = \mathrm{E}[\, Z(\mathbf{x})\,]$ and compute the block variance as $\mathrm{var}(Z_v) = \mathrm{var}(Z(\mathbf{x})) - \overline{\Gamma}(v,v)$ where $\overline{\Gamma}(v,v)$ is the average over v of the variogram $\Gamma(\mathbf{h})$ of $Z(\mathbf{x})$.

1. *The average variogram over v has been computed as $\overline{\Gamma}(v,v) = 8.8$. Use the results obtained in Exercise 32.1 to show that $b_v = 8.1$ and $c_v = 0.4$.*

2. *What value y_c of the standard normal variate Y corresponds to the cut-off z_c when the latter is applied to the block grades Z_v? Compute the values of the recovered ore $T_v(z_c)$, the metal quantity $Q_v(z_c)$ and the average grade $m_v(z_c) = Q_v(z_c)/T_v(z_c)$ for the deposit?*

3. *Compare and comment the results obtained for block and point support.*

EXERCISE 32.4 *[Conditional expectation] There is an area of the iron deposit of Exercise 32.1 (see Figure 32.2) where the exploration drill holes are sufficiently numerous to allow for an individual estimation of small blocs. We wish to estimate the probability that the grade of a small bloc v_3 is above the cut-off $z_c = 58\%$ using the conditional expectation. The change-of-support is taken into account assuming the position of each sample is random inside the block that contains it (this is actually the discrete Gaussian model that will be presented in Chapter 36). The estimation of the bloc v_3 is performed with three data values $z_1 = 49.2$, $z_2 = 65.1$ and $z_3 = 56.8$.*

The conditional expectation is built on the multi-Gaussian hypothesis of the vector

$$\Big(Y(v_1), Y(v_2), Y(v_3), Y_1, Y_2, Y_3 \Big)^{\top} \tag{32.49}$$

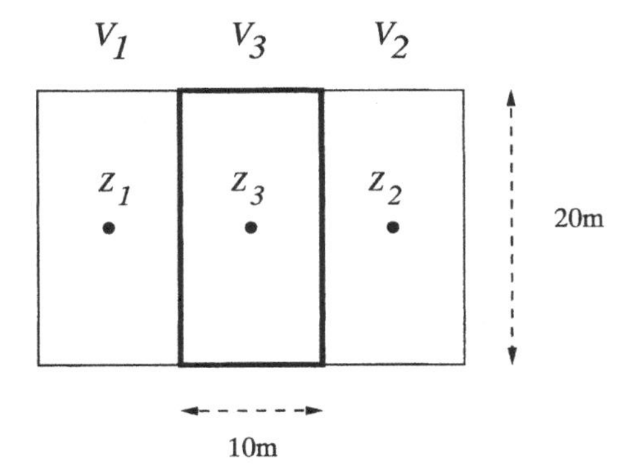

Figure 32.2: Three blocks with three sample values..

where each data value y_α is inside a corresponding bloc v_α. The coefficient of change-of-support is $r = \operatorname{corr}(Y_\alpha, Y(v_\alpha))$ independently of the position of the sample in its block. The covariance function between blocs is $\rho_v(\mathbf{h}) = \rho(v_\mathbf{x}, v_{\mathbf{x}+\mathbf{h}}) = \operatorname{cov}(Y(v_\mathbf{x}), Y(v_{\mathbf{x}+\mathbf{h}}))$, where $\mathbf{h}$ is the distance between the centers of disjoint blocks. The covariance function between points and blocks is $\operatorname{cov}(Y(\underline{\mathbf{x}}), Y(v_{\mathbf{x}+\mathbf{h}})) = r\,\rho_v(\mathbf{h})$ denoting by $\underline{\mathbf{x}}$ the random position of the point. The covariance between two distinct points is $\operatorname{cov}(Y(\underline{\mathbf{x}}), Y(\underline{\mathbf{x}}+\mathbf{h}))) = r^2\,\rho_v(\mathbf{h})$ while the covariance of a point with itself is one.

1. *What are the Gaussian values Y_1, Y_2, Y_3 that correspond to the three data values Z_1, Z_2, Z_3?*

2. *Express the estimator $Y^\star_{\mathrm{KS}}(v_3)$ as a function of the three data values Y_α and of the mean $\mathrm{E}[\,Y(v_\alpha)\,]$. Show that the kriging system can be written:*

$$\begin{pmatrix} 1 + r^2\,\rho_v(20) & r^2\,\rho_v(10) \\ 2\,r^2\,\rho_v(10) & 1 \end{pmatrix} \begin{pmatrix} \lambda_1 \\ \lambda_3 \end{pmatrix} \begin{pmatrix} r\,\rho_v(10) \\ r \end{pmatrix}. \qquad (32.50)$$

3. *The structural analysis has shown that the correlation function is composed of a nugget effect and an exponential model,*

$$\rho_v(\mathbf{h}) = .3\,\mathrm{nug}(\mathbf{h}) + .7\,\exp(-|\mathbf{h}|/8), \qquad (32.51)$$

   that it is isotropic, and that the change-of-support coefficient is $r = 0.8$. Solve the kriging system. What are the values obtained for $Y^\star_{\mathrm{KS}}(v_3)$ and the corresponding simple kriging variance σ^2_{SK}?

4. *What is the distribution of the conditional variable $\Big(Y(v_3) \mid Y_1, Y_2, Y_3\Big)$?*

5. *Compute the value of the conditional expectation $[\mathbf{1}_{Z(v_3)\geq z_c}]^*_{\mathrm{CE}}$.*

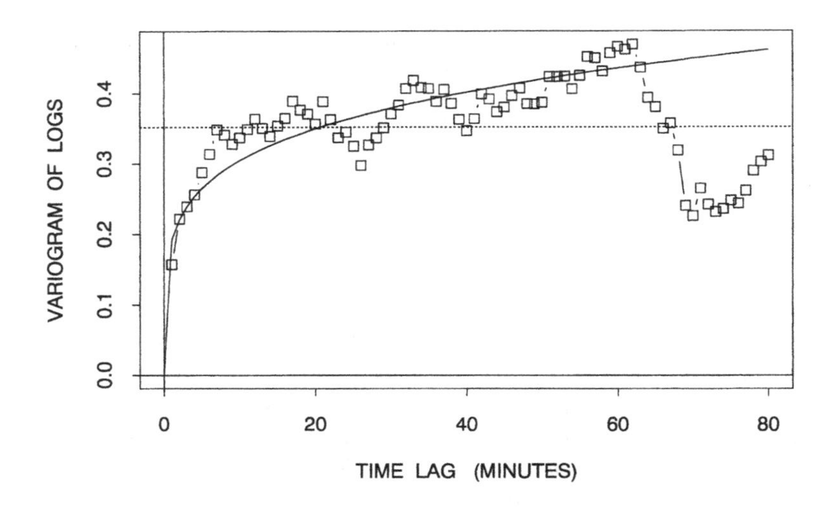

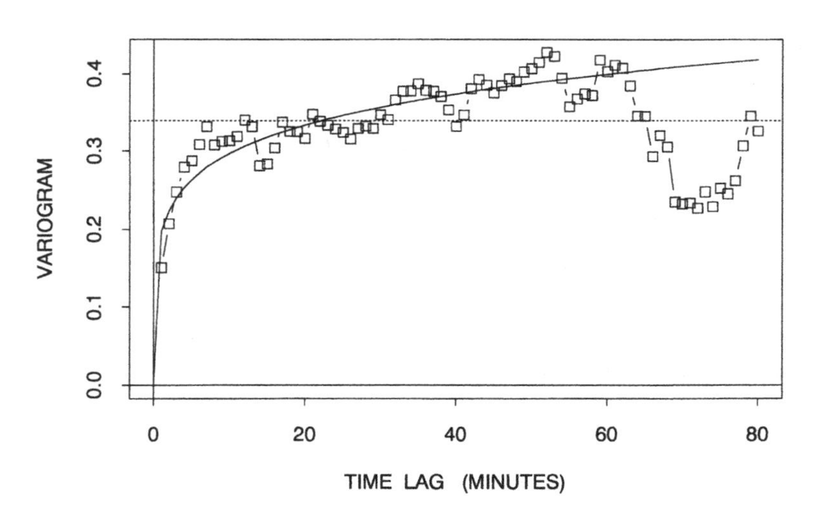

Figure 32.3: Stable model fitting of the variogram of a noise exposure time series (lower graph) and corresponding power variogram model for the logarithm of the series (upper graph).

Stable variogram model

A convenient lognormal framework to study regionalized variables are the locally stationary lognormal random functions [204]. We are not going to develop this topic here, but simply mention the variogram model that naturally arises in this context,

$$\Gamma(\mathbf{h}) \quad = \quad b\,(1-\exp(-\gamma(\mathbf{h}))) \qquad \text{with} \quad b>0, \tag{32.52}$$

where $\Gamma(\mathbf{h})$ is the variogram of the lognormal random function $Z(\mathbf{x})$ (which has a sill b) and $\gamma(\mathbf{h})$ is the variogram of $L(\mathbf{x})$ (which may be unbounded). This is an application of Schoenberg's theorem given on p55.

A particular model of this kind is the stable variogram,

$$\Gamma(\mathbf{h}) \quad = \quad b\left(1-\exp\left(-\frac{|\mathbf{h}|^{\alpha}}{a}\right)\right) \tag{32.53}$$

with $a>0$, $b>0$ and $0<\alpha<2$.

EXAMPLE 32.5 (NOISE EXPOSURE SERIES [341]) *The Figure 32.3 shows the joint modeling of the variograms of a noise exposure time series and of its logarithm by a stable theoretical variogram as defined in Eq. (32.53).*

Lognormal point and block ordinary kriging

We only give the simplest form of ordinary point kriging in the context of a lognormal model and present a corresponding block ordinary kriging.

A lognormal point estimator can be obtained by an ordinary kriging of $L(\mathbf{x})$ and it results in the estimator

$$Z^{\star}(\mathbf{x}_0) \quad = \quad \exp\left(L^{\star}_{\mathrm{LOK}}(\mathbf{x}_0)+\frac{\sigma^2_{\mathrm{LOK}}}{2}-\mu_{\mathrm{OK}}\right) \tag{32.54}$$

where $L^{\star}_{\mathrm{LOK}}(\mathbf{x}_0)$ is the ordinary kriging of $L(\mathbf{x})$ and σ^2_{LOK} is the corresponding kriging variance, as defined in Eqs. (11.1), (11.7) and (11.8) in Chapter 11.

A trivial lognormal block ordinary kriging estimator is obtained by discretizing each block into a substantial number of grid points, performing a lognormal point ordinary kriging at each of the grid points and averaging the result.

EXAMPLE 32.6 (NOISE EXPOSURE [342]) *Industrial noise data have been collected using personal sound exposure meters carried by two operators in different factories. The noise exposure series are shown on Figure 32.4. The upper series, regleur2, is from a worker taking care of different machines and walking in the workshop to check the tuning of the machines. The other series, cyclo41, is from an operator of several machines in the same room; breaks took place from 9h10 to 9h30 and from 11h40 to 11h50. Whereas both variograms (not shown, see [341]) suggest ranges extending*

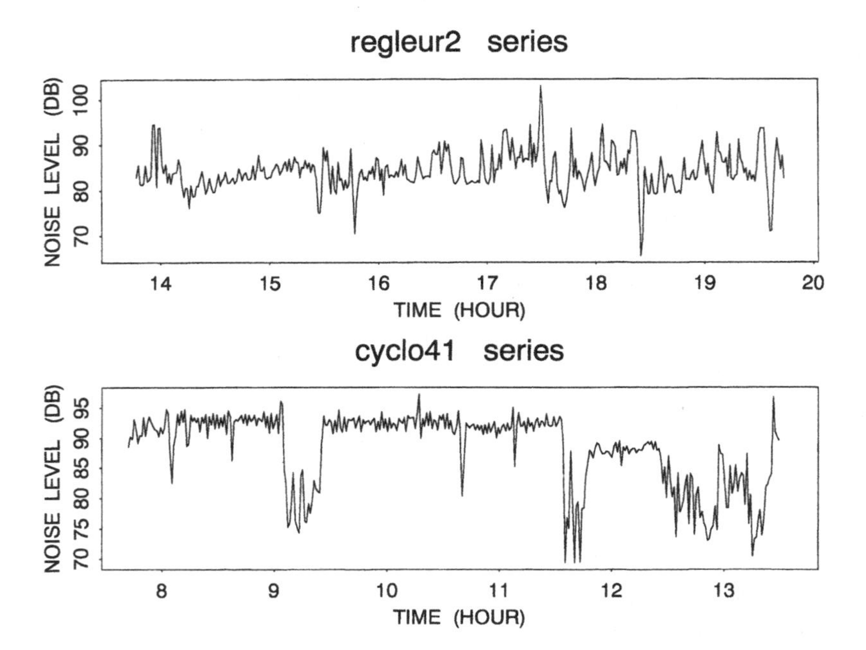

Figure 32.4: Two times series, regleur2 and cyclo41 of 1mn noise level measurements (expressed in decibel).

over several hours, it has to be mentioned that the regleur2 variogram exhibits also a marked structure at the 10mn-scale.

The histograms of the two series are represented on Figures 32.5 and 32.6. For regleur2 the assumption of lognormality seems satisfactory with a fairly symmetric histogram in the decibel scale. The same cannot be said about the histograms of the cyclo41 series.

The classical estimator of the average noise level, termed L_{eq}, consists in computing the arithmetic mean of the sound exposure (which is the additive quantity) and converting the result into the decibel[2] scale [149]. The lognormal kriging is applied in the same way: the 8h-shift mean of sound exposure is computed by lognormal block ordinary kriging as described in this section and the result is then converted into decibels.

For each of the two series, 1000 sets of 36 samples were taken at random. Both estimators were applied to each set of samples and the histograms of the estimated values converted into dB are shown on Figure 32.7.

For regleur2 (upper graphics on Figure 32.7) the standard error of lognormal kriging is downsized almost by half in comparison with the classical L_{eq}. The second mode at 90 dB on the classical L_{eq} histogram, due to an extreme value (in terms of

[2]"In sound we use a logarithmic scale of intensities since the sensitivity of the ear is roughly logarithmic." FEYNMAN et al. [100], p47-4.

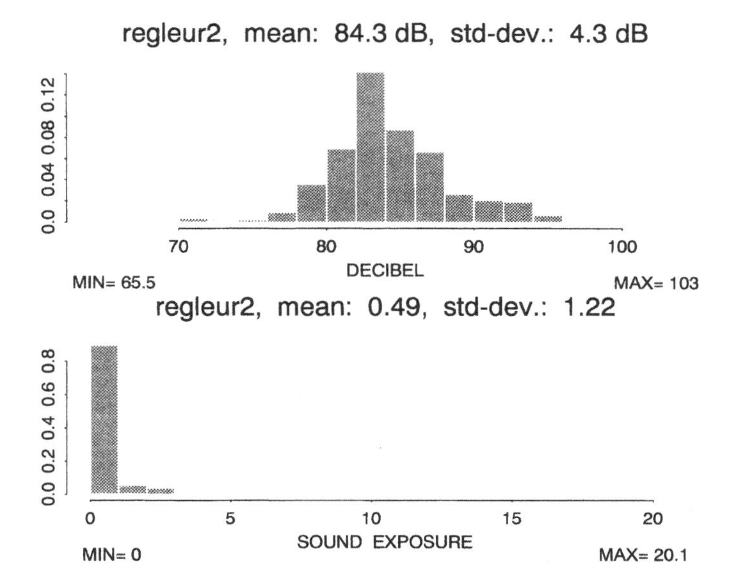

Figure 32.5: Histogram of regleur2 noise time series expressed in terms of both a variable directly proportional to the sound exposure and the logarithm of that variable in decibel scale. The L_{eq} is 86.9 dB.

sound exposure), is not present on the histogram of the lognormal L_{eq}, because the latter estimator is robust to extreme values in the right tail.

For cyclo41 (lower graphics on Figure 32.7) we see that both histograms are fairly symmetrical (with a stronger standard error for the lognormal estimator). The striking feature is the bias for the lognormal estimator. The mean of the classical L_{eq} estimates using 36 samples is equal to the L_{eq} of the 348 values of the series. In comparison, the histogram of the lognormal L_{eq} estimates is shifted to the right by one dB (the mean by .7 dB, the minimum by 1 dB and the maximum by 1.3 dB).

Lognormal estimators are thus only superior in certain situations and cannot be recommended for general use in the statistical treatment of industrial noise data. Geostatistics provides other non-linear estimators which adapt better to an arbitrary distribution of values as we will see in subsequent chapters. However, in present day hearing conservation programs, often less than 30 samples are used in industry to compute 8h-shift L_{eq} values, so that it is difficult to try assessing the shape of an underlying distribution. This situation is likely to change in a near future with the availability of cheap and light personal sound exposure meters, like the one used for measuring the regleur2 and cyclo41 series.

The conclusion is that the lognormal kriging can easily give poor results soon as the lognormal assumption is inadequate.

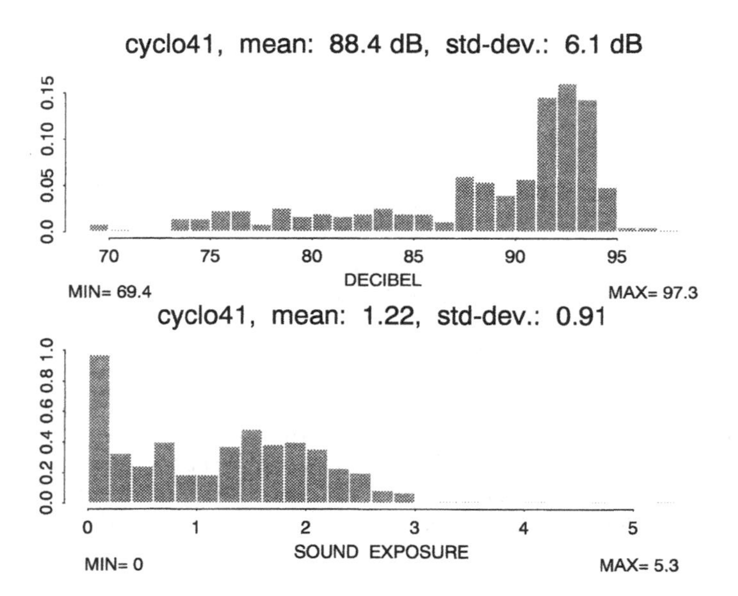

Figure 32.6: Histogram of cyclo41 expressed in terms of both a variable directly proportional to the sound exposure and the logarithm of that variable in decibel scale. The L_{eq} is 90.9 dB.

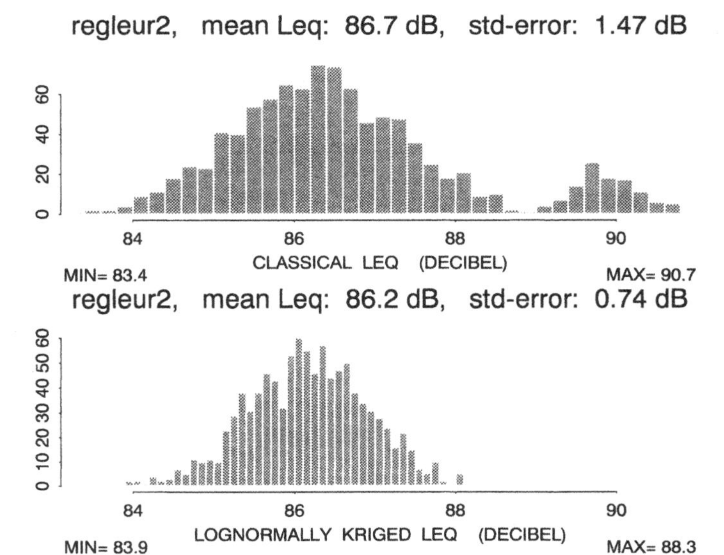

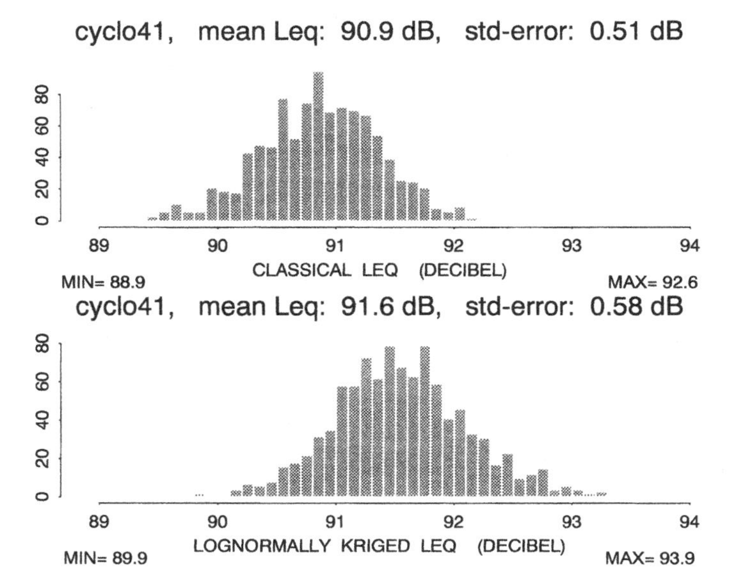

Figure 32.7: Comparison for regleur2 and cyclo41 1mn data of the classical and the lognormally kriged L_{eq} estimator for 1000 sets of 36 randomly selected samples.

33 Gaussian Anamorphosis with Hermite Polynomials

The lognormal model proposes a rather narrow framework for non-linear geostatistical estimation. We examine a more flexible approach which uses Hermite polynomials for transforming a variable with a skewed distribution into a Gaussian variable.

In the presentation of Hermite polynomials we follow closely LANTUÉJOUL [176] where more mathematical detail can be found. We introduce the fitting of the Gaussian anamorphosis in the way of LAJAUNIE [168].

Gaussian anamorphosis

Attempting to generalize the approach of the lognormal model $Z = m\exp(\sigma Y - \sigma^2/2)$ we look for a more flexible model,

$$Z = \varphi(Y) \tag{33.1}$$

in which φ is a class of non-linear functions that establishes a bijective correspondence between a random variable Z and a Gaussian random variable Y. The class of functions for which the problem will be solved is the square-integrable functions.

When applied to a stationary random function, the transformation φ will produce a random function with a Gaussian marginal. However higher-dimensional distributions will not necessarily be multi-Gaussian. In practice an anamorphosis φ will be fitted to the data and a check will have to be made whether at least the bivariate distribution of the Gaussian transform is bi-Gaussian for all spatial distance classes.

Let $F(z) = P(Z > z)$ be a probability distribution function and $G(y) = P(Y < y)$ be the standard normal distribution. If the inverse of F exists, a bijective transformation function φ can be constructed:

$$\varphi(Y) = F^{-1} \circ G(Y) \tag{33.2}$$

Conversely the function φ^{-1} transforms into a normal variable,

$$Y = G^{-1} \circ F(Z) \tag{33.3}$$

so that we have a correspondence between the Gaussian variable and the variable of interest Z as long as F is invertible,

$$\begin{aligned} G(y) &= P(Y < y) = P(\varphi^{-1}(Z) < y) & (33.4)\\ &= P(G^{-1} \circ F(Z) < y) = P(Z < F^{-1} \circ G(y)). & (33.5)\end{aligned}$$

A flexible way of setting up a Gaussian anamorphosis function is by using a development into Hermite polynomials which are now studied in detail.

Hermite polynomials

The basic building block for the Hermite polynomials is the Gaussian density function:

$$g(y) \quad = \quad \frac{1}{\sqrt{2\pi}} \exp(-\frac{x^2}{2}) \tag{33.6}$$

The Hermite polynomials are defined as derivatives of the density function,

$$H_k(y) \quad = \quad \frac{k^{\text{th}} \text{ derivative of Gaussian density}}{\text{Gaussian density}} \tag{33.7}$$

$$\quad = \quad \frac{g^{(k)}(y)}{g(y)} \qquad \text{with} \quad k = 0, 1, \ldots \tag{33.8}$$

Setting $a = 1/\sqrt{2\pi}$, $b = 1/2$ and $g(y) = a \exp(b\,y)$, we compute for example

$$H_0(y) \quad = \quad \frac{g(y)}{g(y)} = 1, \tag{33.9}$$

$$H_1(y) \quad = \quad \frac{g'(y)}{g(y)} = \frac{g(y) \cdot (-2b\,y)}{g(y)} = -y, \tag{33.10}$$

$$H_2(y) \quad = \quad \frac{g''(y)}{g(y)} = \frac{(-y\,g(y))'}{g(y)} = y^2 - 1. \tag{33.11}$$

There is a recurrence formula:

$$H_{k+1}(y) \quad = \quad -y\,H_k(y) - k\,H_{k-1}(y) \qquad \text{for} \quad k \geq 0. \tag{33.12}$$

We calculate for example

$$H_2(y) \quad = \quad -y\,H_1(y) - H_0(y) = y^2 - 1, \tag{33.13}$$

$$H_3(y) \quad = \quad -y\,H_2(y) - H_1(y) = y^3 + 3y. \tag{33.14}$$

As $H_k'(y) = -k\,H_{k-1}(y)$ the recurrence formula can be written

$$H_{k+1}(y) \quad = \quad -y\,H_k(y) + H_k'(y) \qquad \text{for} \quad k \geq 0. \tag{33.15}$$

The key property of Hermite polynomials is their orthogonality with respect to the Gaussian density:

$$\int_{-\infty}^{\infty} H_k(y)\,H_l(y)\,g(y)\,dy \quad = \quad \begin{cases} k! & \text{if } k = l, \\ 0 & \text{if } k \neq l. \end{cases} \tag{33.16}$$

On some occasions normalized Hermite polynomials are used,

$$\eta_k(y) = \frac{1}{\sqrt{k!}} H_k(y), \tag{33.17}$$

so that the orthogonality is written

$$\int_{-\infty}^{\infty} \eta_k(y)\, \eta_l(y)\, g(y)\, dy = \delta_{kl}. \tag{33.18}$$

but not all expressions are more compact, so that for now we shall stick with non-normalized polynomials.

Expanding a function into Hermite polynomials

Any function φ that is square integrable with respect to the Gaussian density, i.e.

$$\int_{-\infty}^{\infty} \varphi^2(y)\, g(y)\, dy < \infty, \tag{33.19}$$

can be expanded into Hermite polynomials:

$$\varphi(y) = \sum_{k=0}^{\infty} \frac{\varphi_k}{k!} H_k(y), \tag{33.20}$$

where the coefficients φ_k are given by,

$$\varphi_k = \int_{-\infty}^{\infty} \varphi(y)\, H_k(y)\, g(y)\, dy. \tag{33.21}$$

The integral (33.19) has the value:

$$\int_{-\infty}^{\infty} \varphi^2(y)\, g(y) = \sum_{k=0}^{\infty} \frac{\varphi_k^2}{k!}, \tag{33.22}$$

and the integral of the cross-product of two functions $\varphi(y)$ and $\psi(y)$ developed into Hermite polynomials is equal to:

$$\int_{-\infty}^{\infty} \varphi(y)\, \psi(y)\, g(y) = \sum_{k=0}^{\infty} \frac{\varphi_k\, \psi_k}{k!}. \tag{33.23}$$

Probabilistic interpretation

We turn to probabilistic questions and examine the mathematical expectation of a Gaussian random variable Y,

$$\mathrm{E}[\,Y\,] = \int_{-\infty}^{\infty} y\, g(y)\, dy \tag{33.24}$$

The orthogonality of the Hermite polynomials in relation (33.16) now comes as,

$$\mathrm{E}[\,H_k(Y)\,H_l(Y)\,] \;=\; \begin{cases} k! & \text{if } k=l, \\ 0 & \text{if } k \neq l. \end{cases} \tag{33.25}$$

Setting $l=0$ we can readily compute the first moment of Hermite polynomials of a Gaussian variable,

$$\mathrm{E}[\,H_k(Y)\,] \;=\; \mathrm{E}[\,H_k(Y)\,H_0(Y)\,] = \begin{cases} 1 & \text{if } k=0, \\ 0 & \text{if } k>0. \end{cases} \tag{33.26}$$

The variance is

$$\mathrm{var}(H_k(Y)) \;=\; \begin{cases} 0 & \text{if } k=0, \\ k! & \text{if } k>0, \end{cases} \tag{33.27}$$

because for $k>0$, using relation (33.25), we have

$$\mathrm{var}(H_k(Y)) \;=\; \mathrm{E}[\,H_k(Y)\,H_k(Y)\,] - 0 = k!, \tag{33.28}$$

while $\mathrm{var}(H_0(Y)) = \mathrm{E}[\,H_0(Y)\,H_0(Y)\,] - 1 = 0$. The fact that the variance of the zero-order polynomial is nil is consistent with the fact that it is deterministic.

The covariance between two Hermite polynomials is therefore

$$\mathrm{cov}(H_k(Y), H_l(Y)) \;=\; \begin{cases} k! & \text{if } k=l>0, \\ 0 & \text{otherwise.} \end{cases} \tag{33.29}$$

COMMENT 33.1 *Hermite polynomials exploit an important fact to be aware of: that different powers of a random variable are linearly uncorrelated. For example, a Gaussian random variable Y and its square Y^2 are uncorrelated as illustrated with 100 samples on Figure 33.1. Even though the shape of the scatter plot clearly reveals the non-linear deterministic dependence, the correlation coefficient, as a measure of linear dependence, indicates a value of only 2%.*

Moments of a function of a Gaussian variable

A function φ of a Gaussian random variable can be expanded into Hermite polynomials, provided its second moment is finite:

$$\mathrm{E}[\,\varphi^2(Y)\,] \;<\; \infty, \tag{33.30}$$

which is the same as relation (33.19).

The expansion of φ is, like in (33.20),

$$\varphi(Y) \;=\; \sum_{k=0}^{\infty} \frac{\varphi_k}{k!}\, H_k(Y), \tag{33.31}$$

with coefficients $\varphi_k = \mathrm{E}[\,\varphi(Y)\,H_k(Y)\,]$.

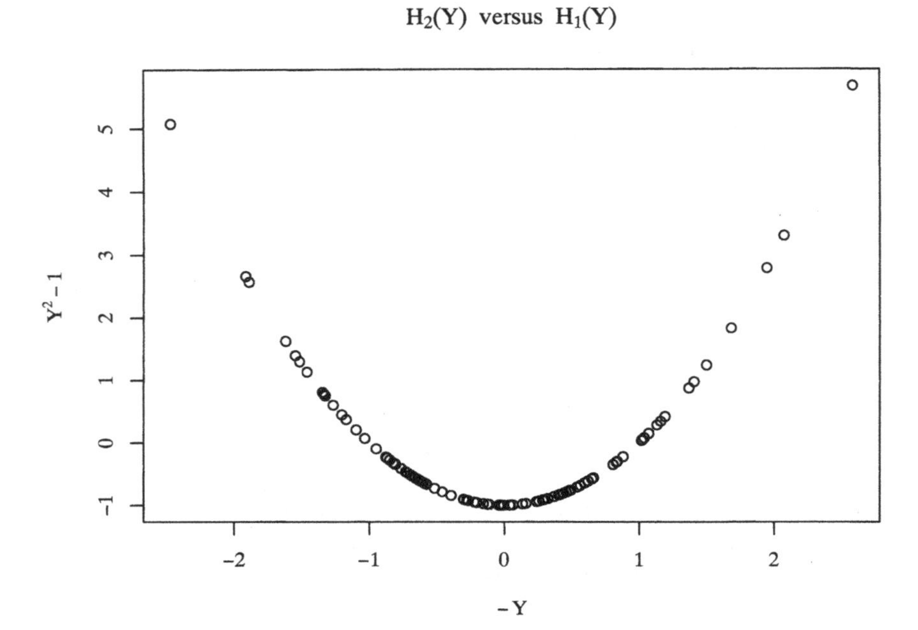

Figure 33.1: Scatter plot between 100 independent samples of a standard normal variable and their squares. The correlation coefficient between $H_1(Y)$ and $H_2(Y)$ is almost zero (value: 0.02).

The first moment of φ is equal to the constant φ_0,

$$\mathrm{E}[\,\varphi(Y)\,] \;=\; \sum_{k=0}^{\infty} \frac{\varphi_k}{k!}\,\mathrm{E}[\,H_k(Y)\,] = \varphi_0, \tag{33.32}$$

as $\mathrm{E}[\,H_k(Y)\,]$ is only non-zero for $k = 0$.

The variance is

$$\mathrm{var}(\varphi(Y)) \;=\; \sum_{k=0}^{\infty}\sum_{l=0}^{\infty} \frac{\varphi_k}{k!}\,\frac{\varphi_l}{l!}\,\mathrm{cov}(H_k(Y), H_l(Y)) \tag{33.33}$$

$$=\; \sum_{k=1}^{\infty} \frac{\varphi_k^2}{k!} \tag{33.34}$$

where the term for $k = 0$ is zero due to Eq. (33.29).

The covariance between two different square integrable functions φ and ψ of a Gaussian variable is,

$$\mathrm{cov}(\varphi(Y), \psi(Y)) \;=\; \sum_{k=0}^{\infty}\sum_{l=0}^{\infty} \frac{\varphi_k}{k!}\,\frac{\psi_l}{l!}\,\mathrm{cov}(H_k(Y), H_l(Y)) \tag{33.35}$$

$$= \sum_{k=1}^{\infty} \frac{\varphi_k \, \psi_k}{k!} \tag{33.36}$$

Conditional expectation of a function of a Gaussian variable

The bi-Gaussian density with correlation coefficient ρ between two random variables U and Y can be expanded using Hermite polynomials:

$$g_\rho(u, y) = \sum_{k=0}^{\infty} \frac{\rho^k}{k!} H_k(u), H_k(y) \, g(u) \, g(y). \tag{33.37}$$

In the same way the conditional density is written:

$$g_\rho(u \mid y) = \sum_{k=0}^{\infty} \frac{\rho^k}{k!} H_k(u), H_k(y) \, g(u). \tag{33.38}$$

Accepting these results the conditional expectation of a Hermite polynomial of U knowing Y is

$$\mathrm{E}[\, H_k(U) \mid Y \,] = \rho^k \, H_k(Y) \tag{33.39}$$

because

$$\begin{aligned} \mathrm{E}[\, H_k(U) \mid Y = y \,] &= \int_{-\infty}^{\infty} H_k(u) \, g_\rho(u \mid y) \, du & (33.40) \\ &= \sum_{l=0}^{\infty} \frac{\rho^l}{l!} H_l(y) \int_{-\infty}^{\infty} H_k(u) \, H_l(u) \, g(u) \, du & (33.41) \\ &= \sum_{k=0}^{\infty} \frac{\rho^k}{k!} H_k(y) \, \mathrm{E}[\, H_k(U) \, H_l(U) \,] & (33.42) \\ &= \rho^k \, H_k(y). & (33.43) \end{aligned}$$

The conditional expectation of a square-integrable function φ of U knowing Y is finally

$$\mathrm{E}[\, \varphi(U) \mid Y \,] = \sum_{k=0}^{\infty} \frac{\varphi_k \, \rho^k}{k!} H_k(Y). \tag{33.44}$$

EXERCISE 33.2 *Show that*

$$\mathrm{cov}(\varphi(U), \psi(Y)) = \sum_{k=1}^{\infty} \frac{\varphi_k \, \psi_k}{k!} \, \rho^k \tag{33.45}$$

using Eq. (33.44).

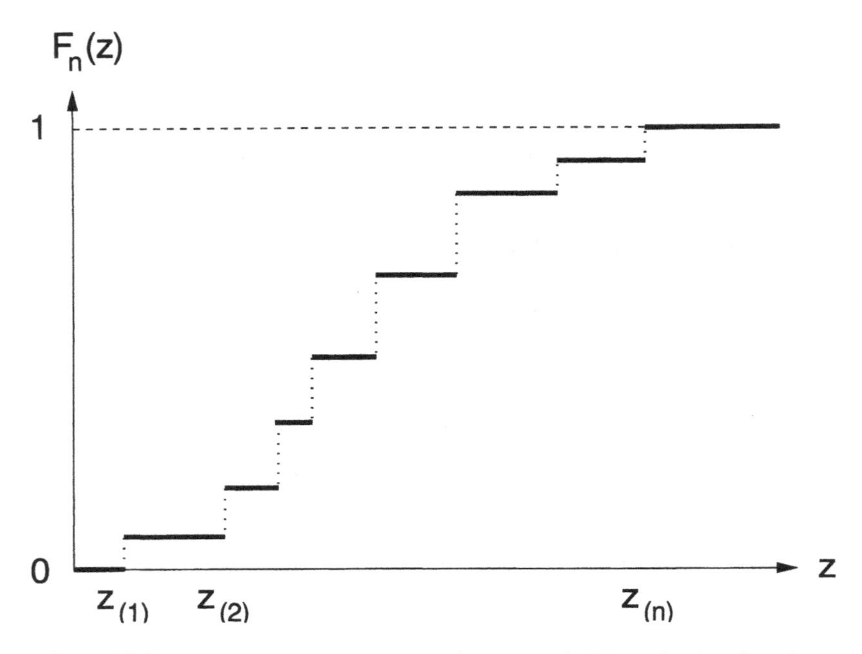

Figure 33.2: Schematic representation of an empirical distribution function.

Empirical Gaussian anamorphosis

In practice we construct from n samples the cumulative histogram, which is an empirical distribution function

$$F_n(z) = \frac{1}{n} \sum_{\alpha=1}^{n} \mathbf{1}_{Z_\alpha < z} \tag{33.46}$$

that provides a discrete representation of the distribution $F(z)= P(z < Z)$ of the variable under study.

Denoting by $Z_{(\alpha)}$ the samples renumbered by increasing value,

$$Z_{(1)} \leq \ldots \leq Z_{(\alpha)} \leq \ldots \leq Z_{(n)} \tag{33.47}$$

we can analyze the behavior of the empirical distribution function. In the case all values $Z_{(\alpha)}$ are different, the value of $F_n(z)$ is zero for z in the interval $]-\infty, Z(1)]$, the value is $(\alpha-1)/n$ in the intervals $]Z(\alpha-1), Z(\alpha)]$ and it is one in the interval $]Z(\alpha), \infty[$. The empirical distribution function is a step-function, as plotted on Figure 33.2. In case that several values, say k, are equal, the step at that level will not be $1/n$ (as for distinct values) but k/n.

In the analogous way we construct an *empirical Gaussian anamorphosis*,

$$\varphi_n(y) = \frac{1}{n} \sum_{\alpha=1}^{n} Z_{(\alpha)} \, \mathbf{1}_{y \in I_\alpha}, \tag{33.48}$$

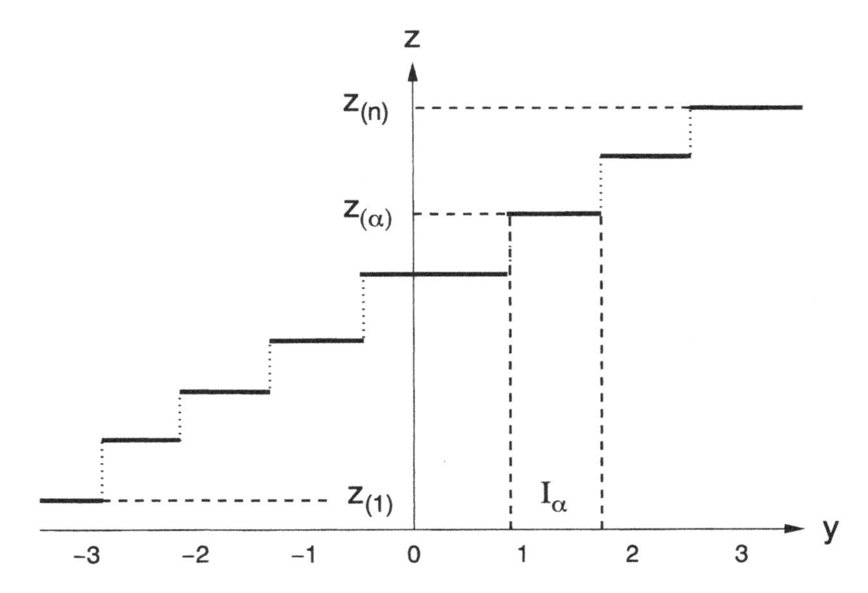

Figure 33.3: Schematic representation of an empirical Gaussian anamorphosis.

where I_α are half-open intervals of the abscissa y defined in terms of the inverse of the Gaussian distribution,

$$I_\alpha = \left]G^{-1}\left(\frac{\alpha-1}{n}\right), G^{-1}\left(\frac{\alpha}{n}\right)\right] \qquad \text{for} \quad \alpha = 1, \ldots, n. \qquad (33.49)$$

In the case of distinct sample values $Z_{(\alpha)}$ the function $\varphi_n(y)$ has the value $Z_{(1)}$ in the interval $I_1 =]-\infty, G^{-1}(1/n)]$, the value $Z_{(n)}$ in the interval $]G^{-1}((n-1)/n), \infty[$ and the value $Z_{(\alpha)}$ for the intervals $]G^{-1}((\alpha-1)/n), G^{-1}(\alpha/n)]$. The corresponding step-function is drawn on Figure 33.3. In the case k subsequent samples have the same value, the step at that level will go to the highest of the k values.

The empirical anamorphosis is not invertible. This can be seen by flipping the axes on Figure 33.3: to each value $Z_{(\alpha)}$ corresponds an interval I_α of y-values whereas the Z-values between sample values have no image.

Smoothing the empirical anamorphosis

The empirical anamorphosis is inadequate for further use and we need to look for a function with better properties. An interesting option for the Gaussian empirical anamorphosis function is to replace it by its development into Hermite polynomials and to truncate the development at an appropriate order. This amounts to smoothen the step function.

The empirical Gaussian anamorphosis is represented with an infinite series of Her-

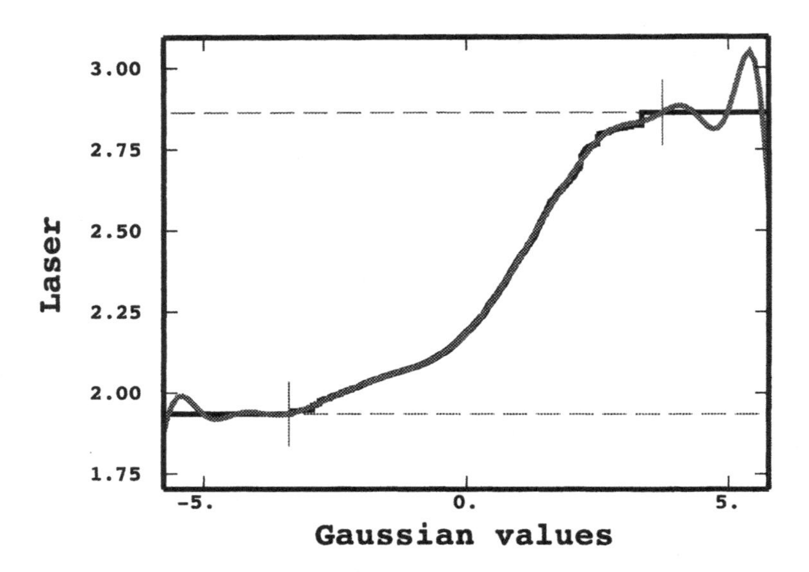

Figure 33.4: Development using 30 Hermite polynomials fitted to the empirical Gaussian anamorphosis of laser measurements.

mite polynomials

$$\varphi_k(y) \;=\; \sum_{k=0}^{\infty} \frac{\varphi_k}{k!}\, H_k(y) \tag{33.50}$$

with coefficients

$$\begin{aligned}
\varphi_k \;&=\; \int_{-\infty}^{\infty} \varphi_n(y)\, H_k(y)\, g(y)\, dy & (33.51)\\
&=\; \sum_{\alpha=1}^{n} Z_{(\alpha)} \int_{I_\alpha} H_k(y)\, g(y)\, dy & (33.52)\\
&=\; \sum_{\alpha=1}^{n} Z_{(\alpha)} \left(H_{k-1}(y_\alpha)\, g(y_\alpha) - H_{k-1}(y_{\alpha-1})\, g(y_{\alpha-1})\right) & (33.53)
\end{aligned}$$

A smooth approximation of the empirical Gaussian anamorphosis is obtained by truncating the development at a fixed value K. This truncated development is what we shall call the *Gaussian anamorphosis*:

$$\varphi^{\star}(y) \;=\; \sum_{k=0}^{K} \frac{\varphi_k}{k!}\, H_k(y) \tag{33.54}$$

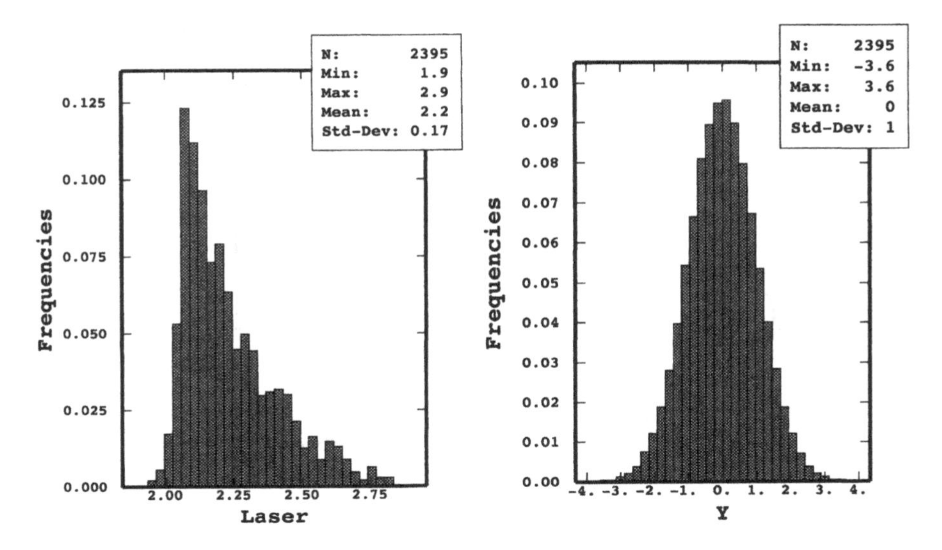

Figure 33.5: Histograms of laser measurements (left) and their Gaussian transformed values (right).

This is actually the best K^{th}-order approximation in the least squares sense and it can be viewed as an eigenvalue problem where the Hermite polynomials are the eigenfunctions (as explained in Chapter 19 when dealing with continuous correspondence analysis).

In practice the development may be truncated somewhere between the 15th and the 50th order. A criterion is to compute the ratio between the variance (33.34) of $\varphi^{\star}(y)$ and the variance s^2 computed from the data. The development is stopped at an order for which this ratio is almost equal to one.

Bijectivity of Gaussian anamorphosis

The limited development tends to be bijective within the interval defined by the minimum and the maximum of the sample values.

Yet the Gaussian anamorphosis function with Hermite polynomials is generally not bijective below the minimum $Y_{\min}= \varphi^{\star-1}(Z_{(1)})$ or above the maximum $Y_{\max}= \varphi^{\star-1}(Z_{(n)})$: the development $\varphi^{\star}(y)$ into polynomials oscillates wildly outside the the interval $[Y_{\min}, Y_{\max}]$. This may be of little importance for non-linear estimation as the results tend to be within the range of the data. However when simulating a Gaussian random function there are a couple of values well outside the interval $[Y_{\min}, Y_{\max}]$ which at some stage need to be converted into the original Z-scale by anamorphosis. A simple practical solution is to define two new points $Y'_{\min}= -10$ and $Y'_{\max}= 10$ for which the Gaussian probabilities can be considered as respectively zero and one (at least numerically, on a computer) and to define two increasing straight line segments

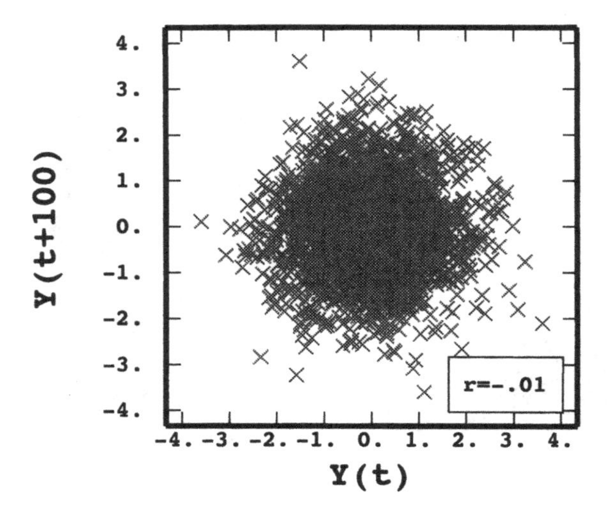

Figure 33.6: Scatter plot between Gaussian anamorphosis values separated by lags of 100 seconds.

as a replacement for the limited development outside the interval $[Y_{\min}, Y_{\max}]$.

LANTUÉJOUL (quoted in [265], p137) proposes to compute a measure of total variation to check whether the anamorphosis is slightly decreasing at some points within the interval $[Y_{\min}, Y_{\max}]$ for a given order K. This provides an additional constraint when selecting the order of the Hermite polynomial: orders K for which this happens should be rejected.

EXAMPLE 33.3 (LASER DATA) *In an oceanographic study [98] a set of laser data of sea level heights (in decimeter) measured during 20 minutes at the Ekofisk oil platform on 1st January 2002 at midnight presented a distribution that was left-skew.*

We perform a Gaussian anamorphosis of this data which is displayed on Figure 33.4. The fit of the empirical anamorphosis with a development into $K= 30$ *Hermite polynomials provides an increasing function within the interval* $[-3.6, 3.6]$ *given by the Gaussian transforms of the minimum* $Z_{(1)}$ *and the maximum* $Z_{(n)}$ *of the laser data, which are shown as two dashed lines parallel to the abscissa. Outside the interval* $[-3.6, 3.6]$ *the anamorphosis function* $\varphi^{\star}(y)$ *wildly oscillates and is no more increasing, so it cannot be used in this form outside that interval as discussed above.*

The histograms of the data before and after anamorphosis are displayed on Figure 33.5. The variogram in this study had shown that the data can be considered as stationary within the 20 minute time domain. We display on Figure 33.6 and Figure 33.7 the scatter plots between values separated by lags of 100 seconds and 1 second. These lagged scatter plots may be used to check the bivariate distributions. It turns out that, although the diagram for a lag of 100 seconds (by the way far beyond

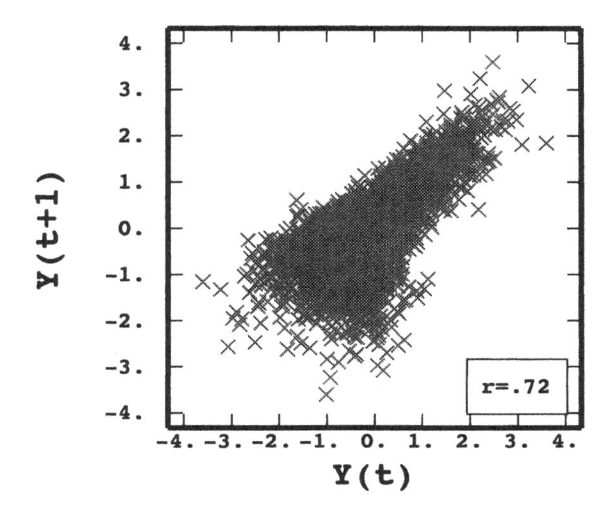

Figure 33.7: Scatter plot between Gaussian values separated by lags of 1 second.

the range of the variogram, which is of only a few seconds) is a good example of a bi-Gaussian dispersion cloud, the scatter plot for a lag of 1 second does not show at all the ellipsoidal shape of a bi-Gaussian density.

So this is a nice example illustrating that if a random function $Y(\mathbf{x})$ *has a Gaussian marginal distribution (like on the histogram of Figure 33.5) this does not necessarily imply that the bivariate distribution for pairs* $Y(\mathbf{x}), Y(\mathbf{x}+\mathbf{h})$ *is bi-Gaussian. Only the reverse is true: bivariate Gaussian distributions imply Gaussian marginal distributions.*

The consequences are twofold. On one hand we need a model to construct bivariate distributions that are not bi-Gaussian when the marginal is Gaussian; this will be provided by the Hermitian isofactorial model. On the other hand we will study models with a marginal distribution having a shape close to that of the data; this will in particular be the case for the gamma distribution with its shape parameter λ. *For the bivariate distribution of two gamma variables we will be able to specify the Laguerre isofactorial model which covers a wide class of bivariate distributions of which the bigamma distribution is a special case.*

34 Isofactorial Models

Isofactorial models are important for modeling the bivariate distribution between a pair of points of a stationary random function. They offer a great variety of possible constructions for modeling random functions between the two extremes of diffusive type and mosaic type.

According to MATHERON [216] isofactorial models stem from quantum mechanics where they have been applied since the nineteen-twenties [174]. Hermite polynomials for example appear as eigenfunctions of the one dimensional harmonic oscillator. In statistics, the Hermitian model is already mentioned by CRAMER [62] and discrete isofactorial models have been used in connection with Markov processes [99, 326]. In geostatistics the work on correspondence analysis by the group around JP BENZÉCRI [234, 24], as described in Chapter 19, has given the impulse for introducing isofactorial models [209].

After a general presentation of isofactorial models the particular cases of the Hermitian and the Laguerre isofactorial models will be examined. These models have been implemented by HU [144, 142] and were successfully used for evaluating reserves of uranium and gold deposits with very skew distributions [142, 143, 182, 52].

Isofactorial bivariate distribution

Let F be a symmetric bivariate distribution,

$$F(du, dv) \quad = \quad F(dv, du) \tag{34.1}$$

with marginal distribution $F(du) = \int_v F(du, dv)$.

A symmetric bivariate distribution is *isofactorial*, if a countable system of orthonormal real functions χ_k exists, complete in $\mathcal{L}^2(F)$, such that

$$\int \chi_k(u)\, \chi_l(u)\, F(du) \quad = \quad \delta_{kl} \tag{34.2}$$

and

$$\int\!\!\int \chi_k(u)\, \chi_l(v)\, F(du, dv) \quad = \quad \delta_{kl}\, T_k, \tag{34.3}$$

where the T_k are real coefficients.

The coefficients T_k represent covariances between factors χ_k,

$$T_k \quad = \quad |\mathrm{E}[\,\chi_k(U)\, \chi_k(V)\,]| \;\leq\; \mathrm{E}[\,(\chi_k(U))^2\,] = 1 \tag{34.4}$$

and are thus bounded by $-1 \leq T_k \leq 1$.

Isofactorial decomposition

We may decompose any function φ that is square integrable with respect to $F(du)$,

$$\int (\varphi(u))^2 \, F(du) \, du < \infty, \tag{34.5}$$

into orthogonal functions χ_k with coefficients φ_k:

$$\varphi(u) = \sum_{k=0}^{\infty} \varphi_k \, \chi_k(u). \tag{34.6}$$

To obtain the value of a coefficient φ_k we multiply both sides of (34.6) by $\chi_l(u) \, F(du)$ and integrate,

$$\int \varphi(u) \, \chi_l(u) \, F(du) = \sum_{k=0}^{\infty} \varphi_k \int \chi_k(u) \, \chi_l(u) \, F(du) \tag{34.7}$$

and taking advantage of orthogonality we get,

$$\varphi_k = \int \varphi(u) \, \chi_k(u) \, F(du). \tag{34.8}$$

For a pair of random variables U and V with an isofactorial bivariate distribution we have

$$\mathrm{E}[\, \chi_k(U) \mid V \,] = T_k \, \chi_k(V). \tag{34.9}$$

EXERCISE 34.1 *Demonstrate the validity of Eq.(34.9) using the fact that the conditional expectation is a square integrable function.*

The conditional expectation of a function φ of one of the variables knowing the other is

$$\mathrm{E}[\, \varphi(U) \mid V \,] = \sum_{k=0}^{\infty} \varphi_k \, T_k \, \chi_k(V). \tag{34.10}$$

Replacing the φ_k by their expression (34.8),

$$\mathrm{E}[\, \varphi(U) \mid V \,] = \int \varphi(u) \sum_{k=0}^{\infty} T_k \, \chi_k(V) \, \chi_k(u) \, F(du), \tag{34.11}$$

it is obvious that the conditional distribution has the form

$$F_v(du) = F(du) \sum_{k=0}^{\infty} T_k \, \chi_k(u) \, \chi_k(v). \tag{34.12}$$

By analogy the decomposition of the isofactorial bivariate distribution is obtained as:

$$F(du, dv) = F(du) \, F(dv) \sum_{k=0}^{\infty} T_k \, \chi_k(u) \, \chi_k(v). \tag{34.13}$$

COMMENT 34.2 *We have already seen the isofactorial decompositions of the Gaussian bivariate and conditional densities in expressions (33.37) and (33.38) on p243. The factors χ_k in that case are normalized Hermite polynomials $\eta_k = H_k/\sqrt{k!}$ and the covariances T_k are equal to ρ^k.*

Isofactorial models

In the geostatistical context U may be a random variable located at one point in the domain $\mathcal{D}$ while V is located at another point. We are interested in modeling the bivariate distribution between any pair of points of the domain for a stationary random function $Y(\mathbf{x})$ and set: $U = Y_{\mathbf{x}}$, $V = Y_{\mathbf{x+h}}$. The *isofactorial model* for the bivariate distribution of the random function now comes as

$$F(dy_{\mathbf{x}}, dy_{\mathbf{x+h}}) = F(dy_{\mathbf{x}})\, F(dy_{\mathbf{x+h}}) \sum_{k=0}^{\infty} T_k(\mathbf{h})\, \chi_k(y_{\mathbf{x}})\, \chi_k(y_{\mathbf{x+h}}) \quad (34.14)$$

where $T_k(\mathbf{h})$ is a correlation function between the pair of points.

A great variety of isofactorial models have been made available to geostatisticians by MATHERON, who developed a passion for the subject and whose work is to a large extent summarized and listed in [14, 15, 16]. CHILÈS & DELFINER [51] provide a comprehensive overview in their chapter on nonlinear methods.

Two general patterns of random functions used in geostatistics can be distinguished [221],

- on one hand a *diffusion type*, with almost surely continuous (but not necessarily differentiable) realizations, which describe phenomena with a diffusive behavior, i.e a gradual change from one location to the other,
- on the other hand a *mosaic type*, with jumps located at surfaces of discontinuity, which represent a geographical space divided into compartments inside which the phenomenon stays constant while it shows sudden change between compartments.

The first type corresponds to the idea of a smooth transition between neighboring values while the second type conveys the image of abrupt change when stepping from one compartment to another.

For diffusion type random functions attention has been focused principally on *isofactorial models with polynomial factors* that can be explicitly computed through recurrence relations. They actually provide a full range of models between the extremes of pure diffusion type and plain mosaic type random functions.

We have already reviewed to some extent the Gaussian model with its associated Hermite polynomials in Chapter 32. There are two other models with a continuous marginal distribution: the gamma model (with Laguerre polynomials) and the beta model (with Jacobi polynomials).

Furthermore, for diffusion type random functions with a discrete marginal distribution, there are five classes of models with polynomial factors [217]. Two of them correspond to processes with an infinite number of states: the Poisson model (with Charlier polynomials) and the negative binomial model (with Meixner polynomials); the latter is a discrete version of the gamma model. Three classes are derived from processes with a finite number of states: the binomial model (with Krawtchouk polynomials), the Jacobi and the anti-Jacobi models (with discrete Jacobi/anti-Jacobi polynomials).

For the specification of a diffusion type isofactorial model with known polynomials two ingredients have to be determined:

- the class of marginal distribution, which will determine the system of polynomial functions χ_k,
- the form of the isofactorial bivariate distribution, which is governed by the $T_k(\mathbf{h})$ correlation functions.

Choice of marginal and of isofactorial bivariate distribution

The choice of the marginal distribution towards which the data will be transformed is guided by the skewness of the data histogram and the possible presence of a large amount of equal values (e.g. a spike at the origin because of many zero values). The data may be considered as the result of either a continuous process or of a discrete process taking a countable finite or infinite number of states.

The choice of a bivariate distribution type constructed with an isofactorial model should be guided by an exploratory analysis of scatter diagrams of pairs of anamorphosed values for different lags.

COMMENT 34.3 HU *[142] discusses the example of a bigamma isofactorial random function (with λ= .1) which is submitted to a Gaussian anamorphosis: the graph of the bivariate density (for $\rho(\mathbf{h})$= .6) of the Gaussian transforms is pear-shaped instead of ellipsoidal. This shows that if the data are transformed to have a Gaussian marginal distribution this does not imply that the bivariate distribution becomes bi-Gaussian. The smaller the value of the parameter λ of the gamma distribution, the less the bivariate density can be considered as bi-Gaussian.*

A different valuable instrument is the first order variogram, defined as the expectation of the absolute value of increments:

$$\gamma_1(\mathbf{h}) \quad = \quad \frac{1}{2}\mathrm{E}[\,|Y(\mathbf{x}+\mathbf{h}) - Y(\mathbf{x})|\,] \tag{34.15}$$

The ratio between the first order and the usual (second order) variograms may help to discriminate between diffusion and mosaic type phenomena, or rather to characterize an intermediate situation between these two extreme types of random function models. Ideally, for a diffusion type the following ratio will take a constant value,

$$\frac{\gamma_1(\mathbf{h})}{\sqrt{\gamma(\mathbf{h})}} \quad = \quad \frac{C_1}{\sqrt{C}}, \tag{34.16}$$

while for the mosaic type we have,

$$\frac{\gamma_1(\mathbf{h})}{\gamma(\mathbf{h})} = \frac{C_1}{C}, \tag{34.17}$$

where C_1 is the sill of the first order variogram $\gamma_1(\mathbf{h})$ and C is the sill of the usual variogram $\gamma(\mathbf{h})$. In practice a plot of the $\gamma_1(\mathbf{h})$ against $\gamma(\mathbf{h})$ model variograms will yield a parabola for a diffusive phenomenon and a straight line for data compliant with the mosaic model.

In a purely diffusion type isofactorial model we will choose covariance functions of the form

$$T_k(\mathbf{h}) = (\rho(\mathbf{h}))^k, \tag{34.18}$$

while they will not depend on k in a mosaic model:

$$T_k(\mathbf{h}) = \rho(\mathbf{h}), \tag{34.19}$$

where for some models the correlation functions $\rho(\mathbf{h})$ will only be allowed with positive values [51].

An important implication is that the two types of random functions have a different coregionalization of indicator functions of level sets $\{Z(\mathbf{x}) \geq z_c\}$. In the case of (34.18) the ranges of the indicator functions will vary with z_c and in particular their coregionalization is not compatible with a linear model of coregionalization. For example, for a left-skew distribution, the ranges of the indicator functions will decrease with increasing z_c because there are fewer values for high cut-off: this is a frequntly observed *destructuration effect*. In the case of (34.19), however, indicator functions of different level sets all have covariance functions with the same range and follow the intrinsic correlation model.

In practice we are likely to be faced with intermediate cases between diffusion type and mosaic type random functions. Specific models will be discussed for the Hermitian and Laguerre isofactorial models.

Hermitian and Laguerre isofactorial distributions

A random function with a Gaussian marginal distribution and with normalized Hermite polynomials η_k has a *Hermitian isofactorial distribution*:

$$F(dy_{\mathbf{x}}, dy_{\mathbf{x}+\mathbf{h}}) = g(y_{\mathbf{x}})\, g(y_{\mathbf{x}+\mathbf{h}}) \sum_{k=0}^{\infty} T_k(\mathbf{h})\, \eta_k(y_{\mathbf{x}})\, \eta_k(y_{\mathbf{x}+\mathbf{h}}) \tag{34.20}$$

In the particular case of a purely diffusive phenomenon the covariance functions take the form $T_k(\mathbf{h})= (\rho(\mathbf{h}))^k$, with $\rho(\mathbf{h}) \in [-1, 1]$ and the bivariate distribution is *bi-Gaussian*. Conversely we see that a random function with a Gaussian marginal distribution is not necessarily Gaussian because its bivariate distributions may not be bi-Gaussian and can have the more general form (34.20).

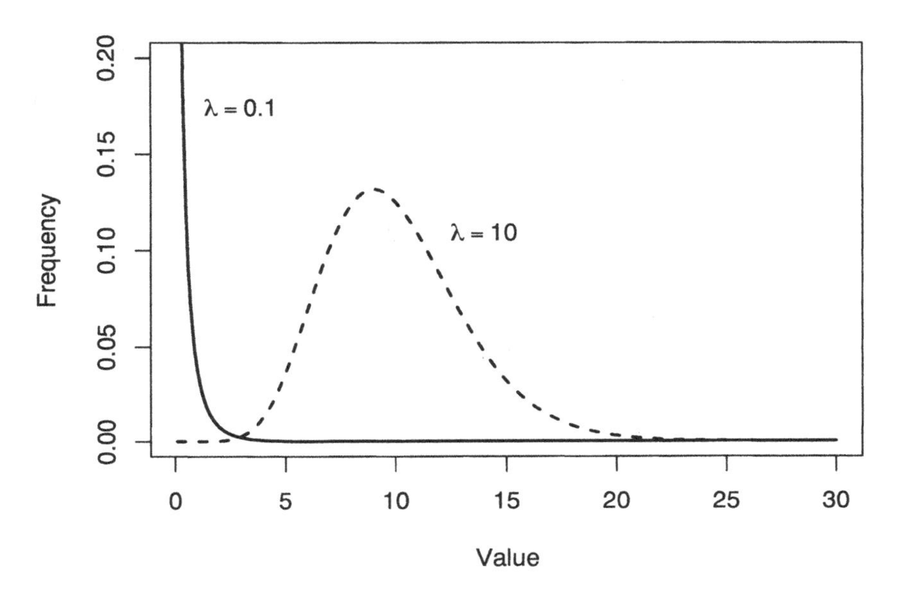

Figure 34.1: Gamma density function for values of λ= .1 and λ= 10.

The gamma model is designed for random functions with positive values. The density function of the gamma distribution is defined as

$$f_\lambda(y) \;=\; \frac{1}{\Gamma(\lambda)}\,\exp(-y)\,y^{\lambda-1} \qquad \text{with} \quad \lambda > 0, y > 0, \tag{34.21}$$

where $\Gamma(\lambda)$ is the classical gamma function and λ is a parameter controlling the shape of the distribution. This is actually the standard gamma density function, obtained by setting the second parameter of the general gamma density function to one.

COMMENT 34.4 *The gamma density function is displayed on Figure 34.2 for values of* λ= .1 *and* λ= 10 *(both the abscissa and the ordinate have been truncated to focus on the shape). With* $\lambda \leq 1$ *the density function is steep at the origin and decreasing, while for* $\lambda > 1$ *a bump appears, and it turns out that for* $\lambda \to \infty$ *the gamma density tends to the bell shape of the Gaussian density.*

The mean and the variance of a gamma distributed random variable take the same value:

$$\mathrm{E}[\,Y\,] \;=\; \lambda, \qquad \mathrm{var}(Y) \;=\; \lambda. \tag{34.22}$$

The Laguerre polynomials are defined as derivatives of the gamma density function

$$L_\lambda^k(y) = \frac{k^{\text{th}} \text{ derivative of gamma density}}{\text{gamma density}} \qquad (34.23)$$

$$= \frac{f_\lambda^{(k)}(y)}{f_\lambda(y)} \qquad \text{with} \quad k = 0, 1, \ldots \qquad (34.24)$$

and the first two polynomials are

$$L_\lambda^0(y) = 1, \qquad L_\lambda^1(y) = \sqrt{\frac{\Gamma(\lambda)}{\Gamma(\lambda+1)}} \left(1 - \frac{y}{\lambda}\right). \qquad (34.25)$$

A recurrence relation allows to compute the others:

$$(\lambda + k)\, L_\lambda^{k+1}(y) = (2k + \alpha - y)\, L_\lambda^k(y) - k\, L_\lambda^{k-1}(y). \qquad (34.26)$$

Normalized Laguerre polynomials are obtained by the relation

$$l_{\lambda k}(y) = \sqrt{\frac{\Gamma(\lambda + k)}{\Gamma(\lambda)\, k!}}\, L_\lambda^k(y). \qquad (34.27)$$

A random function with a gamma marginal distribution has a *Laguerre isofactorial distribution*:

$$F(dy_{\mathbf{x}}, dy_{\mathbf{x}+\mathbf{h}}) = f_\lambda(y_{\mathbf{x}})\, f_\lambda(y_{\mathbf{x}+\mathbf{h}}) \sum_{k=0}^{\infty} T_k(\mathbf{h})\, l_{\lambda k}(y_{\mathbf{x}})\, l_{\lambda k}(y_{\mathbf{x}+\mathbf{h}}) \qquad (34.28)$$

For a purely diffusive random function the covariance functions take the form $T_k(\mathbf{h}) = (\rho(\mathbf{h}))^k$, with $\rho(\mathbf{h}) \in [0, 1]$ and the bivariate distribution is then *bigamma*.

As in (33.39) and (34.9), we have

$$\mathrm{E}\Big[\, l_{\lambda k}(Y_{\mathbf{x}+\mathbf{h}}) \mid Y_{\mathbf{x}} \Big] = T_k(\mathbf{h})\, l_{\lambda k}(Y_{\mathbf{x}}). \qquad (34.29)$$

COMMENT 34.5 *On Figure 34.2 a bigamma density function for the value $\lambda = .1$ is displayed: its swallow-tail shape is very different from the bell-shape for the case $\lambda = 10$ that is shown on Figure 34.3. In both examples a correlation of $\rho(\mathbf{h}) = .6$ was used.*

A gamma anamorphosis $\varphi_\lambda(Y(\mathbf{x}))$ for data of a random function $Z(\mathbf{x})$ can be developed along the same lines as for the Gaussian anamorphosis [144]. The parameter λ will be chosen in such a way that the anamorphosed function most closely resembles the shape of the data cumulative histogram. As criteria for resemblance the coefficient of variation, the selectivity, the median or a normality index may be used. An important criterion is to check the bivariate distribution between points separated by a vector $\mathbf{h}$ and to select a value λ for the model which implies a bivariate distribution with the same shape as that suggested by the data.

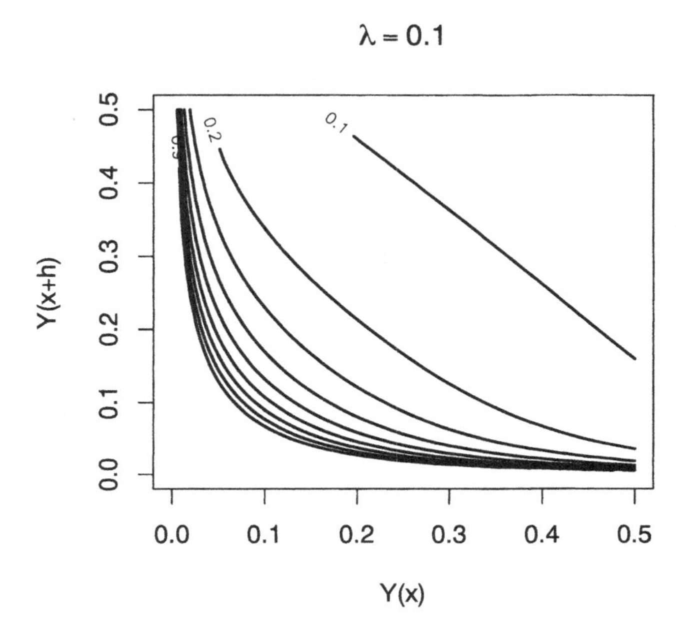

Figure 34.2: Bivariate gamma density for λ= .1 and $\rho(\mathbf{h})$= .6.

EXAMPLE 34.6 (LASER DATA) *The Gauss transformed laser data discussed at in Chapter 32 had an autocorrelation diagram at a lag of 1 second that was not bi-Gaussian (see Figure 33.7 on p249). As can be seen on the 1 second-lag scatter plot of the laser data on Figure 34.4 it seems a better idea to use a bigamma model in such a situation rather than a bi-Gaussian.*

It should be noted that the philosophy of the Gaussian and of the gamma anamorphosis are not the same: while in the former case an arbitrary distribution is brought more or less brutally into a Gaussian setting, in the latter the parameter λ is tuned in such a way that we transform to a distribution that is close to some of the features of the data.

We give the formula for reconstructing the variogram $\gamma_Z(\mathbf{h})$ of $Z(\mathbf{x})$, which comes as

$$\gamma_Z(\mathbf{h}) \quad = \quad \sum_{k=1}^{\infty} \varphi_{\lambda k}^2 \, (1 - T_k(\mathbf{h})). \qquad (34.30)$$

where $\varphi_{\lambda k}$ are the coefficients of the gamma anamorphosis $\varphi_\lambda(Y(\mathbf{x}))$. This formula is useful for fitting the variograms of $Z(\mathbf{x})$ and $Y(\mathbf{x})$ in a coherent manner.

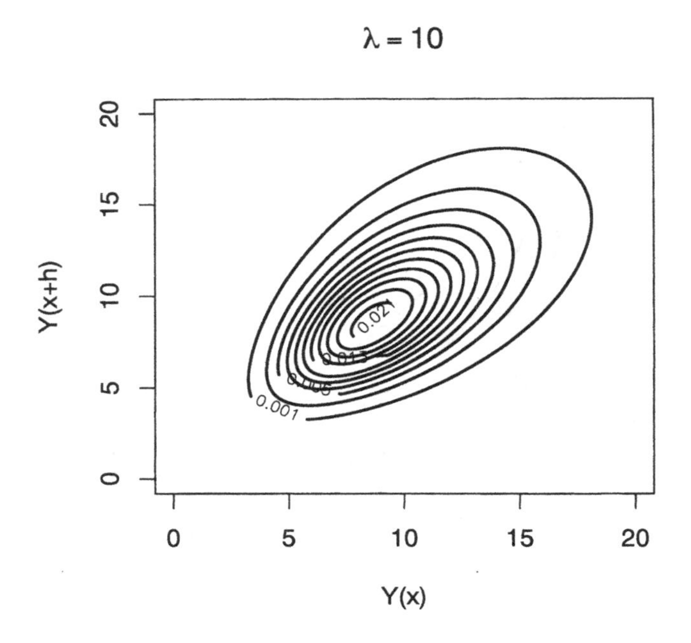

Figure 34.3: Bivariate gamma density for λ= 10 and $\rho(\mathbf{h})$= .6.

Intermediate types between diffusion and mosaic models

We examine only Hermitian and Laguerre isofactorial models in presenting intermediate types between the pure diffusion and the plain mosaic models. In both cases bivariate distributions for intermediate types can be obtained from $T_k(\mathbf{h})$ functions with the general structure [206, 209],

$$T_k(\mathbf{h}) \;=\; \mathrm{E}[\, r^k \,] = \int_{r\in I} r^k \, p(dr) \qquad \text{with} \quad k \geq 0, \tag{34.31}$$

where $I = [-1, 1]$ for Hermitian and $I = [0, 1]$ for Laguerre isofactorial models.

For a pure diffusion type random function the distribution $p(dr)$ is concentrated at a value $r_0 = \rho(\mathbf{h})$,

$$T_k(\mathbf{h}) \;=\; \int_I r^k \, \delta_{r_0}(dr) = r_0^k = (\rho(\mathbf{h}))^k. \tag{34.32}$$

For a plain mosaic type random function the distribution $p(dr)$ will be concentrated at the two values r= 1 and r= 0 with the probabilities,

$$P(r{=}\,1) \;=\; \rho(\mathbf{h}) \qquad \text{and} \qquad P(r{=}\,0) \;=\; (1{-}\rho(\mathbf{h})) \tag{34.33}$$

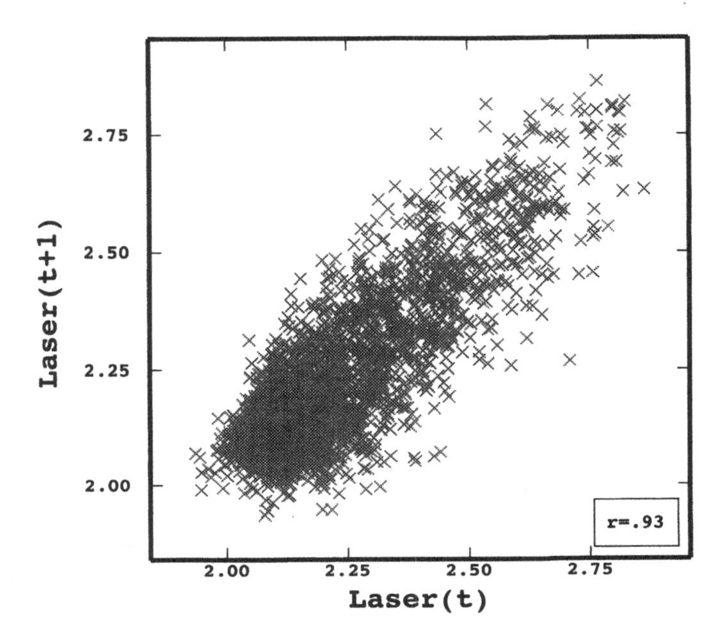

Figure 34.4: Lagged correlation diagram.

so that

$$p(dr) \;=\; \rho(\mathbf{h})\,\delta_1(dr) + (1-\rho(\mathbf{h}))\,\delta_0(dr) \tag{34.34}$$

and

$$T_k(\mathbf{h}) \;=\; \rho(\mathbf{h}) \int_I r^k\,\delta_1(dr) + (1-\rho(\mathbf{h})) \int_I r^k\,\delta_0(dr) \tag{34.35}$$

$$\;=\; \rho(\mathbf{h}) \tag{34.36}$$

Isofactorial models for intermediate types are then almost trivially obtained by taking proportions w and $1-w$ of the two basic types, which is termed the *barycentric* model,

$$T_k(\mathbf{h}) \;=\; w\,(\rho(\mathbf{h}))^k + (1-w)\,\rho(\mathbf{h}) \qquad \text{with} \quad w \in [0,1]. \tag{34.37}$$

COMMENT 34.7 *In terms of variograms, in the barycentric model the normalized first order variogram relates to the normalized variogram in the following way,*

$$\frac{\gamma_1(\mathbf{h})}{C_1} \;=\; w\sqrt{\frac{\gamma(\mathbf{h})}{C}} + (1-w)\,\frac{\gamma(\mathbf{h})}{C} \qquad \textit{with} \quad w \in [0,1], \tag{34.38}$$

where C_1 is the sill of the first order variogram $\gamma_1(\mathbf{h})$ and C is the sill of the usual variogram $\gamma(\mathbf{h})$.

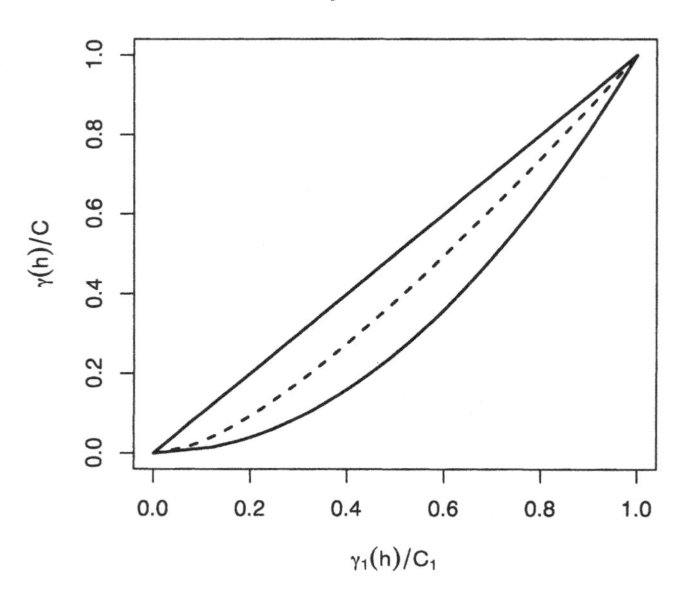

Figure 34.5: Barycentric model.

The Figure 34.5 plots the normalized first order variogram against the normalized second-order variogram for different different values of w. The case $w = 0$ generates a straight line corresponding to a plain mosaic type random function while the case $w = 1$ yields a parabola for a pure diffusion type random function. An intermediate case ($w = .5$) is shown using a dashed curve.

A family of isofactorial models for intermediate random function types, particularly suited for models with $r \in [0,1]$ like the gamma model, can be obtained taking a beta density function for the density $p(dr)$ in (34.31) as suggested by HU [142, 144],

$$p(dr) \;=\; \frac{\Gamma(\mu+\nu)}{\Gamma(\mu)\,\Gamma(\nu)}\, r^{\mu-1}\,(1-r)^{\nu-1} \qquad \text{with} \quad \mu>0, \nu>0. \tag{34.39}$$

By setting $\mu = \beta\,\rho(\mathbf{h})$ and $\nu = \beta\,(1-\rho(\mathbf{h}))$ with a positive parameter β, the following family of correlation functions is obtained

$$T_k(\mathbf{h}) \;=\; \frac{\Gamma(\beta)}{\Gamma(\beta+k)}\,\frac{\Gamma(\beta\,\rho(\mathbf{h})+k)}{\Gamma(\beta\,\rho(\mathbf{h}))} \qquad \text{with} \quad \beta>0. \tag{34.40}$$

COMMENT 34.8 *In terms of variograms, in this model, the first order variogram can be related to the variogram $\gamma(\mathbf{h}) = C(1-\rho(\mathbf{h}))$ by*

$$\frac{\gamma_1(\mathbf{h})}{C_1} \;=\; \frac{\Gamma(\beta)}{\Gamma(\beta+1/2)}\,\frac{\Gamma(\beta\,\gamma(\mathbf{h})/C\,+\,1/2)}{\Gamma(\beta\,\gamma(\mathbf{h})/C)} \qquad \textit{with} \quad \beta>0, \tag{34.41}$$

where C_1 *is the sill of the first order variogram* $\gamma_1(\mathbf{h})$.

An different form given by HU [142] *is,*

$$\frac{\gamma_1(\mathbf{h})}{C_1} = \frac{\Gamma(\beta+1/2)}{\Gamma(\beta)} \frac{\Gamma(\beta\, C/\gamma(\mathbf{h}))}{\Gamma(\beta\, C/\gamma(\mathbf{h}) + 1/2)} \quad \textit{with} \quad \beta > 0, \qquad (34.42)$$

which provides, when plotted, a sensibly differently shaped family of intermediate curves between the pure diffusion and the plain mosaic cases.

35 Isofactorial Change of Support

Considering the position of a sample as random within a block and making use of Cartier's relation, isofactorial change of support support models are gained in the form of discrete point-block models [205, 207]. A case study on a lead-silver deposit is provided in an exercise (with solution).

The point-block-panel problem

The point-block-panel problem arose in mining, but it may easily be encountered in other fields. In mining exploration samples of a few cm^3 are taken at hectometric spacing and analyzed. Production areas of are termed *panels* and are typically of hectometric size (say $100\times100\times5\text{m}^3$). The panels are subdivided into basic production units, the blocks (say $10\times10\times5\text{m}^3$). These blocks correspond in mining to the quantity of material an engine can carry and a decision will be taken whether the engine is directed to the plant for processing the material or else to the waste dump. During production several samples are taken in each block and only blocks with an average grade superior to a cut-off value z_c, above which they are profitable, will be kept.

The point-block-panel problem consists in anticipating before production, on the basis of the initial exploration data, what will be the proportion of profitable blocks within each panel, so a decision can be taken whether or not to start extraction of a given panel. Isofactorial change-of-support models and corresponding disjunctive kriging will make it possible to estimate different block selectivity statistics for individual panels.

The Figure 35.1 sketches a typical point-block-panel estimation problem. The three different supports involved are points $\mathbf{x}$ denoting the exploration samples, blocks v and panels V. In the model, the sample points $\mathbf{x}$ will be considered as a randomly located in the blocks.

Cartier's relation and point-block correlation

Suppose $\underline{\mathbf{x}}$ is a point located randomly inside a block v. Then the conditional expectation of the randomly located random function value knowing the block value is the block value,

$$\mathrm{E}[\,Z(\underline{\mathbf{x}})\mid Z(v)\,] \;=\; Z(v), \tag{35.1}$$

which is known as *Cartier's relation.*

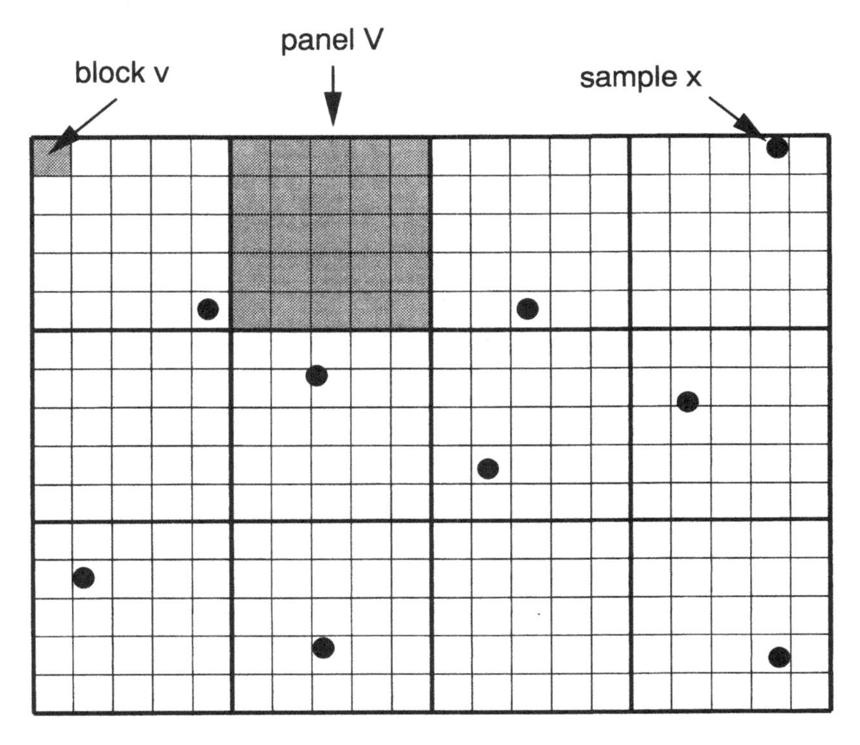

Figure 35.1: Panels V partitioned into blocks v with points $\mathbf{x}$ considered as randomly located within the blocks that contain them.

To understand this relation we rewrite the conditional expectation making the two sources of randomness more explicit,

$$\mathrm{E}[\, Z(\underline{\mathbf{x}}) \mid Z(V)\,] = \mathrm{E}_{\underline{\mathbf{x}}}[\, \mathrm{E}_Z[\, Z(\underline{\mathbf{x}}) \mid Z(v)]] \tag{35.2}$$

$$= \frac{1}{|v|} \int\limits_{\mathbf{x}\in v} \mathrm{E}_Z[\, Z(\mathbf{x}) \mid Z(v)]\, d\mathbf{x} \tag{35.3}$$

Now the relation is easily demonstrated:

$$\mathrm{E}[\, Z(\underline{\mathbf{x}}) \mid Z(v)\,] = \mathrm{E}_Z\left[\frac{1}{|v|}\int Z(\mathbf{x})\, d\mathbf{x} \mid Z(v)\right] \tag{35.4}$$

$$= \mathrm{E}[\, Z(v) \mid Z(v)\,] = Z(v). \tag{35.5}$$

For a point Gaussian anamorphosis $Z(\mathbf{x})= \varphi(Y(x))$ and a corresponding block anamorphosis $Z(v)= \varphi_v(Y(v))$ we have by Cartier's relation,

$$\mathrm{E}[\, \varphi(Y(\underline{\mathbf{x}})) \mid Z(v)\,] = \mathrm{E}[\, \varphi(Y(\underline{\mathbf{x}})) \mid \varphi_v(Y(v))\,] = \varphi_v(Y(v)). \tag{35.6}$$

Using the fact that $\mathrm{E}[\,\varphi(Y(\underline{\mathbf{x}}))\mid Z(v)\,]$ is equivalent to $\mathrm{E}[\,\varphi(Y(\underline{\mathbf{x}}))\mid Y(v)\,]$ and using (34.9) we have for the Hermitian isofactorial model

$$\varphi_v(Y(v)) = \sum_{k=0}^{\infty} \frac{\varphi_k}{k!}\,\mathrm{E}[\,H_k(Y(\underline{\mathbf{x}}))\mid Y_v\,] \qquad (35.7)$$

$$= \sum_{k=0}^{\infty} \frac{\varphi_k}{k!}\,\rho(\cdot,v)\,H_k(Y(v)), \qquad (35.8)$$

where $\rho(\cdot,v)$ is the correlation coefficient between $H_k(Y(\underline{\mathbf{x}}))$ and $H_k(Y(v)$.

In the particular case of a pure diffusion this becomes

$$\varphi_v(Y(v)) = \sum_{k=0}^{\infty} \frac{\varphi_k}{k!}\,r^k\,H_k(Y(v)), \qquad (35.9)$$

and r is called the *point-block correlation coefficient.*

The mean of the block variable $Z(v)$ is equal to the mean of the point variable,

$$\mathrm{E}[\,Z(v)\,] = \mathrm{E}[\,Z(x)\,] = \varphi_0 \qquad (35.10)$$

The variance of $Z(v)$ in terms of the block anamorphosis is

$$\mathrm{var}(Z(v)) = \mathrm{var}(\varphi_v(Y(v))) = \sum_{k=1}^{\infty} \frac{\varphi_k^2}{k!}\,r^{2k}, \qquad (35.11)$$

where the point-block correlation coefficient r is yet to be determined.

The variance of $Z(v)$ can be computed using the variogram,

$$\mathrm{var}(Z(v)) = \overline{C}(v,v) = \gamma(\infty) - \overline{\gamma}(v,v), \qquad (35.12)$$

and then the point-block correlation coefficient is easily computed by inverting relation (35.11).

EXERCISE 35.1 *The development of an exponential function into Hermite polynomials is:*

$$\exp(a\,y) = \exp\left(\frac{a^2}{2}\right)\sum_{k=0}^{\infty} \frac{(-a)^k}{k!}\,H_k(y). \qquad (35.13)$$

1. *Show that the coefficients φ_k of the anamorphosis function,*

$$Z = \varphi(Y) = \sum_{k=0}^{\infty} \frac{\varphi_k}{k!}\,H_k(Y), \qquad (35.14)$$

of a lognormal variable $Z = \exp(\nu + \sigma\,Y)$ with a mean $\mathrm{E}[\,Z\,] = m$ are:

$$\varphi_k = (-\sigma)^k\,m. \qquad (35.15)$$

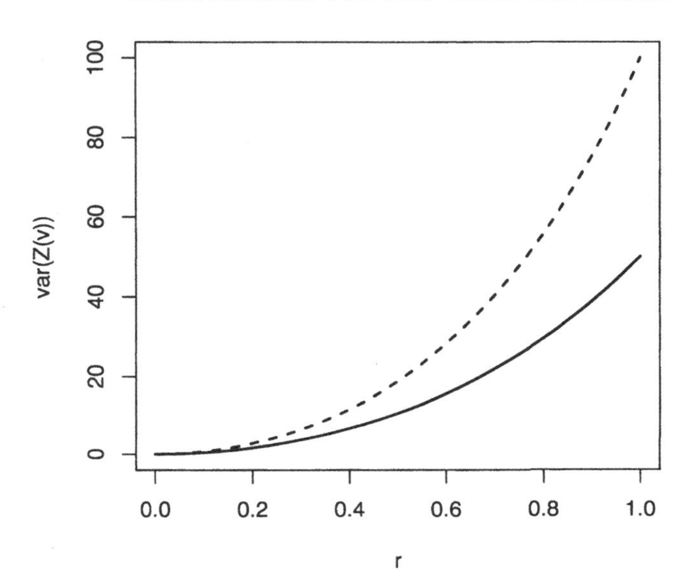

Figure 35.2: Relation between block variance $\text{var}(Z(v))$ and the point-block correlation coefficient r for two lognormal random functions with the same mean, point variances $\text{var}(Z(\mathbf{x}))$ of 50 (solid line) and of 100 (dashed line).

2. *A lognormal variable Z has a mean m= 10 and a variance $\text{var}(Z)$= 50. At which order K is the variance of Z reconstructed up to 99.99%?*

3. *Let $\varphi_v(Y)= m\,\exp(\sigma\, r\, Y + \sigma^2\, r^2/2)$ be the lognormal block anamorphosis under assumption of permanence of lognormality. Show that the coefficient r is identical with the correlation coefficient of the discrete Gaussian point-bloc model.*

EXAMPLE 35.2 *The plot of the relation between the block variance $\text{var}(Z(v))$ and the point-block correlation coefficient r is shown for two lognormal random functions with a mean m= 10 on Figure 35.2. One has a point variance $\text{var}(Z(\mathbf{x}))$= 50 and the relation (35.12) is used to display the block variances for different block sizes v with a solid line; the other has a point variance of $\text{var}(Z(\mathbf{x}))$= 100 and the relation (35.12) is displayed with a dashed line. In practice the block variance is known from the variogram and the coefficient r can be read from the graph.*

The Laguerre isofactorial point-block model involves a point gamma anamorphosis φ_λ and a block anamorphosis φ_μ,

$$Z(\mathbf{x}) \;=\; \varphi_\lambda(Y(\mathbf{x})) \qquad Z(v) \;=\; \varphi_\mu(Y(v)) \tag{35.16}$$

where $\mu \geq \lambda$ as the block distribution is less skew than the point distribution.

The isofactorial model for the generally asymmetric point-block density is

$$f_{\lambda\mu}(y_{\mathbf{x}}, y_v) = f_{\lambda}(y_{\mathbf{x}})\, f_{\mu}(y_v) \sum_{k=0}^{\infty} \rho_k(\mathbf{x}, v)\, C_k(\lambda, \mu)\, l_{\lambda k}(y_{\mathbf{x}})\, l_{\mu k}(y_v) \quad (35.17)$$

with

$$C_k(\lambda, \mu) \doteq \sqrt{\frac{\Gamma(\lambda + k)\, \Gamma(\mu)}{\Gamma(\mu + k)\, \Gamma(\lambda)}} \quad (35.18)$$

The block anamorphosis is linked to the point anamorphosis coefficients $\varphi_{\lambda k}$ by

$$Z(v) = \sum_{k=0}^{\infty} \varphi_{\lambda k}\, \rho_k(\cdot, v)\, C_k(\lambda, \mu)\, l_{\mu k}(y_v) \quad (35.19)$$

and the point-block correlation coefficients $\rho_k(\cdot, v)$ are determined through the relation

$$\operatorname{var}(Z(v)) = \sum_{k=1}^{\infty} \frac{\varphi_{\lambda k}^2}{k!}\, \rho_k^2(\cdot, v)\, C_k^2(\lambda, \mu). \quad (35.20)$$

Discrete Gaussian point-block model

In a point-block-panel problem we may have a vector of n samples $Y(\underline{\mathbf{x}})$ each randomly located inside one of N different blocks v. We are interested in a simple description of the correlation structure of the multi-Gaussian vector of the sample and the block values,

$$(\, Y(\underline{\mathbf{x}}_1), \ldots, Y(\underline{\mathbf{x}}_n), Y(v_1), \ldots, Y(v_N)\,). \quad (35.21)$$

We shall suppose that each sample $Y(\underline{\mathbf{x}})$ in a block v is conditionally independent on $Y(v)$ and the conditional distribution has a mean $r_{\underline{\mathbf{x}}v}\, y_v$ and a variance $1 - r_{\underline{\mathbf{x}}v}^2$. So for two samples with $\underline{\mathbf{x}}$ and $\underline{\mathbf{x}}'$ in distinct blocks we can write

$$Y(\underline{\mathbf{x}}) = r_{\underline{\mathbf{x}}v}\, Y(v) + \sqrt{1 - r_{\underline{\mathbf{x}}v}^2}\, U_1 \quad (35.22)$$

$$Y(\underline{\mathbf{x}}') = r_{\underline{\mathbf{x}}'v'}\, Y(v') + \sqrt{1 - r_{\underline{\mathbf{x}}'v'}^2}\, U_2, \quad (35.23)$$

where U_1 and U_2 are two uncorrelated Gaussian variables independent of $Y(v)$. The coefficients $r_{\underline{\mathbf{x}}v}$ and $r_{\underline{\mathbf{x}}'v}$ express the correlation of the two sample values with the block value $Y(v)$. The conditional independence of the samples entails that U_1 and U_2 are independent so that we get

$$\operatorname{cov}(Y(\mathbf{x}), Y(\mathbf{x}')) = r_{\underline{\mathbf{x}}\underline{\mathbf{x}}'} = r_{\underline{\mathbf{x}}v}\, r_{\underline{\mathbf{x}}'v'}\, r_{vv'} \quad (35.24)$$

We also have

$$r_{\underline{\mathbf{x}}v'} = r_{\underline{\mathbf{x}}v}\, r_{vv'} \quad (35.25)$$

so that

$$r_{\underline{\mathbf{x}}\,\underline{\mathbf{x}}'} \;=\; r_{\underline{\mathbf{x}}v}\, r_{\underline{\mathbf{x}}v'} = r_{\underline{\mathbf{x}}v}^2\, r_{vv'}\,. \tag{35.26}$$

The coefficient $r_{\underline{\mathbf{x}}v}$ is the point-block coefficient r computed from (35.11), so that

$$r_{\underline{\mathbf{x}}\,\underline{\mathbf{x}}'} \;=\; r^2\, r_{vv'} \tag{35.27}$$

$$r_{\underline{\mathbf{x}}v'} \;=\; r\, r_{vv'}\,. \tag{35.28}$$

For specifying the discrete Gaussian point-block model we still need to determine the block-block correlations $r_{vv'}$. This can be done using the relation

$$\overline{C}(v,v') \;=\; \sum_{k=1}^{\infty} \frac{\varphi^2}{k!}\left(r^2\, r_{vv'}\right)^k, \tag{35.29}$$

but by solving this equation for each pair of blocks separately the matrix is in general not positive definite; the way out is to fit a covariance function to the coefficients $r_{vv'}$.

General structure of isofactorial change-of-support

The structure of isofactorial change of support models has been set up by MATHERON [215, 216]. A generally asymmetric bivariate distribution $F(du,dv)$ with marginal distributions $F_1(du)$ and $F_2(dv)$ is isofactorial if there exist two corresponding orthonormal systems χ_k^i, $i=1,2$, complete in $\mathcal{L}^2(F_i)$,

$$\int \chi_k^i(u)\,\chi_l^i(u)\,F_i(du) \;=\; \delta_{kl}, \tag{35.30}$$

whose orthogonality extends to the bivariate distribution:

$$\int \chi_k^1(u)\,\chi_l^2(v)\,F(du,dv) \;=\; \delta_{kl}\,T_k. \tag{35.31}$$

In the context of a stationary random function $Z(\mathbf{x})=\varphi(Y(\mathbf{x}))$ with anamorphosed distribution $F(dy)$ and the block variable $Z_v(\mathbf{x})=\varphi_v(Y_v(\mathbf{x}))$ we have

$$\varphi(Y(\mathbf{x})) \;=\; \sum_{k=0}^{\infty} \varphi_k\,\chi_k(Y(\mathbf{x})) \tag{35.32}$$

$$\varphi_v(Y_v(\mathbf{x})) \;=\; \sum_{k=0}^{\infty} \varphi_k^v\,\chi_k^v(Y_v(\mathbf{x})) \tag{35.33}$$

The point-point and block-block distributions are

$$F(dy_{\mathbf{x}},dy_{\mathbf{x}'}) \;=\; F(dy_{\mathbf{x}})\,F(dy_{\mathbf{x}'})\sum_{k=0}^{\infty} T_k(\mathbf{x},\mathbf{x}')\,\chi_k(y_{\mathbf{x}})\,\chi_k(y_{\mathbf{x}'}) \tag{35.34}$$

$$F(dy_v,dy_{v'}) \;=\; F_v(dy_v)\,F_v(dy_{v'})\sum_{k=0}^{\infty} T_k(v,v')\,\chi_k^v(y_v)\,\chi_k^v(y_{v'}). \tag{35.35}$$

The point-block distribution between two locations is

$$F(dy_{\mathbf{x}}, dy_{v'}) = F(dy_{\mathbf{x}})\, F_v(dy_{v'}) \sum_{k=0}^{\infty} T_k(\mathbf{x}, v')\, \chi_k(y_{\mathbf{x}})\, \chi_k^v(y_{v'}). \tag{35.36}$$

The inference of the different T_k coefficients for a discrete point-block model is analogous to the discrete Gaussian point-block model. The example of the Laguerre isofactorial model has received much attention and is treated in detail in [142, 182, 51].

EXERCISE 35.3 *[Reserve estimation in a lead/silver deposit] In this 2D bivariate study we shall estimate reserves of lead (Pb) and silver (Ag). The exercise is due to Jacques* RIVOIRARD. *The parameters are about those of a real case study performed by Peter* DOWD *on a Pb/Zn/Ag ore body in Broken Hill, Australia (described in [156] on p256).*

The Z_1 variable (Pb) has a mean grade of 13 % with a variance of 50. The variogram consists of a nugget-effect and a spherical model with a range of 60m:

$$\gamma_1(\mathbf{h}) \quad = \quad 10 + 40\, sph(60) \qquad \textit{for} \quad |\mathbf{h}| \neq 0. \tag{35.37}$$

The Z_2 variable (Ag) has a mean grade of 4 (expressed in ounces/ton), a variance of 3 and the variogram model:

$$\gamma_2(\mathbf{h}) \quad = \quad 1.2 + 1.8\, sph(60) \qquad \textit{for} \quad |\mathbf{h}| \neq 0. \tag{35.38}$$

The cross variogram between lead and silver has no nugget-effect:

$$\gamma_{12}(\mathbf{h}) \quad = \quad 5\, sph(60). \tag{35.39}$$

RECOVERABLE RESERVES OF LEAD

The histogram of point values of lead has been modeled using a Gaussian anamorphosis truncated at the second order:

$$Z_1(\mathbf{x}) \quad = \quad \varphi(Y(\mathbf{x})) = \sum_{k=0}^{2} \frac{\varphi_k}{k!}\, H_k(Y(\mathbf{x})) \tag{35.40}$$

with coefficients $\varphi_0 = 13$, $\varphi_1 = -6.928$ and $\varphi_2 = 2$. Note that the Hermite polynomials are not normalized,

$$H_0(Y(\mathbf{x})) = 1; \qquad H_1(Y(\mathbf{x})) = -Y(\mathbf{x}); \qquad H_2(Y(\mathbf{x})) = (Y(\mathbf{x}))^2 - 1. \tag{35.41}$$

The selection mining units are blocks of size 24×24 m^2. The grades of these blocks are assumed to be perfectly known at the time of the selection (no information effect). The cut-off grade used for selection on grades $Z_1(v)$ is $z_{1c} = 15\%$.

Knowing the point variogram, the variance of the blocks has been computed as $\mathrm{var}(Z_1(v)) = \sigma_{1v}^2 = 28$.

1. *Make a graphical representation of the anamorphosis function. What is its domain of validity?*

2. *In the framework of the discrete Gaussian model the anamorphosis of the block values is*

$$Z_1(v) \;=\; \varphi_{r_1}(Y_v) = \sum_{k=0}^{2} \frac{\varphi_k}{k!}\, r_1^k\, H_k(Y_v) \tag{35.42}$$

Knowing that the variance of the blocks is

$$\sigma_{1v}^2 \;=\; \sum_{k=1}^{2} \frac{(\varphi_k)^2}{k!}\, r_1^{2k}, \tag{35.43}$$

compute the value of the change-of-support coefficient r_1.

[Hint: to solve this equation, first find a solution for $(r_1)^2$ *and then take its square root. This avoids manipulating a fourth order equation.]*

3. *Compute the cut-off* y_c *applied to* Y_v *which corresponds to the cut-off* $z_{1c} = 15\%$.

4. *Using a table of* $G(y) = \int_{-\infty}^{y} g(t)\, dt$ *compute the tonnage of recovered ore:*

$$T(z_{1c}) \;=\; \mathrm{E}[\, \mathbf{1}_{Y_v > y_c}\,] = P[Y_v > y_c] \tag{35.44}$$

5. *The quantity of lead recovered can be written:*

$$Q_1(z_{1c}) \;=\; \mathrm{E}[\, Z(v)\, \mathbf{1}_{Y_v > y_c}\,] = \int_{y_c}^{\infty} \varphi_{r_1}(Y_v)\, g(y)\, dy \tag{35.45}$$

$$\;=\; \sum_{k=0}^{2} \frac{\varphi_k}{k!}\, r_1^k \int_{y_c}^{\infty} H_k(y)\, g(y)\, dy \tag{35.46}$$

$$\;=\; \varphi_0 \int_{y_c}^{\infty} g(y)\, dy - \sum_{k=1}^{2} \frac{\varphi_k}{k!}\, r_1^k\, H_{k-1}(y_c)\, g(y_c) \tag{35.47}$$

Compute $Q_1(z_{1c})$ *using tables of* $G(y)$ *and* $g(y)$. *Calculate* $m_1(z_{1c})$ *and compare with overall mean of the deposit.*

ANAMORPHOSIS OF SILVER

The anamorphosis of the point grades of silver is written

$$Z_2(\mathbf{x}) \;=\; \psi(U(\mathbf{x})) = \sum_{k=0}^{2} \frac{\psi_k}{k!}\, H_k(U(\mathbf{x})) \tag{35.48}$$

with coefficients $\psi_0 = 4$, $\psi_1 = -1.71$ *and* $\psi_2 = 0.384$. *The Gaussian transform of the silver values is denoted* $U(\mathbf{x})$.

1. *Compute and plot the anamorphosis function. What is its domain of validity?*

2. *The variance of silver on block support is* $\mathrm{var}(Z_2(v)) = 1.26$. *Compute the value of the change-of-support coefficient* r_2.

   [Hint: first solve for $(r_2)^2$ *and then take the square root of the value.]*

RECOVERABLE RESERVES OF SILVER WHEN SELECTING ON LEAD

The blocks are selected on the basis of the lead values as data about silver grades is scarce. We shall therefore attempt to estimate the recovered quantity of silver for a selection performed on the basis of lead. The quantity to estimate is thus:

$$Q_2(z_{1c}) \quad = \quad \mathrm{E}[\, Z_2(v)\, \mathbf{1}_{Z_1(v)>z_{1c}}\,]. \tag{35.49}$$

The quantity of silver conditioning on the Gaussian transform of lead is,

$$Q_2(z_{1c}) \quad = \quad \mathrm{E}[\, Z_2(v)\, \mathbf{1}_{Y_v>y_c}\,] = \mathrm{E}[\, \mathrm{E}[\, Z_2(v)|Y_v\,]\, \mathbf{1}_{Y_v>y_c}\,]. \tag{35.50}$$

In the last expression we need to compute the conditional expectation of silver knowing lead:

$$\mathrm{E}[\, Z_2(v)|Y_v\,] \quad = \quad \mathrm{E}[\sum_{k=0}^{2} \frac{\psi_k}{k!}\, r_2^k H_k(U_v)|Y_v\,] \tag{35.51}$$

$$\quad = \quad \sum_{k=0}^{2} \frac{\psi_k}{k!}\, r_2^k\, \mathrm{E}[\, H_k(U_v)|Y_v\,] \tag{35.52}$$

$$\quad = \quad \sum_{k=0}^{2} \frac{\psi_k}{k!}\, r_2^k\, \rho^k\, H_k(Y_v) = \psi_{r_2\rho}(Y_v) \tag{35.53}$$

where the anamorphosis $\psi_{r_2\rho}(Y_v)$ *depends on both the silver change-of-support coefficient* r_2 *and the correlation coefficient* ρ *between the Gaussian variables* Y_v *and* U_v.

1. *Knowing that the value of the cross-covariance between* $Z_1(v)$ *and* $Z_2(v)$ *is*

$$\mathrm{cov}(Z_1(v), Z_2(v)) \quad = \quad 3.5 \tag{35.54}$$

   and using the relation

$$\mathrm{cov}(Z_1(v), Z_2(v)) \quad = \quad \sum_{k=1}^{2} \frac{\varphi_k\, \psi_k}{k!}\, r_1^k\, r_2^k\, \rho^k \tag{35.55}$$

   compute the value of the correlation coefficient ρ.

2. *Compute the recovered quantity of silver*

$$Q_2(z_{1c}) \quad = \quad \mathrm{E}[\, \psi_{r_2\rho}(Y_v)\, \mathbf{1}_{Y_v>y_c}\,] = \int_{y_c}^{\infty} \psi_{r_2\rho}(y)\, g(y)\, dy. \tag{35.56}$$

   Calculate $m_2(z_{1c})$.

LOCAL ESTIMATION OF LEAD

The aim of this study is to estimate recoverable lead reserves for panels V of size 72×72 m^2, that is:

- *the proportion P of blocks having a grade superior to $z_{1c} = 15\%$,*
- *the quantity Q of lead for these blocks.*

The method used will be a "uniform conditioning" (presented in detail in Section 36) by the grade $Z_1(V)$ of the panel. From the variogram of lead we know that the variance of the panels is $\mathrm{var}(Z_1(V)) = 12$.

1. *Assuming that the anamorphosis is valid at the level of the panels,*

$$Z_1(V) = \varphi_{r_1'}(Y_V) = \sum_{k=0}^{2} \frac{\varphi_k}{k!}\,(r_1')^k\,H_k(Y_V) \tag{35.57}$$

compute the value of the panel change-of-support coefficient r_1'.

2. *The estimation will be performed for a particular panel with a lead grade of $z_1(V) = 12.76\%$. What is the corresponding value y_V?*

3. *The notation $\underline{v}$ indicates a block located randomly in the panel V. Assuming the pair $(Y_{\underline{v}}, Y_V)$ to be bi-Gaussian, the correlation between $Y_{\underline{v}}$ and Y_V is then given by the correlation $R = r_1'/r_1$. What is the expression for the estimator of the proportion of blocks above cut-off within a panel with a given value $z_1(V)$:*

$$P^\star(z_{1c}) = \mathrm{E}[\,\mathbf{1}_{Z_1(\underline{v})>z_{1c}} | Z_1(V) = z_1(V)\,] \quad ? \tag{35.58}$$

4. *What is the value of $P^\star(z_{1c})$ for a panel value of $z_1(V) = 12.76\%$?*

5. *The quantity of lead within a block is written:*

$$Q_{\underline{v}}(z_{1c}) = \sum_{p=0}^{\infty} \frac{Q_p}{p!}\,H_p(Y_{\underline{v}}) \quad \text{with} \quad Q_p = \sum_{k=0}^{2} \frac{\varphi_k}{k!}\,r^k\,J_{pk}(y_c) \tag{35.59}$$

where

$$J_{pk}(y_c) = \int_{y_c}^{\infty} H_p(y)\,H_k(y)\,g(y)\,dy. \tag{35.60}$$

A table of precomputed values of Q_p coefficients will be given below.

Give the expression for $\mathrm{E}[\,H_p(Y_{\underline{v}})|Y_V\,]$ and the expression for the estimator of the quantity of metal,

$$Q_V^\star(z_{1c}) = \mathrm{E}[\,Q_{\underline{v}}(z_{1c})|Y_V\,]. \tag{35.61}$$

p	Q_p	$\frac{Q_p}{\sqrt{p!}}$	$\frac{Q_p}{\sqrt{p!}}\, R^p$
0	6.074	6.074	6.074
1	-7.385	-7.385	-4.852
2	4.887	3.456	1.492
3	2.828	1.155	0.327
4	-8.382	-1.711	-0.319
5	-6.281	-0.573	-0.070
6	35.275	1.315	0.106
7	18.079	0.255	0.014
8	-217.692	-1.084	-0.038
9	-37.782	-0.063	-0.001
10	1752.361	0.920	0.014

Table 35.1: List of coefficients Q_p.

6. *In numerical computations it can be preferable to use normalized Hermite polynomials,*

$$\eta_p(Y_V) \quad = \quad \frac{H_p(Y_V)}{\sqrt{p!}} \tag{35.62}$$

to avoid large numbers for the polynomials.

The expression (35.59) is then

$$Q_{\underline{v}}(z_{1c}) \quad = \quad \sum_{p=0}^{\infty} \frac{Q_p}{\sqrt{p!}}\, \eta_p(Y_{\underline{v}}) \tag{35.63}$$

Rewrite the expression obtained in the previous question in terms of normalized Hermite polynomials.

7. *On Table 35.1 we list the Q_p coefficients up to the order 10. The table also lists the coefficients normalized by $\sqrt{p!}$ and furthermore when multiplied by R^p. The last column of Table 35.1 shows that $Q_V^\star$ should not be computed for a too low order of p.*

   Compute $Q_V^\star(z_{1c})$ for a panel with grade $z_1(V) = 12.76\%$ truncating at the order $p = 6$. Calculate the mean grade of the selected blocs.

   [Hint: use the recurrence formula $H_{k+1}(y) = -y\, H_k(y) - k\, H_{k-1}(y)$.]

36 Kriging with Discrete Point-Bloc Models

Disjunctive kriging is the classical method to estimate non-linear functions of the data. Several geostatistical methods for local estimation of non linear functions of block variables are presented, including conditional expectation, disjunctive kriging and uniform conditioning.

Non-linear function of a block variable

In selection problems we are interested in estimating some non-linear function ψ of $Z(v)$ like for example the indicator function

$$\mathbf{1}_{Z(v)\geq z_c} \tag{36.1}$$

which tells whether or not a block variable $Z(v)$ is above some prescribed level z_c.

Having data about the point variable $Z(\mathbf{x})$ the steps are the following for setting up a discrete Gaussian point-bloc model:

1. compute the point anamophosis $Z(\mathbf{x})= \varphi(Y(\mathbf{x}))$,
2. knowing the variogram $\gamma(\mathbf{h})$ of $Z(\mathbf{x})$, compute the point-block coefficient r of the bloc anamorphosis $Z(v)= \varphi_v(Y(v))$ in (35.9), by first evaluating the variance of $Z(v)$ with (35.12) and then computing r using (35.11),
3. again with the variogram $\gamma(\mathbf{h})$ of $Z(\mathbf{x})$, compute the block covariances $\overline{C}(v,v')$ and then corresponding Gaussian block covariances $\rho(v,v')$ using the relation (35.29), taking care that the matrix of these covariances is positive definite.

The anamorphosis of a non-linear function ψ can be specified on the basis of the anamorphosis of $Z(v)$,

$$\psi(Z(v)) \quad = \quad \psi\circ\varphi_v(Y(v)) = \sum_{k=0}^{\infty} \frac{\psi_k}{k!}\, H_k(Y_v). \tag{36.2}$$

Thus any non-linear function of $Z(v)= \varphi_v(Y(v)$ can be represented as a linear function of the orthogonal polynomials $H_k(Y_v)$ just by using specific coefficients ψ_k. We now present different estimation methods of $\psi(Z(v))$ from data $Z(\mathbf{x}_\alpha)$ which build on this decomposition.

Conditional expectation and disjunctive kriging of a bloc

A first approach, which requires an assumption of multinormality, is to calculate the *conditional expectation* (CE) in the following steps:

1. compute $Y(\mathbf{x}_\alpha) = \varphi^{-1}(Z(\mathbf{x}_\alpha))$ and identify it with $Y(\underline{\mathbf{x}}_\alpha)$, as in the model the samples are considered as being randomly located within corresponding blocks v_α;

2. perform simple kriging of the block value

$$Y^{\mathrm{SK}}(v_0) \;=\; \sum_{\alpha=1}^{n} w_\alpha \, Y(\underline{\mathbf{x}}_\alpha) \tag{36.3}$$

   solving the system

$$\sum_{\beta=1}^{n} w_\beta \, \rho(\underline{\mathbf{x}}_\alpha, \underline{\mathbf{x}}_\beta) \;=\; \rho(\underline{\mathbf{x}}_\alpha, v_0) \qquad \text{for all } \alpha, \tag{36.4}$$

   where $\rho(\underline{\mathbf{x}}_\alpha, \underline{\mathbf{x}}_\beta)$, $\rho(\underline{\mathbf{x}}_\alpha, v_0)$ are obtained by the formulas (35.27) and (35.28).

   The simple block kriging variance is

$$\sigma^2_{\mathrm{SK}} \;=\; 1 - \sum_{\alpha=1}^{n} w_\alpha \, \rho(\underline{\mathbf{x}}_\alpha, v_0) = 1 - s^2, \tag{36.5}$$

   where s^2 represents the variance of $Y^{\mathrm{SK}}(v_0)$;

3. the estimate of the non-linear function ψ for a specific block v_0 is

$$[\psi(Z(v_0))]^{\star}_{\mathrm{CE}} \;=\; \sum_{k=0}^{K} \frac{\psi_k}{k!} \, s^k \, H_k\left(\frac{Y^{\mathrm{SK}}(v_0)}{s}\right). \tag{36.6}$$

A second approach for estimating $\psi(Z(v_0)$ is by *disjunctive kriging* (DK) which estimates directly the terms of the orthogonal decomposition and thus merely requires an assumption of binormality instead of multinormality. It is implemented in the following steps:

1. compute the first K polynomials $H_k(Y(\mathbf{x}_\alpha))$ considering again the location of each sample as random within the block v_α that contains it;

2. perform for $k = 1, \ldots, K$ the bloc simple kriging

$$H^{\star}_k(Y(v_0)) \;=\; \sum_{\alpha=1}^{n} w_{\alpha k} \, H_k(Y(\underline{\mathbf{x}}_\alpha)), \tag{36.7}$$

where the weights $w_{\alpha k}$ are solution of

$$\sum_{\beta=1}^{n} w_{\alpha k} \left(\rho(\underline{\mathbf{x}}_\alpha, \underline{\mathbf{x}}_\beta)\right)^k = \left(\rho(\underline{\mathbf{x}}_\alpha, v_0)\right)^k \qquad \text{for all } \alpha, \tag{36.8}$$

with the kriging variance

$$\sigma_k^2 = k! \left[(1 - \sum_{\alpha=1}^{n} w_{\alpha k} \left(\rho(\underline{\mathbf{x}}_\alpha, v_0)\right)^k \right] \tag{36.9}$$

and with correlations ρ computed from the discrete Gaussian point-bloc model;

3. the disjunctive kriging estimate then comes as

$$[\psi(Z(v_0))]^{\star}_{\mathrm{DK}} = \psi_0 + \sum_{k=1}^{K} \frac{\psi_k}{k!} H_k^{\star}(Y(v_0)) \tag{36.10}$$

and the corresponding disjunctive kriging variance is

$$\sigma^2_{\mathrm{DK}} = \sum_{k=1}^{K} \left(\frac{\psi_k}{k!}\right)^2 \sigma_k^2. \tag{36.11}$$

Disjunctive kriging of a panel

We come back to the point-block-panel problem exposed in Section 35 (Figure 35.1, p263). We now consider the problem of predicting the average of a non-linear function of block values for a given panel V_0 as shown on Figure 36.1. The panel V_0 is partitioned into N blocks v, each of the samples being located at random in a different block (inside or outside the panel) and we denote $M(V_0)$ the average of a linear function over the panel,

$$M(V_0) = \frac{1}{N} \sum_{u=1}^{N} \psi(Z(v_u)) \tag{36.12}$$

$$= \sum_{k=0}^{\infty} \frac{\psi_k}{k!} \underbrace{\frac{1}{N} \sum_{u=1}^{N} H_k(Y(v_u))}_{M_k}. \tag{36.13}$$

The estimator for the panel is simply the disjunctive kriging,

$$M^{\star}_{\mathrm{DK}}(V_0) = \psi_0 + \sum_{k=1}^{K} \frac{\psi_k}{k!} M_k^{\star} \tag{36.14}$$

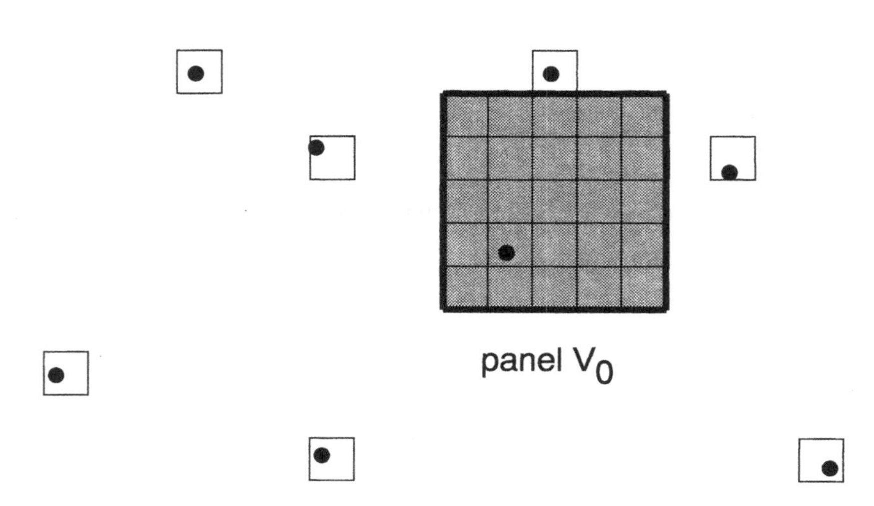

Figure 36.1: Panel V_0 partitioned into blocks v and samples points $\mathbf{x}$ considered as randomly located within blocks v that contain them.

with

$$M_k^\star = \sum_{\alpha=1}^{n} w_{\alpha k}\, H_k(Y(\underline{\mathbf{x}}_\alpha)), \tag{36.15}$$

where the kriging weights are obtained from

$$\sum_{\beta=1}^{n} w_{\alpha k}\, (\rho(\underline{\mathbf{x}}_\alpha, \underline{\mathbf{x}}_\beta))^k = \frac{1}{N}\sum_{u=1}^{N}(\rho(\underline{\mathbf{x}}_\alpha, v_u))^k \qquad \text{for all } \alpha, \tag{36.16}$$

with the kriging variance

$$\sigma_k^2 = \frac{k!}{N^2}\sum_{u=1}^{N}\sum_{u'=1}^{N}(\rho(v_u, v_{u'}))^k - \frac{k!}{N}\sum_{u=1}^{N}\sum_{\alpha=1}^{n} w_{\alpha k}\, (\rho(\underline{\mathbf{x}}_\alpha, v_u))^k \tag{36.17}$$

and the disjunctive kriging variance is again

$$\sigma_{\mathrm{DK}}^2 = \sum_{k=1}^{K}\left(\frac{\psi_k}{k!}\right)^2 \sigma_k^2. \tag{36.18}$$

Uniform conditioning

In some point-block-panel problems panel averages $Z(V)$ are available. For example a general circulation model (GCM) provides values for panels of 100×100 km^2 size while a station network provides point samples $\mathbf{x}_\alpha$. We may be interested in a technique that provides estimates of statistics for smaller spatial units obtained by partitioning the panels, say blocks of size 5×5 km^2. This is called a *downscaling* problem in statistical climatology [335].

The geostatistical analysis of the point data $Z(\mathbf{x})$ provides a variogram $\gamma(\mathbf{h})$, point anamorphosis coefficients φ_k and a point-block coefficient r using the relation (35.11). In the same way a point-panel correlation coefficient r',

$$Z(V) = \sum_{k=0}^{\infty} \frac{\varphi_k}{k!}\,(r')^k\, H_k(Y(V)), \tag{36.19}$$

is computed from

$$\operatorname{var}(Z(V)) = \operatorname{var}(\varphi_V(Y(V))) = \sum_{k=1}^{\infty} \frac{\varphi_k^2}{k!}\,(r')^{2k}. \tag{36.20}$$

and a block-panel coefficient is then obtained as $r_{vV} = r'/r$.

Considering a block $\underline{v}$ randomly located in V and writing,

$$Z(\underline{v}) = \sum_{k=0}^{\infty} \frac{\varphi_k}{k!}\, r^k\, H_k(Y(\underline{v})), \tag{36.21}$$

we have

$$\mathrm{E}[\,Z(\underline{v}) \mid Z(V)\,] = \sum_{k=0}^{\infty} \frac{\varphi_k}{k!}\, r^k\, \mathrm{E}[\,H_k(Y(\underline{v})) \mid Y(V)\,] \tag{36.22}$$

$$= \sum_{k=0}^{\infty} \frac{\varphi_k}{k!}\, r^k\, (r_{vV})^k\, H_k(Y(V)) \tag{36.23}$$

The *uniform conditioning* technique [205, 270] consists in taking the conditional expectation of some non-linear function of the blocks with respect to known panel values. For example, the proportion of blocks within a panel V_0 that are above a certain rainfall level z_c may be of interest: this is calculated by uniform conditioning on the GCM value $Z(V)$

$$\mathrm{E}[\,\mathbf{1}_{Z(\underline{v})\geq z_c} \mid Z(V_0)\,] = \mathrm{E}[\,\mathbf{1}_{Y(\underline{v})\geq z_c} \mid Y(V_0)\,] \tag{36.24}$$

$$= 1 - G\left(\frac{y_c - r_{vV}\, Y(V_0)}{\sqrt{1 - r_{vV}^2}}\right) \tag{36.25}$$

assuming that the conditional variable $Y(\underline{v}) \mid Y(V_0)$ is normally distributed with mean $r_{vV}\, Y(V_0)$ and variance $1 - r_{vV}^2$.

In many applications the panel value $Z(V)$ is not available and the trick is to generate it by ordinary kriging. This has been viewed as a way of introducing a large scale non-stationarity in the model and has been applied in mining reserves assessment in this manner [258, 188, 259].

EXERCISE 36.1 *[Iron ore deposit] This is a continuation of Exercise 32.4 on p230 in which the conditional expectation method was applied in a lognormal framework. The discrete Gaussian model for three samples and blocs displayed on Figure 32.2 was discussed there. Now we apply disjunctive kriging.*

Normalized Hermite polynomials are defined as

$$\eta_k(y) \quad = \quad \frac{H_k(y)}{\sqrt{k}} \tag{36.26}$$

with $\eta_1(y) = -y$ *and* $\eta_2(y) = (y^2-1)/\sqrt{2}$.

The development of the indicator function of a block Gaussian variable can be written:

$$\mathbf{1}_{Y(v_3)\geq y_c} \quad = \quad \sum_{k=0}^{\infty} \psi_k\, \eta_k(Y(v_3)) \tag{36.27}$$

$$\quad = \quad 1 - G(y_c) - \sum_{k=1}^{\infty} \frac{\eta_{k-1}(y_c)\, g(y_c)}{\sqrt{k}}\, \eta_k(Y(v_3)). \tag{36.28}$$

1. *What are the values for the coefficients* ψ_0, ψ_1 *and* ψ_2 *in this development?*

2. *What are the values of the Hermite polynomials* $\eta_1(Y_\alpha)$ *and* $\eta_2(Y_\alpha)$ *for the three sample locations?*

3. *What is the value of* $\Big[\eta_1(Y(v_3))\Big]^{\star}_{\text{SK}}$*?*

4. *Express the estimator of the second Hermite polynomial* $\Big[\eta_2(Y(v_3))\Big]^{\star}_{\text{SK}}$ *as a function of the three data values* $\eta_2(Y_\alpha)$ *and of the mean* $\mathrm{E}[\,\eta_2(Y(v_3))\,]$. *Develop its kriging system and solve it. What is the value of* $\Big[\eta_2(Y(v_3))\Big]^{\star}_{\text{SK}}$*?*

5. *Supposing that the terms* $\Big[\eta_k(Y(v_3))\Big]^{\star}_{\text{SK}} \cong 0$ *for* $k>2$ *and can be neglected, what is the value of the disjunctive kriging estimate* $\Big[\mathbf{1}_{Z(v_3)\geq z_c}\Big]^{\star}_{\text{DK}}$*?*

EXERCISE 36.2 *[Global Miner and Smart Invest] The company* Global Miner *has performed an exploration campaign of a deposit. The production will be based on blocs* v *with a surface of* 12×12 m^2 *which will serve as a selection support. The evaluation of the deposit was done with a discrete Gaussian point-block model and 2D disjunctive kriging.*

The point anamorphosis of the deposit is:

$$Z(\mathbf{x}) = \varphi(Y(\mathbf{x})) = \sum_{k=0}^{2} \frac{\varphi_k}{k!} H_k(Y(\mathbf{x})) \tag{36.29}$$

$$= 10 - 6.368\, H_1(Y(\mathbf{x})) + \frac{4.055}{2} H_2(Y(\mathbf{x})) \tag{36.30}$$

where $H_1(Y(\mathbf{x}))= -Y(\mathbf{x})$ *et* $H_2(Y(\mathbf{x}))= Y^2(\mathbf{x})-1$ *are Hermite polynomials.*

- *What are the value of the mean and of the variance of* $Z(\mathbf{x})$*?*

The fitted variogram $\gamma(\mathbf{h})$ *was used to compute* $\overline{\gamma}(v,v) = 22$. *In a discrete Gaussian point-bloc model the block anamorphosis is:*

$$Z_v(\mathbf{x}) = \varphi_v(Y_v(\mathbf{x})) = \sum_{k=0}^{K} \frac{\varphi_k}{k!} (r)^k H_k(Y_v(\mathbf{x})) \tag{36.31}$$

when stopping the development at an order $K = 2$.

- *Determine the value of the point-bloc coefficient* r.
- *What is the value* y_c *corresponding to a cut-off grade* z_c*=11 used for selecting the blocks?*

During a presentation of the reserves of the deposit by Global Miner *for potential investors a consultant of* Smart Invest *took note of the following three parameters:* $m = 10$, $\mathrm{var}(Z) = 50$, *as well as the value of the point-block correlation* r. *The consultant wants to get quickly a rough idea of global reserves and uses a lognormal model with change-of-support under assumption of permanence of lognormality using these parameters.*

Global Miner *had announced a proportion of 42% of the deposit that could go into production:*

- *is this figure compatible with the quick assessment of global reserves by* Smart Invest *'s consultant?*

Part F

Non-Stationary Geostatistics

37 External Drift

We start the study of non-stationary methods with a multivariate method that is applicable to auxiliary variables that are densily sampled over the whole domain and linearly related to the principal variable. Such auxiliary variables can be incorporated into a kriging system as *external drift* functions.

Two applications are discussed: one is about kriging temperature using elevation as external drift, while the other is a continuation of the Ebro case study from Chapter 27 using the ouput of a numerical model as external drift.

Depth measured with drillholes and seismic

It may happen that two variables measured in different ways reflect the same phenomenon and that the primary variable is precise, but known only at few locations, while the secondary variable cannot be accurately measured, but is available everywhere in the spatial domain.

The classical example is found in petroleum exploration, where the top of a reservoir has to be mapped. The top is typically delimited by a smooth geologic layer because petroleum is usually found in sedimentary formations. Data is available from two sources

- the precise measurements of depth stemming from drillholes. They are not a great many because of the excessively high cost of drilling. We choose to model the drillholes with a second-order stationary random function $Z(\mathbf{x})$ with a known covariance function $C(\mathbf{h})$.

- inaccurate measurements of depth, deduced from seismic travel times and covering the whole domain at a small scale. The seismic depth data provides a smooth image of the shape of the layer with some inaccuracy due to the difficulty of converting seismic reflection times into depths. This second variable is represented as a regionalized variable $s(\mathbf{x})$ and is considered as deterministic.

As $Z(\mathbf{x})$ and $s(\mathbf{x})$ are two ways of expressing the phenomenon "depth of the layer" we assume that $Z(\mathbf{x})$ is on average equal to $s(\mathbf{x})$ up to a constant a_0 and a coefficient b_1,

$$\mathrm{E}[\,Z(\mathbf{x})\,] \quad = \quad a_0 + b_1\, s(\mathbf{x}). \tag{37.1}$$

The deterministic function $s(\mathbf{x})$ describes the overall shape of the layer in inaccurate depth units while the data about $Z(\mathbf{x})$ give at a few locations an information about the exact depth. The method merging both sources of information uses $s(\mathbf{x})$ as an external drift function for the estimation of $Z(\mathbf{x})$.

Estimating with a shape function

We consider the problem of improving the estimation of a second order stationary random function $Z(\mathbf{x})$ by taking into account a function $s(\mathbf{x})$ describing its average shape. The estimator is the linear combination

$$Z^{\star}(\mathbf{x}_0) = \sum_{\alpha=1}^{n} w_\alpha Z(\mathbf{x}_\alpha) \tag{37.2}$$

with weights constrained to unit sum, so that

$$\mathrm{E}[\, Z^{\star}(\mathbf{x}_0)\,] = \mathrm{E}[\, Z(\mathbf{x}_0\,], \tag{37.3}$$

which can be developed into

$$\begin{aligned} \mathrm{E}[\, Z^{\star}(\mathbf{x}_0)\,] &= \sum_{\alpha=1}^{n} w_\alpha \, \mathrm{E}[\, Z(\mathbf{x}_\alpha)\,] \\ &= a_0 + b_1 \sum_{\alpha=1}^{n} w_\alpha \, s(\mathbf{x}_\alpha) \\ &= a_0 + b_1 \, s(\mathbf{x}_0). \end{aligned} \tag{37.4}$$

This last equation implies that the weights should be consistent with an exact interpolation of $s(\mathbf{x})$

$$s(\mathbf{x}_0) = \sum_{\alpha=1}^{n} w_\alpha \, s(\mathbf{x}_\alpha). \tag{37.5}$$

The objective function to minimize in this problem consists of the estimation variance σ_{E}^2 and of two constraints

$$\phi = \sigma_{\mathrm{E}}^2 - \mu_0 \Big(\sum_{\alpha=1}^{n} w_\alpha - 1 \Big) - \mu_1 \Big(\sum_{\alpha=1}^{n} w_\alpha \, s(\mathbf{x}_\alpha) - s(\mathbf{x}_0) \Big). \tag{37.6}$$

The result is the kriging system

$$\begin{cases} \displaystyle\sum_{\beta=1}^{n} w_\beta \, C(\mathbf{x}_\alpha - \mathbf{x}_\beta) - \mu_0 - \mu_1 \, s(\mathbf{x}_\alpha) = C(\mathbf{x}_\alpha - \mathbf{x}_0) \qquad \text{for } \alpha = 1, \dots, n \\ \displaystyle\sum_{\beta=1}^{n} w_\beta = 1 \\ \displaystyle\sum_{\beta=1}^{n} w_\beta \, s(\mathbf{x}_\beta) = s(\mathbf{x}_0). \end{cases} \tag{37.7}$$

The mixing of a second-order stationary random function with a (non stationary) mean function looks surprising in this example. There may be situations where this makes sense. Stationarity is a concept that depends of scale: the data can suggest stationarity at a large scale (for widely spaced data points on $Z(\mathbf{x})$) while at a smaller scale things look non-stationary (when inspecting the fine detail provided by a function $s(\mathbf{x})$).

Estimating external drift coefficients

The external drift method consists in integrating into the kriging system supplementary universality conditions about one or several external drift variables $s_i(\mathbf{x})$, $i = 1, \ldots, N$ measured exhaustively in the spatial domain. Actually the functions $s_i(\mathbf{x})$ need to be known at all locations $\mathbf{x}_\alpha$ of the samples as well as at the nodes of the estimation grid.

The conditions

$$\sum_{\alpha=1}^{n} w_\alpha\, s_i(\mathbf{x}_\alpha) \;=\; s_i(\mathbf{x}_0) \qquad i = 1, \ldots, N \tag{37.8}$$

are added to the kriging system independently of the inference of the covariance function, hence the qualificative *external*.

The kriging system with multiple external drift is

$$\begin{cases} \displaystyle\sum_{\beta=1}^{n} w_\beta\, C(\mathbf{x}_\alpha - \mathbf{x}_\beta) - \mu_0 - \sum_{i=1}^{N} \mu_i\, s_i(\mathbf{x}_\alpha) \;=\; C(\mathbf{x}_\alpha - \mathbf{x}_0) & \\ \qquad\qquad\qquad\qquad\qquad\qquad \text{for } \alpha = 1, \ldots, n & \\ \displaystyle\sum_{\beta=1}^{n} w_\beta \;=\; 1 & \\ \displaystyle\sum_{\beta=1}^{n} w_\beta\, s_i(\mathbf{x}_\beta) \;=\; s_i(\mathbf{x}_0) & \text{for } i = 1, \ldots, N. \end{cases} \tag{37.9}$$

When applying the method with a moving neighborhood it is interesting to map the coefficients b_i of each external drift to measure the influence of each external variable on the principal variable in different areas of the region. An estimate $b^\star_{i_0}$ of a particular coefficient indexed i_0 is obtained by modifying the right hand side of the kriging system in the following way

$$\begin{cases} \displaystyle\sum_{\beta=1}^{n} w_\beta\, C(\mathbf{x}_\alpha - \mathbf{x}_\beta) - \mu_0 - \sum_{i=1}^{N} \mu_i\, s_i(\mathbf{x}_\alpha) = \boxed{0} & \text{for } \alpha = 1, \ldots, n \\ \displaystyle\sum_{\beta=1}^{n} w_\beta = \boxed{0} & \\ \displaystyle\sum_{\beta=1}^{n} w_\beta\, s_i(\mathbf{x}_\beta) = \boxed{\delta_{i i_0}} & \text{for } i = 1, \ldots, N, \end{cases} \tag{37.10}$$

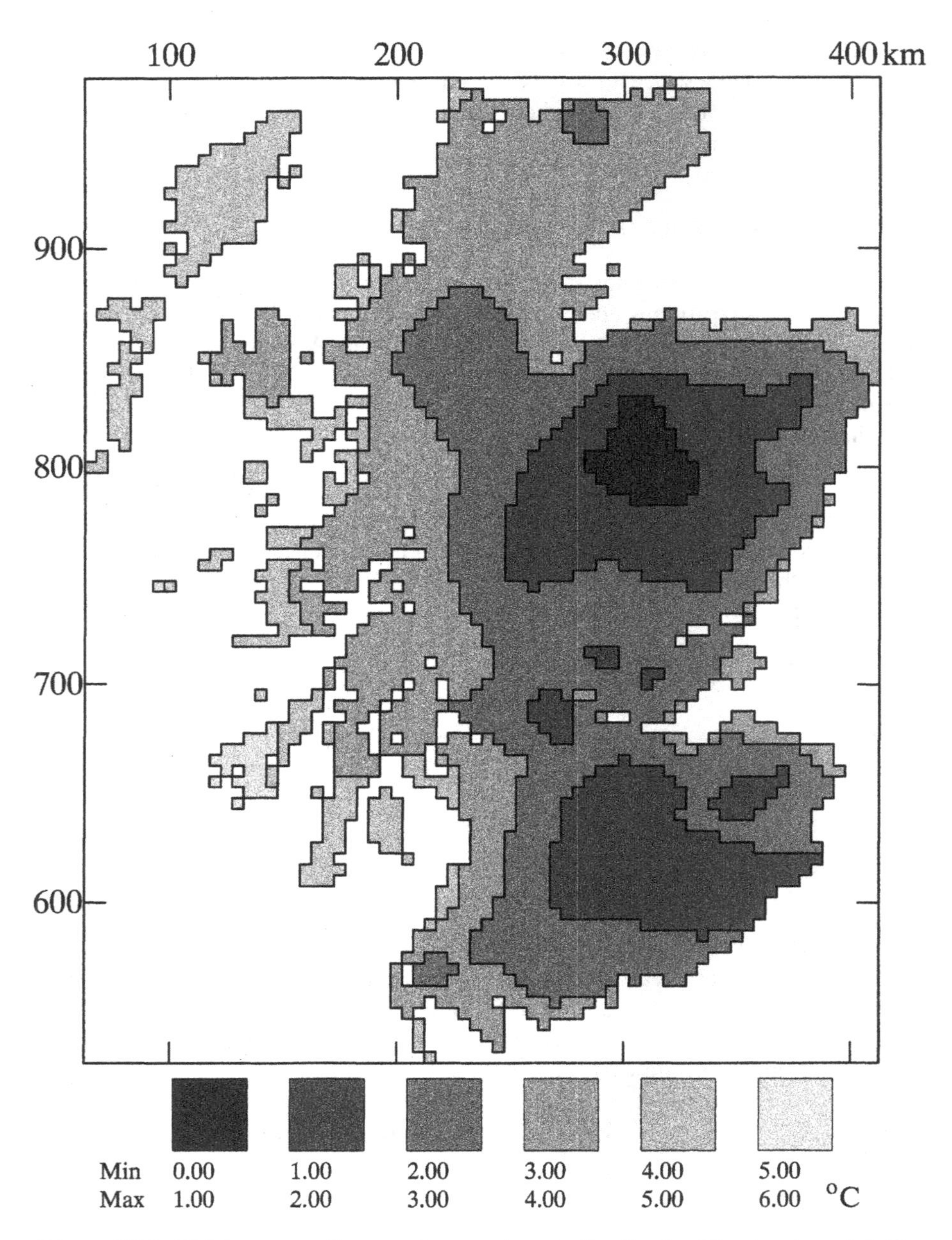

Figure 37.1: Kriging mean January temperature in Scotland without external drift.

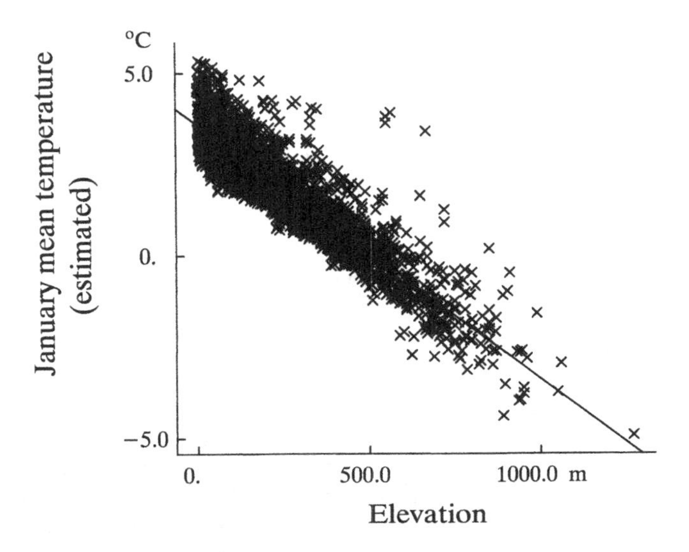

Figure 37.2: Estimated values of kriging with external drift plotted against elevation: the estimated values above 400m extrapolate out of the range of the data.

where δ_{ii_0} is a Kronecker symbol which is one for $i = i_0$ and zero otherwise.

EXAMPLE 37.1 *(see [146] for the full details of this case study) We have data on mean January temperature at 146 stations in Scotland which are all located below 400m altitude, whereas the Scottish mountain ranges rise up to 1344m (Ben Nevis). The variogram of this data will be discussed later (Figure 38.2, p306) .*

Figure 37.1 displays a map obtained by kriging without the external drift. The estimated temperatures stay within the range of the data (0 to 5 degrees Celsius). Temperatures below the ones observed can however be expected at altitudes above 400m.

It could be verified on scatter plots that temperature depends on elevation in a reasonably linear way. The altitude of the stations is known and elevation data, digitized from a topographic map, is available at the nodes of the estimation grid. The map of mean January temperature on Figure 37.3 was obtained by incorporating into the kriging system the elevation data as an external drift $s(\mathbf{x})$. *The kriging was performed with a neighborhood of 20 stations and the estimated values now extrapolate outside the range of the data at higher altitudes as seen on the diagram of Figure 37.2.*

The map of the estimated coefficient $b_1^\star$ *of the external drift is displayed on Figure 37.4. The coefficient is more important in absolute value in the west, in areas where stations are scarcer: the kriging estimator relies more heavily on the secondary variable when few data on the primary variable are available in the vicinity.*

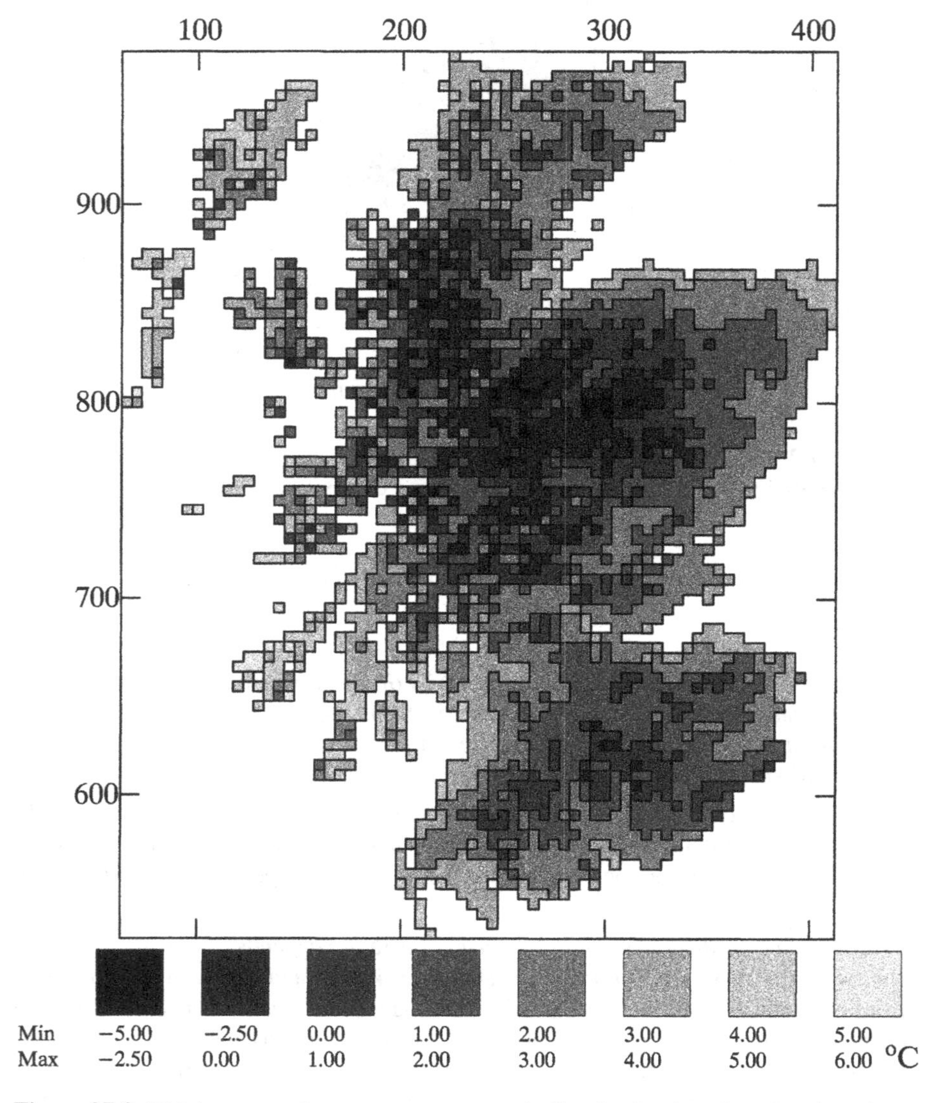

Figure 37.3: Kriging mean January temperature in Scotland using elevation data from a digitized topographic map as external drift.

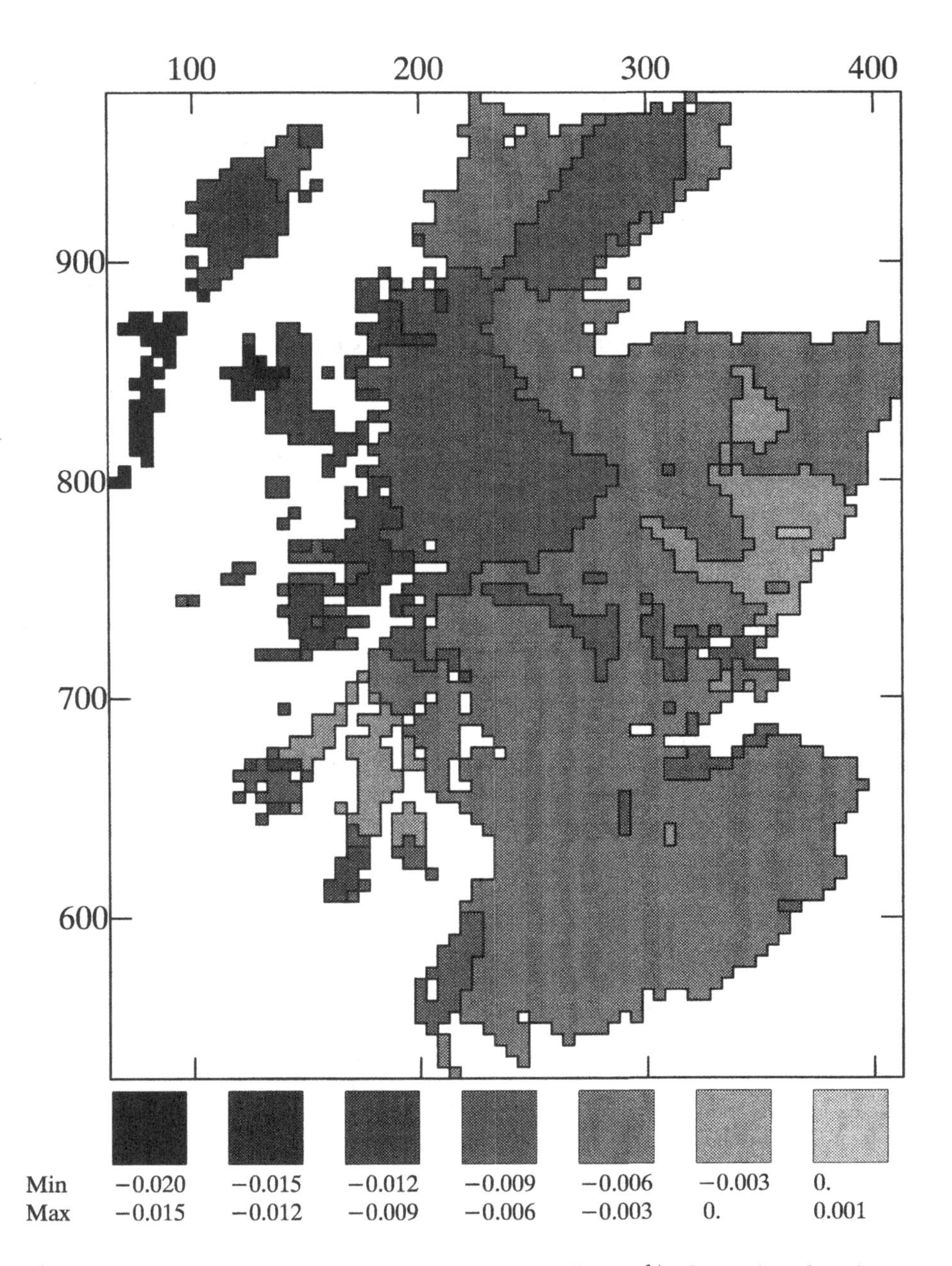

Figure 37.4: Map of the estimated external drift coefficient $b_1^{\star}$ when using elevation as the external drift for kriging temperature with a moving neighborhood.

Cross validation with external drift

The subject of cross validation with the external drift method was examined in detail by CASTELIER [37, 38] and is presented in the form of three exercises. The random function $Z(\mathbf{x})$ is second order stationary and only one drift function $s(\mathbf{x})$ is taken into account.

EXERCISE 37.2 *We are interested in pairs of values (z_α, s_α) of a variable of interest $Z(\mathbf{x})$ with a covariance function $C(\mathbf{h})$ and an external drift variable $s(\mathbf{x})$ at locations $\mathbf{x}_\alpha$, $\alpha = 1, \ldots, n$ in a domain $\mathcal{D}$. The values are arranged into vectors* $\mathbf{z}$ *and* $\mathbf{s}$ *of dimension n.*

To clarify notation we write down the systems of simple kriging, ordinary kriging, kriging with external drift and kriging of the mean.

Simple Kriging (SK):

$$\mathbf{C}\,\mathbf{w}_{\mathrm{SK}} = \mathbf{c}_0 \quad \textit{and} \quad \mathbf{w}_{\mathrm{SK}} = \mathbf{R}\,\mathbf{c}_0 \qquad \textit{with} \quad \mathbf{R} = \mathbf{C}^{-1}. \tag{37.11}$$

Ordinary Kriging (OK):

$$\begin{pmatrix} \mathbf{C} & \mathbf{1} \\ \mathbf{1}^\top & 0 \end{pmatrix} \begin{pmatrix} \mathbf{w}_{\mathrm{OK}} \\ -\mu_{\mathrm{OK}} \end{pmatrix} = \begin{pmatrix} \mathbf{c}_0 \\ 1 \end{pmatrix} \quad \textit{i.e.}\ \mathbf{K} \begin{pmatrix} \mathbf{w}_{\mathrm{OK}} \\ -\mu_{\mathrm{OK}} \end{pmatrix} = \mathbf{k}_0, \tag{37.12}$$

and

$$\begin{pmatrix} \mathbf{w}_{\mathrm{OK}} \\ -\mu_{\mathrm{OK}} \end{pmatrix} = \mathbf{K}^{-1}\,\mathbf{k}_0 \qquad \textit{with} \quad \mathbf{K}^{-1} = \begin{pmatrix} \mathbf{A} & \mathbf{v} \\ \mathbf{v}^\top & u \end{pmatrix}. \tag{37.13}$$

Kriging with external drift (KE):

$$\begin{pmatrix} \mathbf{C} & \mathbf{1} & \mathbf{s} \\ \mathbf{1}^\top & 0 & 0 \\ \mathbf{s}^\top & 0 & 0 \end{pmatrix} \begin{pmatrix} \mathbf{w}_{\mathrm{KE}} \\ -\mu_{\mathrm{KE}} \\ -\mu'_{\mathrm{KE}} \end{pmatrix} = \begin{pmatrix} \mathbf{c}_0 \\ 1 \\ s_0 \end{pmatrix} \quad \textit{i.e.}\ \mathbf{F} \begin{pmatrix} \mathbf{w}_{\mathrm{KE}} \\ -\mu_{\mathrm{KE}} \\ -\mu'_{\mathrm{KE}} \end{pmatrix} = \mathbf{f}_0. \tag{37.14}$$

Kriging of the mean (KM):

$$\mathbf{K} \begin{pmatrix} \mathbf{w}_{\mathrm{KE}} \\ -\mu_{\mathrm{KE}} \end{pmatrix} = \begin{pmatrix} \mathbf{o} \\ 1 \end{pmatrix}, \tag{37.15}$$

and

$$m^\star = \sum_{\alpha=1}^{n} w_\alpha^{\mathrm{SK}}\, z_\alpha = (\mathbf{z}^\top, 0)\, \mathbf{K}^{-1} \begin{pmatrix} \mathbf{o} \\ 1 \end{pmatrix}. \tag{37.16}$$

a) Show that the variance of ordinary kriging is

$$\sigma^2_{\mathrm{OK}} = c_{00} - \mathbf{k}_0^\top\, \mathbf{K}^{-1}\, \mathbf{k}_0. \tag{37.17}$$

b) Using the following relations between matrices and vectors of OK,

$$\begin{array}{llll} \mathbf{C}\,\mathbf{A} & + & \mathbf{1}\,\mathbf{v}^{\top} & = \mathbf{I}, \\ \mathbf{C}\mathbf{v} & + & \mathbf{1}\,u & = \mathbf{0}, \\ \mathbf{1}^{\top}\mathbf{A} & & & = \mathbf{0}^{\top}, \\ \mathbf{1}^{\top}\mathbf{v} & & & = 1, \end{array} \tag{37.18--37.21}$$

show that the kriging of the mean can be expressed as

$$m^{\star} = \frac{\mathbf{z}^{\top}\mathbf{R}\,\mathbf{1}}{\mathbf{1}^{\top}\mathbf{R}\,\mathbf{1}}. \tag{37.22}$$

c) We define Castelier's scalar product

$$<\mathbf{x},\mathbf{y}> = \frac{\mathbf{x}^{\top}\mathbf{R}\,\mathbf{y}}{\mathbf{1}^{\top}\mathbf{R}\,\mathbf{1}} \tag{37.23}$$

and a corresponding mean and covariance

$$\mathrm{E}_n[\mathbf{z}] = <\mathbf{z},\mathbf{1}>, \tag{37.24}$$

$$\mathrm{E}_n[\mathbf{z},\mathbf{s}] = <\mathbf{z},\mathbf{s}>, \tag{37.25}$$

$$\mathrm{cov}_n(\mathbf{z},\mathbf{s}) = \mathrm{E}_n[\mathbf{z},\mathbf{s}] - \mathrm{E}_n[\mathbf{z}]\,\mathrm{E}_n[\mathbf{s}]. \tag{37.26}$$

Defining Castelier's pseudo-scalar product using the inverse $\mathbf{K}^{-1}$ of OK (instead of the inverse $\mathbf{R}$ of SK),

$$\mathcal{K}_n(\mathbf{z},\mathbf{s}) = (\mathbf{z}^{\top},0)\,\mathbf{K}^{-1}\begin{pmatrix}\mathbf{s}\\0\end{pmatrix}, \tag{37.27}$$

show that

$$\mathcal{K}_n(\mathbf{z},\mathbf{s}) = \mathbf{z}^{\top}\mathbf{A}\,\mathbf{s} = \mathbf{1}^{\top}\mathbf{R}\,\mathbf{1}\,\mathrm{cov}_n(\mathbf{z},\mathbf{s}). \tag{37.28}$$

d) As the external drift is linked to the expectation of the variable of interest by a linear relation,

$$\mathrm{E}[Z(\mathbf{x})] = a + b\,s(\mathbf{x}), \tag{37.29}$$

and as the coefficient b and the constant a of the external drift are estimated by

$$b^{\star} = (\mathbf{z}^{\top},0,0)\,\mathbf{F}^{-1}\begin{pmatrix}0\\0\\1\end{pmatrix} \quad \text{and} \quad a^{\star} = (\mathbf{z}^{\top},0,0)\,\mathbf{F}^{-1}\begin{pmatrix}0\\1\\0\end{pmatrix}, \tag{37.30}$$

show that $b^{\star}$ and $a^{\star}$ are the coefficients of a linear regression in the sense of Castelier's scalar product

$$b^{\star} = \frac{\mathrm{cov}_n(\mathbf{z},\mathbf{s})}{\mathrm{cov}_n(\mathbf{s},\mathbf{s})} \quad \text{and} \quad a^{\star} = \mathrm{E}_n[\mathbf{z}] - b^{\star}\,\mathrm{E}_n[\mathbf{s}]. \tag{37.31}$$

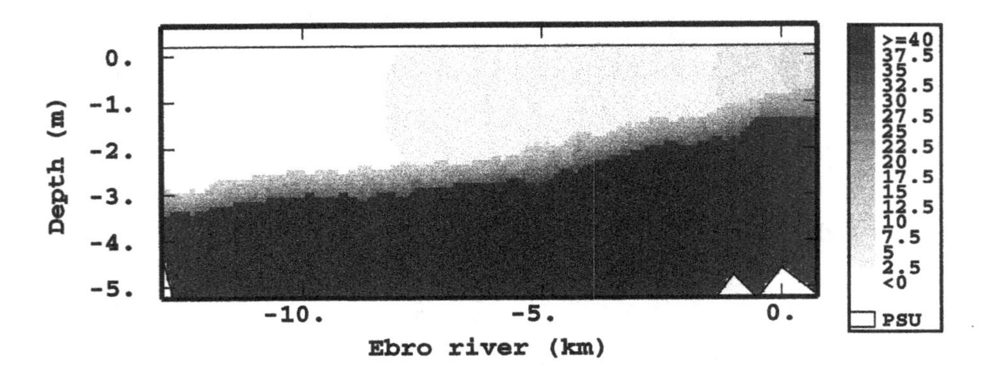

Figure 37.5: Interpolated salinity computed by the MIKE12 numerical model.

EXERCISE 37.3 *Assume, in a first step, that we have an additional data value z_0 at the point $\mathbf{x}_0$. Let us rewrite the left hand matrix of OK with $n+1$ informations,*

$$\mathbf{K}_0 = \begin{pmatrix} c_{00} & \mathbf{k}_0^\top \\ \mathbf{k}_0 & \mathbf{K} \end{pmatrix} \quad \text{and} \quad \mathbf{K}_0^{-1} = \begin{pmatrix} u_{00} & \mathbf{v}_0^\top \\ \mathbf{v}_0 & \mathbf{A}_0 \end{pmatrix}, \tag{37.32}$$

where $c_{00} = C(\mathbf{x}_0 - \mathbf{x}_0)$, $\mathbf{K}$, $\mathbf{k}_0$ are defined as in the previous exercise.

As $\mathbf{K}_0\,\mathbf{K}_0^{-1} = \mathbf{I}$, we have the relations

$$c_{00}\, u_{00} + \mathbf{k}_0^\top \mathbf{v}_0 = 1, \tag{37.33}$$

$$\mathbf{k}_0\, u_{00} + \mathbf{K}\, \mathbf{v}_0 = \mathbf{0}, \tag{37.34}$$

$$c_{00}\, \mathbf{v}_0^\top + \mathbf{k}_0^\top \mathbf{A}_0 = \mathbf{0}^\top, \tag{37.35}$$

$$\mathbf{k}_0\, \mathbf{v}_0^\top + \mathbf{K}\, \mathbf{A}_0 = \mathbf{I}. \tag{37.36}$$

a) Show that

$$u_{00} = \frac{1}{\sigma^2_{\mathrm{OK}}}. \tag{37.37}$$

b) Let $\mathbf{z}_0^\top = (z_0, \mathbf{z}^\top)$ and $\mathbf{s}_0^\top = (s_0, \mathbf{s}^\top)$ be vectors of dimension $n+1$, and

$$\mathcal{K}_{n+1}(\mathbf{z}_0, \mathbf{s}_0) = (\mathbf{z}_0^\top, 0)\, \mathbf{K}_0^{-1} \begin{pmatrix} \mathbf{s}_0 \\ 0 \end{pmatrix}. \tag{37.38}$$

Show that

$$\begin{aligned} \mathcal{K}_{n+1}(\mathbf{z}_0, \mathbf{s}_0) &= \\ \mathcal{K}_n(\mathbf{z}, \mathbf{s}) &+ u_{00}\,(z_0 - (\mathbf{z}^\top, 0)\,\mathbf{K}^{-1}\mathbf{k}_0) \cdot (s_0 - (\mathbf{s}^\top, 0)\,\mathbf{K}^{-1}\mathbf{k}_0). \end{aligned} \tag{37.39}$$

c) Let $\Delta(z_0|\mathbf{z})$ be the OK error

$$\Delta(z_0|\mathbf{z}) = z_0 - z_0^\star = z_0 - (\mathbf{z}^\top, 0)\, \mathbf{K}^{-1}\mathbf{k}_0. \tag{37.40}$$

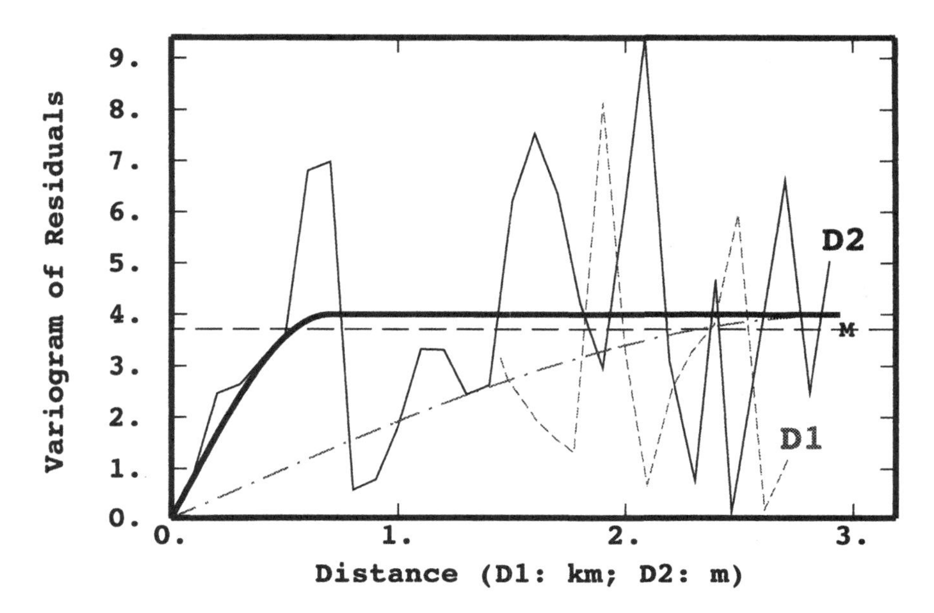

Figure 37.6: Variogram of residuals between chlorophyll water samples and drift where the interpolated MIKE12 salinity intervenes as an external drift. The experimental variogram was computed in two directions (D1: horizontal, D2: vertical), fitted with a spherical model using a geometric anisotropy. The abscissa should be read in kilometers for D1 and in meters for D2.

Show that

$$\mathcal{K}_{n+1}(\mathbf{z}_0, \mathbf{s}_0) = z_0 \, \frac{\Delta(s_0|\mathbf{s})}{\sigma^2_{\mathrm{OK}}} - (\mathbf{z}^\top, 0) \, \mathbf{K}^{-1} \, \mathbf{k}_0 \, \frac{\Delta(s_0|\mathbf{s})}{\sigma^2_{\mathrm{OK}}} + \mathcal{K}_n(\mathbf{z}, \mathbf{s}), \tag{37.41}$$

and that

$$\frac{\Delta(s_0|\mathbf{s})}{\sigma^2_{\mathrm{OK}}} = (u_{00}, \mathbf{v}_0^\top) \cdot \begin{pmatrix} \mathbf{s}_0 \\ 0 \end{pmatrix}. \tag{37.42}$$

EXERCISE 37.4 *In a second step, we are interested in an OK with $n{-}1$ points. We write $\Delta(z_\alpha|\mathbf{z}_{[\alpha]})$ for the error at a location $\mathbf{x}_\alpha$ with a kriging using only the $n{-}1$ data contained in the vector,*

$$\mathbf{z}_{[\alpha]} = (z_1, \ldots, z_{\alpha-1}, z_{\alpha+1}, \ldots, z_n)^\top, \tag{37.43}$$

and we denote $\sigma^2_{[\alpha]}$ the corresponding ordinary kriging variance.

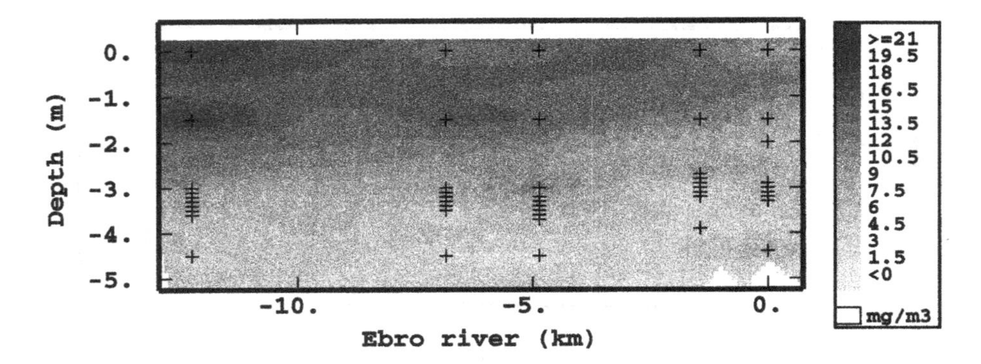

Figure 37.7: Kriging of chlorophyll water samples using the interpolated MIKE12 salinity as an external drift.

a) Establish the following elegant relation

$$\mathbf{K}^{-1}\begin{pmatrix}\mathbf{s}\\0\end{pmatrix} = \begin{pmatrix}\vdots\\ \dfrac{\Delta(s_\alpha|\mathbf{s}_{[\alpha]})}{\sigma^2_{[\alpha]}}\\ \vdots\\ m^\star\end{pmatrix}. \tag{37.44}$$

b) As the coefficient of the drift is estimated by

$$b^\star = \sum_{\alpha=1}^{n} w_b^\alpha\, z_\alpha = \frac{\mathcal{K}_n(\mathbf{z},\mathbf{s})}{\mathcal{K}_n(\mathbf{s},\mathbf{s})}, \tag{37.45}$$

show that

$$w_b^\alpha = \frac{\Delta(s_\alpha|\mathbf{s}_{[\alpha]})}{\sigma^2_{[\alpha]}\,\mathcal{K}_n(\mathbf{s},\mathbf{s})}. \tag{37.46}$$

The computation of the errors $\Delta(s_\alpha|\mathbf{s}_{[\alpha]})$ can be interesting in applications. They can serve to explore the influence of the values s_α on the KE estimator in a given neighborhood.

Regularity of the external drift function

CASTELIER [37, 38] has used the description of the influence of the external drift data on the kriging weights to show that the external drift should preferably be a smoother function than realizations of the primary variable $Z(\mathbf{x})$. This question can be checked on data by examining the behavior of the experimental variogram of $s(\mathbf{x})$ and comparing it with the variogram model adopted for $Z(\mathbf{x})$. The behavior at the

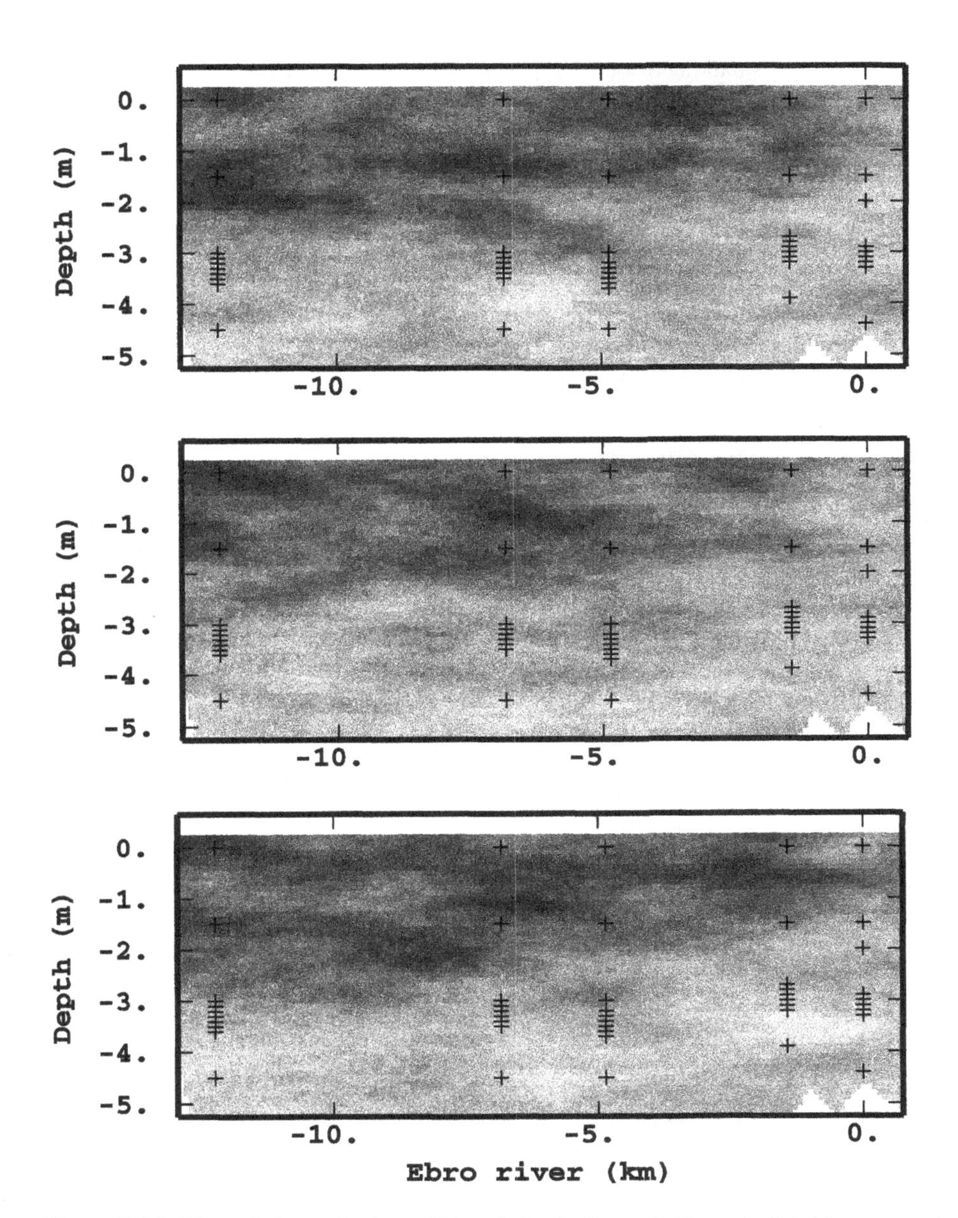

Figure 37.8: Three independent conditional simulations of chlorophyll taking account of MIKE12 salinity. The greyscale is the same as on Figure 37.7.

origin of the experimental variogram of $s(\mathbf{x})$ should be more regular than that of the variogram model of $Z(\mathbf{x})$.

For example, using the same setting as in the previous section, the ratio of the squared cross-validation error of the external drift kriging with respect to the variance $\sigma^2_{[\alpha]}$ of ordinary kriging (omitting the sample at the location $\mathbf{x}_\alpha$),

$$\frac{\Delta(s_\alpha|\mathbf{s}_{[\alpha]})^2}{\sigma^2_{[\alpha]}}, \tag{37.47}$$

could in an ideal situation be constant for all $\mathbf{x}_\alpha$. In this particular case $s(\mathbf{x})$ could be considered as a realization (up to a constant factor) of the random function $Z(\mathbf{x})$ and the covariance function of the external drift would be proportional to the covariance function of $Z(\mathbf{x})$. This ratio is worth examining in practice as it explains the structure of the weights used for estimating the external drift coefficient $b^\star$ (see Exercise 37.4).

Cokriging with multiple external drift

The last problem to consider is how to set up a cokriging with multiple external drift. In practice, however, it may be quite tricky to formulate the corresponding coregionalization model.

We consider a set of external drifts s_k, $k=1\dots K$ that are available both at the sample locations and at the prediction locations (typically the nodes of a grid spanned over the domain). Assuming the variables $Z_i(\mathbf{x})$ are related to the second set of variables s_k by the relations:

$$\mathrm{E}[Z_i(\mathbf{x})] = \sum_{k=1}^{K} a_k^i + b_k^i\, s_k(\mathbf{x}) \qquad \text{for} \quad i=1,\dots,N, \tag{37.48}$$

the samples of the second set can be included into the ordinary cokriging system using additional constraints,

$$\sum_{\alpha=1}^{n_i} w_\alpha^k\, s_k(\mathbf{x}_\alpha) = \delta_{ii_0}\, s_k(\mathbf{x}_0) \qquad \text{for all } i. \tag{37.49}$$

The cokriging system with multiple external drift finally comes as

$$\begin{cases} \displaystyle\sum_{j=1}^{N}\sum_{\beta=1}^{n_j} w_\beta^j\, \gamma_{ij}(\mathbf{x}_\alpha-\mathbf{x}_\beta) + \mu_i + \sum_{k=1}^{K} \mu_k\, s_k(\mathbf{x}_\alpha) = \gamma_{ii_0}(\mathbf{x}_\alpha-\mathbf{x}_0) & \text{for } \forall\alpha, i, \\ \displaystyle\sum_{\beta=1}^{n_i} w_\beta^i = \delta_{ii_0} & \text{for } \forall i, \\ \displaystyle\sum_{\beta=1}^{n_i} w_\beta^i\, s_k(\mathbf{x}_\beta) = \delta_{ii_0}\, s_k(\mathbf{x}_0) & \text{for } \forall i, k, \end{cases} \tag{37.50}$$

where μ_i, μ_k are Lagrange multipliers.

The external drift variables s_k on one hand need to be linearly related to the different variables Z_i in conjunction with which they are used. On the other hand the different drift variables have to be linearly independent among themselves. A constant external drift for example is redundant with the condition that the weights of the principal variable should satisfy $\sum w_\alpha^{i_o} = 1$ and will thus cause the cokriging system to be singular.

Ebro estuary: numerical model output as external drift

We pursue the case study of Chapter 27 using the ouput of a hydrodynamical model as external drift. The software package MIKE12 [89, 255] provides a two-layer model which enables computation of the average concentration, average temperature, average salinity and water level for the top and the bottom layers of the estuary.

We display on Figure 37.5 an interpolated map based on MIKE12 salinity output for the Ebro estuary on the 5th October 1999. It is interesting to note that contrarily to the kriged map of conductivity on Figure 27.3 on p186 the MIKE12 model output suggests that the transition zone between fresh water and salt wedge may not be horizontal (as assumed when defining the anisotropy in the different variogram models used up to this point). Its inclination reflects an upstream downward slope of the upper limit of the salt wedge resulting from the dynamics incorporated into the model and does not seem incompatible with the Hydrolab data displayed on Figure 27.1 on p184. So the model provides a new perspective that can be used in a kriging of the chlorophyll values.

The variogram of residuals between chlorophyll water samples and drift (where the interpolated MIKE12 salinity intervened as an external drift) is shown on Figure 37.6. The external drift kriging based on this model is displayed on Figure 37.7. Comparing it with the cokriging on Figure 27.7 we can see the effect of the MIKE12 model output which results in an inclined transition zone throughout the estuary.

Comparing results of conditional simulations and kriging

Stochastic conditional simulations using the external drift geostatistical model were performed and three independent realizations are displayed on Figure 37.8, which are all compatible with the chlorophyll water samples. To get an idea of how much the different estimations of the distribution of chlorophyll within the analyzed cross section of the Ebro river differ we summarize the results using selectivity curves $T(z)$ and $m(z)$ which were presented in Chapter 31.

On Figure 37.9 we see curves $T(z)$ that represent the proportion of chlorophyll above cutoff z (in the Ebro river section). At cutoff 15 mg/m3 the four displayed curves (starting from the lowest) are:

- the kriged chlorophyll values using the MIKE12 external drift, as displayed on Figure 37.7 (thick curve);

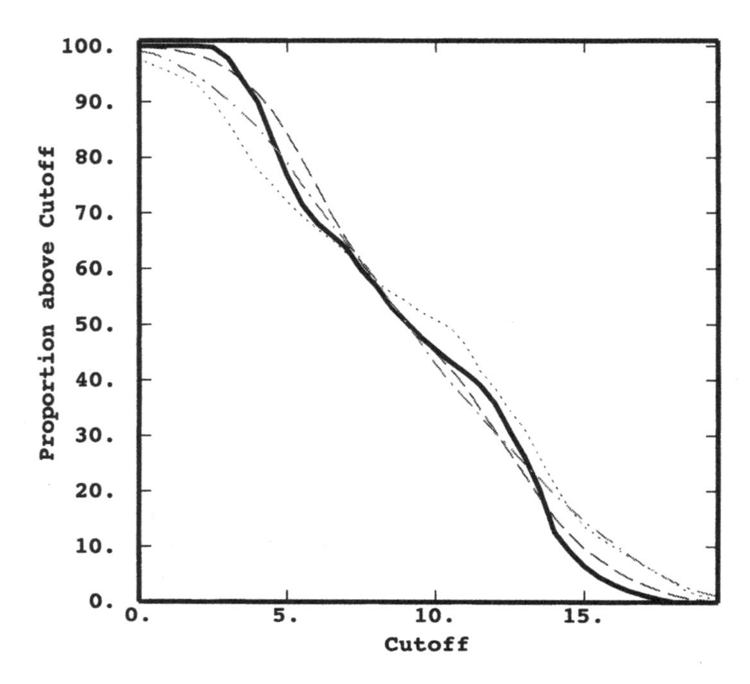

Figure 37.9: Proportion $T(z)$ of chlorophyll values above cutoff z. Comparison of external drift kriging (thick curve) with different simulations.

- one of the realizations of a corresponding conditional simulation, as displayed on Figure 37.8 (dashed curve);
- a realization of a conditional simulation using a cubic variogram model, as displayed on the upper graph of Figure 27.11 (dotted curve);
- a realization of a conditional simulation using an exponential variogram model, as displayed on the lower graph of Figure 27.11 (dotted curve);

The proportion of chlorophyll decreases when the cutoff is increased, so all four curves decrease. The kriging can be viewed as the average of a great many realizations: that is why it is smoother as individual realizations. At one end, the corresponding proportion of values for large values, say above 14 mg/m^3, is lower for kriging than for the conditional simulations; at the other end, say below 3 mg/m^3, the reverse occurs, i.e. the proportion of kriged values above cutoff is larger than for the simulations. Among the different realizations the one from a simulation incorporating external drift has a lower proportion for cutoffs above 14 mg/m^3. For problems in which the cutoff is equal to the median it does not make a difference whether kriged or simulated values are used to compute the proportion of chlorophyll above cutoff.

The Figure 37.10 plots the curves $m(z)$, representing the mean of the values above cutoff, for the same four calculations (kriging with external drift, conditional simula-

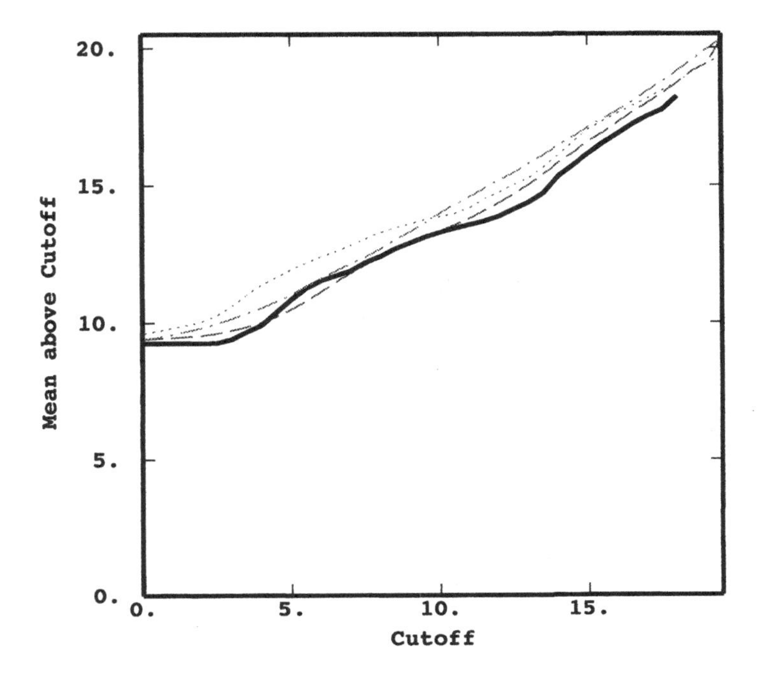

Figure 37.10: Mean $m(z)$ of chlorophyll values above cutoff z. Comparison of external drift kriging (thick curve) with different simulations.

tions with or without external drift). The kriging is smoother than the simulations and yields systematically lower estimates of mean values above 11 mg/m^3 or below 4 mg/m^3. The simulations that are not constrained by external drift yield larger estimated values than the simulation (dashed line) incorporating MIKE12 output as external drift.

38 Universal Kriging

The external drift approach has left questions about the inference of the variogram unanswered. The theory of universal kriging, i.e. kriging with several *universality conditions*, will help deepen our understanding. The universal kriging model splits the random function into a linear combination of deterministic functions, known at any point of the region, and a random component, the residual random function. It turns out that this model is in general difficult to implement, because the underlying variogram of the random component can only be inferred in exceptional situations.

Universal kriging system

On a domain $\mathcal{D}$ we have functions f_l of the coordinates $\mathbf{x}$, which are considered deterministic, because they are known at any location of the domain.

The random function $Z(\mathbf{x})$ is composed of a deterministic component $m(\mathbf{x})$, the *drift*, and a second-order stationary random function $Y(\mathbf{x})$,

$$Z(\mathbf{x}) = m(\mathbf{x}) + Y(\mathbf{x}). \tag{38.1}$$

Assuming in this section that $Y(\mathbf{x})$ is second-order stationary with a mean zero and a covariance function $C(\mathbf{h})$, we have

$$\mathrm{E}[\,Z(\mathbf{x})\,] = m(\mathbf{x}). \tag{38.2}$$

We suppose that the drift $m(\mathbf{x})$ can be represented as a linear combination of the deterministic f_l functions with non-zero coefficients a_l,

$$m(\mathbf{x}) = \sum_{l=0}^{L} a_l\, f_l(\mathbf{x}). \tag{38.3}$$

The function $f_0(\mathbf{x})$ is defined as constant

$$f_0(\mathbf{x}) = 1. \tag{38.4}$$

For kriging we use the linear combination

$$Z^{\star}(\mathbf{x}_0) = \sum_{\alpha=1}^{n} w_\alpha\, Z(\mathbf{x}_\alpha). \tag{38.5}$$

We want no bias

$$\mathrm{E}[\,Z(\mathbf{x}_0) - Z^{\star}(\mathbf{x}_0)\,] \;=\; 0, \tag{38.6}$$

which yields

$$m(\mathbf{x}_0) - \sum_{\alpha=1}^{n} w_\alpha\, m(\mathbf{x}_\alpha) \;=\; 0, \tag{38.7}$$

and

$$\sum_{l=0}^{L} a_l\,\big(f_l(\mathbf{x}_0) - \sum_{\alpha=1}^{n} w_\alpha\, f_l(\mathbf{x}_\alpha)\big) \;=\; 0. \tag{38.8}$$

As the a_l are non zero, the following set of constraints on the weights w_α emerges

$$\sum_{\alpha=1}^{n} w_\alpha\, f_l(\mathbf{x}_\alpha) \;=\; f_l(\mathbf{x}_0) \qquad \text{for} \quad l = 0, \dots, L, \tag{38.9}$$

which are called ***universality conditions***.

For the constant function $f_0(\mathbf{x})$ this is the usual condition

$$\sum_{\alpha=1}^{n} w_\alpha \;=\; 1. \tag{38.10}$$

Developing the expression for the estimation variance, introducing the constraints into the objective function together with Lagrange parameters μ_l and minimizing, we obtain the ***universal kriging*** (UK) system

$$\begin{cases} \displaystyle\sum_{\beta=1}^{n} w_\beta\, C(\mathbf{x}_\alpha - \mathbf{x}_\beta) - \sum_{l=0}^{L} \mu_l\, f_l(\mathbf{x}_\alpha) = C(\mathbf{x}_\alpha - \mathbf{x}_0) & \text{for } \alpha = 1, \dots, n \\ \displaystyle\sum_{\beta=1}^{n} w_\beta\, f_l(\mathbf{x}_\beta) = f_l(\mathbf{x}_0) & \text{for } l = 0, \dots, L, \end{cases} \tag{38.11}$$

and in matrix notation

$$\begin{pmatrix} \mathbf{C} & \mathbf{F} \\ \mathbf{F}^{\top} & \mathbf{0} \end{pmatrix} \begin{pmatrix} \mathbf{w} \\ -\boldsymbol{\mu} \end{pmatrix} \;=\; \begin{pmatrix} \mathbf{c} \\ \mathbf{f} \end{pmatrix}. \tag{38.12}$$

For this system to have a solution, it is necessary that the matrix

$$\mathbf{F} \;=\; (\,\mathbf{f}_0, \dots, \mathbf{f}_L\,) \tag{38.13}$$

is of full column rank, i.e. the column vectors $\mathbf{f}_l$ have to be linearly independent. This means in particular that there can be no other constant vector besides the vector $\mathbf{f}_0$. Thus the functions $f_l(\mathbf{x})$ have to be selected with care.

Estimation of the drift

The drift can be considered random if the coefficients multiplying the deterministic functions are assumed random

$$M(\mathbf{x}) = \sum_{l=0}^{L} A_l f_l(\mathbf{x}) \qquad \text{with} \qquad \mathrm{E}[A_l] = a_l. \tag{38.14}$$

A specific coefficient A_{l_0} can be estimated from the data using the linear combination

$$A^\star_{l_0} = \sum_{\alpha=1}^{n} w^\alpha_{l_0} Z(\mathbf{x}_\alpha). \tag{38.15}$$

In order to have no bias we need the following L constraints

$$\sum_{\alpha=1}^{n} w^\alpha_{l_0} f_l(\mathbf{x}_\alpha) = \delta_{ll_0} = \begin{cases} 1 & \text{if } l = l_0 \\ 0 & \text{if } l \neq l_0. \end{cases} \tag{38.16}$$

Computing the estimation variance of the random coefficient and minimizing with the constraints we obtain the system

$$\begin{cases} \displaystyle\sum_{\beta=1}^{n} w^\beta_{l_0} C(\mathbf{x}_\alpha - \mathbf{x}_\beta) - \sum_{l=0}^{L} \mu_{ll_0} f_l(\mathbf{x}_\alpha) = 0 & \text{for } \alpha = 1, \ldots, n \\ \displaystyle\sum_{\beta=1}^{n} w^\beta_{l_0} f_l(\mathbf{x}_\beta) = \delta_{ll_0} & \text{for } l = 0, \ldots, L. \end{cases} \tag{38.17}$$

What is the meaning of the Lagrange multipliers $\mu_{l_0 l_1}$ (for two fixed indices l_0 and l_1)? We start from

$$\mathrm{cov}(A^\star_{l_1}, A^\star_{l_0}) = \sum_{\alpha=1}^{n} \sum_{\beta=1}^{n} w^\alpha_{l_1} w^\beta_{l_0} C(\mathbf{x}_\alpha - \mathbf{x}_\beta). \tag{38.18}$$

Taking into account the kriging equations,

$$\sum_{\alpha=1}^{n} w^\alpha_{l_1} \sum_{\beta=1}^{n} w^\beta_{l_0} C(\mathbf{x}_\alpha - \mathbf{x}_\beta) = \sum_{\alpha=1}^{n} w^\alpha_{l_1} \sum_{l=0}^{L} \mu_{ll_0} f_l(\mathbf{x}_\alpha), \tag{38.19}$$

and the constraints on the weights,

$$\sum_{l=0}^{L} \mu_{ll_0} \sum_{\alpha=1}^{n} w^\alpha_{l_1} f_l(\mathbf{x}_\alpha) = \sum_{l=0}^{L} \mu_{ll_0} \delta_{ll_1}, \tag{38.20}$$

we see that

$$\mathrm{cov}(A^\star_{l_0}, A^\star_{l_1}) = \mu_{l_0 l_1}. \tag{38.21}$$

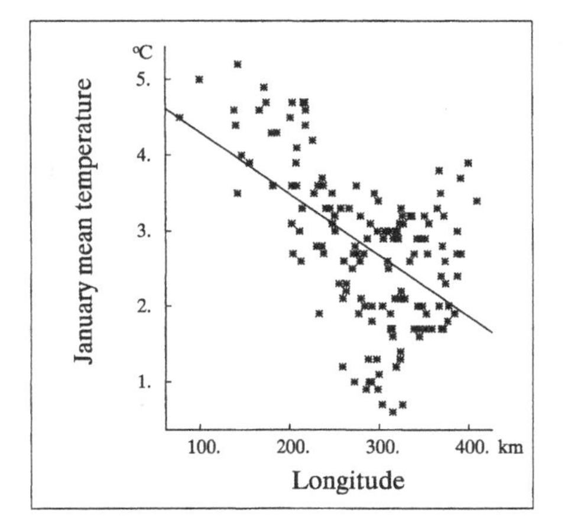

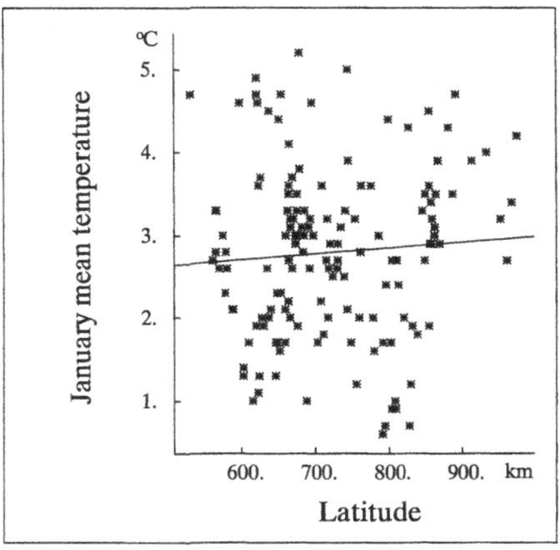

Figure 38.1: Scatter diagrams of temperature with longitude and latitude: there is a systematic decrease from west to east, while there is no such trend in the north-south direction.

A Lagrange parameter $\mu_{l_0 l_1}$ represents the covariance between two drift coefficients indexed l_0 and l_1.

The estimated drift at a specific location is obtained by combining the estimated coefficients with the values of the deterministic functions at that location,

$$m^{\star}(\mathbf{x}_0) \quad = \quad \sum_{l=0}^{L} a_l^{\star} \, f_l(\mathbf{x}_0). \tag{38.22}$$

It can be shown that the same solution is obtained without explicitly estimating the drift coefficients, simply by a kriging of the mean function

$$M^{\star}(\mathbf{x}_0) \quad = \quad \sum_{\alpha=1}^{n} w_{\alpha}^{\mathrm{KM}} \, Z(\mathbf{x}_\alpha). \tag{38.23}$$

The corresponding system is

$$\begin{cases} \displaystyle\sum_{\beta=1}^{n} w_{\beta}^{\mathrm{KM}} \, C(\mathbf{x}_\alpha - \mathbf{x}_\beta) - \sum_{l=0}^{L} \mu_{\mathrm{KM}}^{l} \, f_l(\mathbf{x}_\alpha) = 0 & \text{for } \alpha = 1, \ldots, n \\ \displaystyle\sum_{\beta=1}^{n} w_{\beta}^{\mathrm{KM}} \, f_l(\mathbf{x}_\beta) = f_l(\mathbf{x}_0) & \text{for } l = 0, \ldots, L. \end{cases} \tag{38.24}$$

Underlying variogram and estimated residuals

We have studied the case of a *known* covariance function (which is equivalent to a bounded variogram). In practice we need to infer the variogram in the presence of drift. In the decomposition of $Z(\mathbf{x})$ into drift and a residual random function,

$$Y(\mathbf{x}) \quad = \quad Z(\mathbf{x}) - m(\mathbf{x}), \tag{38.25}$$

we have not discussed how the *underlying variogram* $\gamma(\mathbf{h})$ associated with Y can be inferred, knowing that both quantities, drift and residual, need to be estimated. The underlying variogram is defined as

$$\gamma(\mathbf{h}) \quad = \quad \frac{1}{2}\operatorname{var}(Z(\mathbf{x}+\mathbf{h}) - Z(\mathbf{x})) = \frac{1}{2}\operatorname{E}[\,(Y(\mathbf{x}+\mathbf{h}) - Y(\mathbf{x}))^2\,]. \tag{38.26}$$

Experimentally we have access to an estimated residual $R^\star(\mathbf{x})$ by forming the difference between an estimated drift $M^\star(\mathbf{x})$ and $Z(\mathbf{x})$ at data locations $\mathbf{x}_\alpha$,

$$R^\star(\mathbf{x}_\alpha) \quad = \quad Z(\mathbf{x}_\alpha) - M^\star(\mathbf{x}_\alpha) = Z(\mathbf{x}_\alpha) - \sum_{l=1}^{L} A_l^\star f_l(\mathbf{x}_\alpha). \tag{38.27}$$

In the drift, the term $A_0^\star$ associated with the $f_0 = 1$ monomial has been dropped right away as it will vanish in the variogram expressions, because increments will be taken, which are zero for this monomial. By the way the estimation of the term $A_0^\star$ is problematic within the framework of an intrinsic hypothesis: this question is dwelled at length in [200, 45, 44].

The variogram of the estimated residuals between two data locations is

$$\begin{aligned}
\gamma^\star(\mathbf{x}_\alpha, \mathbf{x}_\beta) \quad &= \quad \frac{1}{2}\operatorname{E}[\,(R^\star(\mathbf{x}_\alpha) - R^\star(\mathbf{x}_\beta))^2\,] \\
&= \quad \gamma(\mathbf{x}_\alpha, \mathbf{x}_\beta) + \frac{1}{2}\operatorname{var}(M^\star(\mathbf{x}_\alpha) - M^\star(\mathbf{x}_\beta)) \\
&\quad - \operatorname{cov}\Big((Z(\mathbf{x}_\alpha) - Z(\mathbf{x}_\beta)), (M^\star(\mathbf{x}_\alpha) - M^\star(\mathbf{x}_\beta))\Big).
\end{aligned} \tag{38.28}$$

The variogram of the estimated residuals is composed of the underlying variogram and two terms representing bias.

If $\mathcal{S}$ is the set of sample points, we can define the geometric covariogram of the sample points as

$$\mathfrak{K}_{\mathcal{S}}(\mathbf{h}) \quad = \quad \int_{\mathcal{D}} \mathbf{1}_{\mathcal{S}}(\mathbf{x})\, \mathbf{1}_{\mathcal{S}}(\mathbf{x}+\mathbf{h})\, d\mathbf{x}, \tag{38.29}$$

where $\mathcal{D}$ is the domain of interest and $\mathbf{1}_{\mathcal{S}}(\mathbf{x})$ is the indicator function of the sample set.

The experimental regional variogram of the estimated residuals is then computed as

$$G^\star(\mathbf{h}) \quad = \quad \frac{1}{2\,\mathfrak{K}_{\mathcal{S}}(\mathbf{h})} \int_{\mathcal{D}} \mathbf{1}_{\mathcal{S}}(\mathbf{x})\, \mathbf{1}_{\mathcal{S}}(\mathbf{x}+\mathbf{h})\, (R^\star(\mathbf{x}+\mathbf{h}) - R^\star(\mathbf{x}))\, d\mathbf{x}. \tag{38.30}$$

Taking the expectation of $G^{\star}(\mathbf{h})$ gives a three terms expression, which is however not equal to the underlying variogram

$$\mathrm{E}[\,G^{\star}(\mathbf{h})\,] \;=\; \mathrm{T}_1 - 2\,\mathrm{T}_2 + \mathrm{T}_3 \neq \gamma(\mathbf{h}), \tag{38.31}$$

where

$$\mathrm{T}_1 \;=\; \frac{1}{2\,\mathfrak{K}_S(\mathbf{h})} \int_{\mathcal{D}} \mathbf{1}_S(\mathbf{x})\, \mathbf{1}_S(\mathbf{x}{+}\mathbf{h})\, \mathrm{E}[\,(Z(\mathbf{x}{+}\mathbf{h}) - Z(\mathbf{x}))^2\,]\, d\mathbf{x}, \tag{38.32}$$

$$\begin{aligned}\mathrm{T}_2 \;=\; & \frac{1}{2\,\mathfrak{K}_S(\mathbf{h})} \int_{\mathcal{D}} \mathbf{1}_S(\mathbf{x})\, \mathbf{1}_S(\mathbf{x}{+}\mathbf{h}) \sum_{l=1}^{L} \Big(\mathrm{E}[\,A_l^{\star}\,(Z(\mathbf{x}{+}\mathbf{h}) - Z(\mathbf{x}))\,] \\ & \times (f_l(\mathbf{x}{+}\mathbf{h}) - f_l(\mathbf{x})) \Big)\, d\mathbf{x},\end{aligned} \tag{38.33}$$

$$\begin{aligned}\mathrm{T}_3 \;=\; & \frac{1}{2\,\mathfrak{K}_S(\mathbf{h})} \int_{\mathcal{D}} \mathbf{1}_S(\mathbf{x})\, \mathbf{1}_S(\mathbf{x}{+}\mathbf{h}) \sum_{l=1}^{L} \sum_{s=1}^{L} \mathrm{cov}(A_l^{\star}, A_s^{\star}) \\ & \times (f_l(\mathbf{x}{+}\mathbf{h}) - f_l(\mathbf{x}))\, (f_s(\mathbf{x}{+}\mathbf{h}) - f_s(\mathbf{x}))\, d\mathbf{x}.\end{aligned} \tag{38.34}$$

The first term is the underlying variogram while the two other terms represent the bias, which is generally considerable. If we assume to know the drift, replacing $A_l^{\star}$ by a_l, the second term vanishes, but not the third,

$$\mathrm{E}[\,G^{\star}(\mathbf{h})\,] \;=\; \mathrm{T}_1 - \mathrm{T}_3 \neq \gamma(\mathbf{h}). \tag{38.35}$$

Thus even with known drift we can only hope to obtain the underlying variogram from the variogram of the residuals in the vicinity of the origin. Due to its smoothness, the increments of the drift have a small value at very short distances, making T_3 negligible.

For regularly gridded data two methods exist for inferring the underlying variogram (see [199, 48]). For data irregularly scattered in space, in few exceptional situations the underlying variogram is directly accessible:

- when the drift occurs only in one part of a domain which is assumed homogeneous, then the variogram model can be inferred in the other, stationary part of the domain and transferred to the non stationary part;
- when the drift is not active in a particular direction of space, the variogram inferred in that direction can be extended to the other directions under an assumption of isotropic behavior of the underlying variogram.

To these important exceptions we can add the case of a weak drift, when the variogram can reasonably well be inferred at short distances, for which the bias is assumed not to be strong. This situation is very similar to the case when ordinary kriging is applied in a moving neighborhood with a locally stationary model. The necessity for including drift terms into the kriging system may be subject to discussion in that case and is usually sorted out by cross-validation.

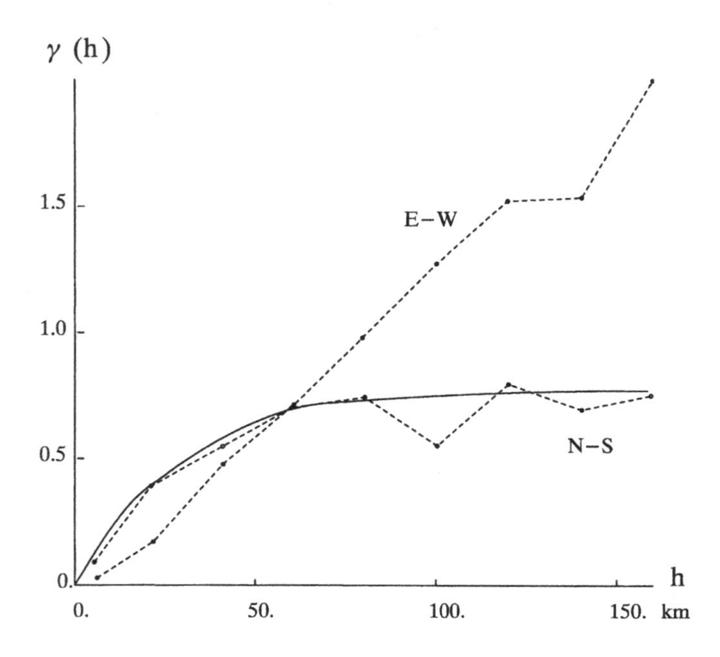

Figure 38.2: Variograms of average January temperature data in the east-west and north-south directions: a model is fitted in the direction with no drift.

EXAMPLE 38.1 *The Scottish temperature data [146] provides an example for drift being absent in a particular direction of space. On Figure 38.1 we see the scatter diagrams of average January temperature in Scotland plotted against longitude and latitude . Clearly there is a systematic decrease of mean temperature from west to east (longitudes) due to the effect of the Gulf stream which warms up the western coast of Scotland in the winter. Along latitudes there is no trend to be seen.*

The variograms along longitude and latitude are shown on Figure 38.2. The east-west experimental variogram is influenced by the drift and grows without bounds while the shape of the north-south variogram suggests a sill: the latter has been modeled using three spherical functions with ranges of 20, 70 and 150km. By assuming that the underlying variogram is isotropic, we can use the north-south variogram model also in the east-west direction. The interpretation of the east-west experimental variogram is that drift masks off the underlying variogram in that direction.

From universal to intrinsic kriging

In general it is difficult and cumbersome to infer the underlying variogram (assuming it existence). Two steps will lead us to a more sound and workable model:

1. The class of deterministic functions is restricted to functions which are *translation-invariant* and pairwise orthogonal, i.e. the class of exponentials-polynomials.

2. A specific structural analysis tool is defined, the *generalized covariance function* $K(\mathbf{h})$, which filters these translation-invariant functions.

As the variogram represents only a particular subclass of generalized covariances, we understand that the quest for the underlying variogram was doomed to be vain: the residuals may have a structure which is simply not compatible with a variogram and which can only be captured by a generalized covariance.

Instead of building a model by pasting together a second-order stationary (or intrinsic) random function and a given drift which will generate universality conditions in the kriging system, we will have a different approach. We take a particular class of deterministic functions and infer generalized covariance functions filtering them, what leads to *authorization* constraints instead of universality conditions in the kriging system. Thus we will speak of *intrinsic kriging* as its formulation and meaning changes within the theory of generalized intrinsic random functions. We provide a very elementary sketch of that powerful theory of in the next chapter.

39 Translation Invariant Drift

The characterization of the drift by a linear combination of translation-invariant deterministic functions opens the gate to the powerful theory of the intrinsic random functions of order k. A more general tool of structural analysis than the variogram can be defined in this framework: the generalized covariance function. The constraints on the weights in the corresponding intrinsic kriging appear as conditions for the positive-definiteness of the generalized covariance. Parallels between this formulation and the spline approach to interpolation can be drawn.

Exponential-polynomial basis functions

With estimation problems in mind we consider sets weights w_α which interpolate exactly some basis functions $f_l(\mathbf{x})$,

$$\sum_{\alpha=1}^{n} w_\alpha \, f_l(\mathbf{x}_\alpha) \quad = \quad f_l(\mathbf{x}_0), \tag{39.1}$$

for any $l = 0, \ldots, L$.

By introducing a weight $w_0 = -1$ we can rewrite the expression as

$$\sum_{\alpha=0}^{n} w_\alpha \, f_l(\mathbf{x}_\alpha) \quad = \quad 0 \qquad \text{for} \quad l = 0, \ldots, L, \tag{39.2}$$

and we call this an *authorized linear combination*, which filters the functions $f_l(\mathbf{x})$ for a set of $n{+}1$ points $\mathbf{x}_\alpha$.

The authorized linear combination should be translation-invariant for all l and for any vector $\mathbf{h}$:

$$\sum_{\alpha=0}^{n} w_\alpha \, f_l(\mathbf{x}_\alpha) \; = \; 0 \qquad \Rightarrow \qquad \sum_{\alpha=0}^{n} w_\alpha \, f_l(\mathbf{x}_\alpha{+}\mathbf{h}) \; = \; 0. \tag{39.3}$$

To achieve this, the $L + 1$ basis functions $f_l(\mathbf{x})$ should generate a *translation-invariant* vector space. It can be shown [210] that only sets of functions belonging to the class of exponentials-polynomials fulfill this condition. Families in this class are polynomials of degree $\leq k$, trigonometric functions and exponential functions.

COMMENT 39.1 *In two spatial dimensions with coordinate vectors* $\mathbf{x} = (x_1, x_2)^\top$ *the following monomials are often used*

$$\begin{aligned} f_0(\mathbf{x}) &= 1, \quad f_1(\mathbf{x}) = x_1, \quad f_2(\mathbf{x}) = x_2, \\ f_3(\mathbf{x}) &= (x_1)^2, \quad f_4(\mathbf{x}) = x_1 \cdot x_2, \quad f_5(\mathbf{x}) = (x_2)^2, \end{aligned} \tag{39.4}$$

which generate a translation invariant vector space. The actual number $L+1$ *of basis functions of the drift depends on the degree* k *of the drift in the following way:*

for $k = 0$ *we have one basis function and* $L = 0$,
for $k = 1$ *we have three basis functions and* $L = 2$,
for $k = 2$ *we have six basis functions and* $L = 5$.

Intrinsic random functions of order k

Let w be a measure $\sum w_\alpha \, \delta_{\mathbf{x}_\alpha}$ which attributes weights w_α to points $\mathbf{x}_\alpha$ and let us denote by

$$Z(w) = \sum_{\alpha=0}^{n} w_\alpha \, Z(\mathbf{x}_\alpha) \tag{39.5}$$

the linear combination of these weights with random variables at locations $\mathbf{x}_\alpha$ in a domain $\mathcal{D}$.

The vector space of authorized linear combinations W_k contains the measures for which the corresponding weights filter the basis functions, i.e. $w \in W_k$ if the w_α fulfill equation (39.2) for any l.

A non-stationary random function $Z(\mathbf{x})$ is called an *intrinsic random function of order k* (IRF-k) if for any measure $w \in W_k$ the linear combination

$$\sum_{\alpha=1}^{n} w_\alpha \, Z(\mathbf{x}_\alpha + \mathbf{h}) \tag{39.6}$$

is zero-mean second-order stationary for any vector $\mathbf{h}$.

COMMENT 39.2 *We want to mention that a more abstract definition is possible [203]. The linear combination (39.6) can be denoted by* $Z(\tau_\mathbf{h}\, w)$, *where* $\tau_\mathbf{h}$ *is a translation operator. We can then call an abstract IRF-*k *the linear application* $\mathcal{Z}$ *of the space* W_k *into a Hilbert space of random variables of zero-expectation and finite variance, such that for any* $w \in W_k$ *the random function* $Z(\tau_\mathbf{h}\, w)$ *is second-order stationary in* $\mathbf{h}$. *The abstract* $\mathcal{Z}$ *describes the class of all IRF-*k *having the same* k*-order increments as a given* $Z(\mathbf{x})$, *which is called a representation of* $\mathcal{Z}$.

Generalized covariance function

A symmetric function, $K(\mathbf{h}) = K(-\mathbf{h})$, is a *generalized covariance function* of an IRF-k Z if

$$\operatorname{var}(Z(w)) = \sum_{\alpha=1}^{n}\sum_{\beta=1}^{n} w_\alpha\, w_\beta\, K(\mathbf{x}_\alpha - \mathbf{x}_\beta) \tag{39.7}$$

for any $w \in W_k$. The existence and uniqueness of $K(\mathbf{h})$ can be demonstrated, see [203].

It is interesting to note that in analogy to the property (7.10) of the variogram we have

$$\lim_{|\mathbf{h}|\mapsto\infty} \frac{K(\mathbf{h})}{|\mathbf{h}|^{2k+2}} = 0. \tag{39.8}$$

Actually a generalized covariance is a *k-th order conditionally positive definite* function

$$\sum_{\alpha=0}^{n}\sum_{\beta=0}^{n} w_\alpha\, w_\beta\, K(\mathbf{x}_\alpha - \mathbf{x}_\beta) \geq 0 \tag{39.9}$$

$$\text{for} \quad \sum_{\alpha=0}^{n} w_\alpha\, f_l(\mathbf{x}_\alpha) = 0, \qquad l = 0, \ldots, L.$$

From the definition (39.9) we see that the negative of the variogram is the generalized covariance of an intrinsic random function of order zero

$$K(\mathbf{h}) = -\gamma(\mathbf{h}) \qquad \text{for} \quad k = 0. \tag{39.10}$$

The theory of intrinsically stationary random functions of order k provides a generalization of the random functions with second order stationary increments, which themselves were set in a broader framework than the second order stationary functions.

The inference of the highest order of the drift polynomial to be filtered and the selection of a corresponding optimal generalized covariance is performed using automatic algorithms, which we shall not describe here (see for example [80, 158, 242, 51]).

EXAMPLE 39.3 *A possible model for a k-th order generalized covariance is*

$$K_\theta(\mathbf{h}) = \Gamma\left(-\frac{\theta}{2}\right) |\mathbf{h}|^\theta \qquad \textit{with } 0 < \theta < 2k{+}2, \tag{39.11}$$

where $\Gamma(\cdot)$ is the gamma function.

EXAMPLE 39.4 *From the preceding model a generalized covariance can be derived which has been (unfortunately) called the* polynomial model

$$K_{pol}(\mathbf{h}) \;=\; \sum_{u=0}^{k} b_u\,(-1)^{u+1}\,|\mathbf{h}|^{2u+1} \qquad \text{with} \quad b_u \geq 0. \tag{39.12}$$

The conditions on the coefficients b_u are sufficient. Looser bounds on these coefficients are given in MATHERON *[203].*

The polynomial generalized covariance model is a nested model which has the striking property that it is built up with several structures having a different behavior at the origin. For k= 1 (linear drift) the term for u= 0 is linear at the origin and is adequate for the description of a regionalized variable which is continuous but not differentiable. The term for u= 1 is cubic and thus appropriate for differentiable regionalized variables.

If a nugget-effect component is added to the polynomial generalized covariance model, discontinuous phenomena are also covered. This extended polynomial generalized covariance model is flexible with respect to the behavior at the origin and well suited for automatic fitting: inessential structures get zero weights b_u, provided the fitting is correct.

The polynomial generalized covariance model can be used with a moving kriging neighborhood.

Intrinsic kriging

To estimate a value at a location $\mathbf{x}_0$, using a linear combination of weights w_α with data, the error

$$\sum_{\alpha=1}^{n} w_\alpha\, Z(\mathbf{x}_\alpha) - Z(\mathbf{x}_0) \tag{39.13}$$

should be an authorized linear combination, i.e. the weights should satisfy (39.2).

The estimation variance in terms of the generalized covariance comes as

$$\begin{aligned} &\mathrm{E}\Big[\Big(\sum_{\alpha=1}^{n} w_\alpha\, Z(\mathbf{x}_\alpha) - Z(\mathbf{x}_0) \Big)^2 \Big] \qquad\qquad (39.14) \\ &= K(\mathbf{x}_0-\mathbf{x}_0) - 2\sum_{\alpha=1}^{n} w_\alpha\, K(\mathbf{x}_0-\mathbf{x}_\alpha) + \sum_{\alpha=1}^{n}\sum_{\beta=1}^{n} w_\alpha\, w_\beta\, K(\mathbf{x}_\alpha-\mathbf{x}_\beta). \end{aligned}$$

Minimizing this variance respecting the constraints (39.2) we get the *intrinsic kriging* system

$$\begin{cases} \displaystyle\sum_{\beta=1}^{n} w_\beta\, K(\mathbf{x}_\alpha-\mathbf{x}_\beta) ds - \sum_{l=0}^{L} \mu_l\, f_l(\mathbf{x}_\alpha) \;=\; K(\mathbf{x}_\alpha-\mathbf{x}_0) & \text{for } \alpha = 1,\ldots,n \\ \displaystyle\sum_{\beta=1}^{n} w_\beta\, f_l(\mathbf{x}_\beta) \;=\; f_l(\mathbf{x}_0) & \text{for } l = 0,\ldots,L. \end{cases} \tag{39.15}$$

We have exactly the same system as for universal kriging, see expression (38.11), page 301. The meaning of the constraints has changed however: they show as *authorization constraints* in intrinsic kriging, being conditions for the existence of the generalized covariance function $K(\mathbf{h})$ in the IRF-k model.

Trigonometric temporal drift

To motivate the problem of space-time drift let us take the example of earth magnetism measured from a traveling ship (see SÉGURET & HUCHON [297]). When studying the magnetism of the earth near the equator, a strong variation in time is observed which is due to the 24 hour rotation of the earth. Measurements taken from a ship which zigzags during several days in an equatorial zone need to be corrected from the effect of this periodic variation.

The magnetism is modeled with a space-time random function $Z(\mathbf{x},t)$ using two uncorrelated intrinsic random functions of order k, $Z_{k,\omega}(\mathbf{x})$ and $Z_{k,\omega}(t)$, as well as spatial polynomial drift $m_k(\mathbf{x})$ and temporal periodic drift $m_\omega(t)$

$$Z(\mathbf{x},t) \;=\; Z_{k,\omega}(\mathbf{x}) + m_k(\mathbf{x}) + Z_{k,\omega}(t) + m_\omega(t), \tag{39.16}$$

where $\mathbf{x}$ is a point of the spatial domain $\mathcal{D}$ and t is a time coordinate.

The variation in time of earth magnetism is known from stations on the land. Near the equator a trigonometric drift is dominant,

$$m_{\omega,\varphi}(t) \;=\; \sin(\omega\, t + \varphi), \tag{39.17}$$

with a period of ω=24 hours and a phase φ. Using a well-known relation this can be rewritten as

$$m_\omega(t) \;=\; a\,\sin(\omega\, t) + b\,\cos(\omega\, t), \tag{39.18}$$

where $a = \cos\varphi$ and $b = \sin\varphi$ are multiplicative constants which will be expressed implicitly in the weights, so that the phase parameter φ does not need to be provided explicitly.

Filtering trigonometric temporal drift

In an equatorial region the time component $Z_{k,\omega}(t)$ of magnetism is usually present as weak noise while the temporal drift dominates the time dimension. Neglecting the random variation in time, the model can be expressed as the approximation

$$Z(\mathbf{x},t) \;\cong\; Z_{k,\omega}(\mathbf{x}) + m_k(\mathbf{x}) + m_\omega(t). \tag{39.19}$$

For mapping magnetism without the bias of the periodic variation in time, the generalized covariance is inferred taking into account the space-time drift. Subsequently the temporal drift is filtered by setting the corresponding equations to zero on the right hand side of the following intrinsic kriging system

$$\begin{cases} \sum_{\beta=1}^{n} w_\beta\, K(\mathbf{x}_\alpha - \mathbf{x}_\beta) & -\sum_{l=0}^{L} \mu_l\, f_l(\mathbf{x}_\alpha) - \nu_1 \sin(\omega\, t_\alpha) - \nu_2 \cos(\omega\, t_\alpha) \\ & = K(\mathbf{x}_\alpha - \mathbf{x}_0) \qquad \text{for } \alpha = 1,\dots,n \\ \sum_{\beta=1}^{n} w_\beta\, f_l(\mathbf{x}_\beta) = f_l(\mathbf{x}_0) & \qquad\qquad \text{for } l = 0,\dots,L \\ \sum_{\beta=1}^{n} w_\beta \sin(\omega\, t_\beta) = \boxed{0} & \\ \sum_{\beta=1}^{n} w_\beta \cos(\omega\, t_\beta) = \boxed{0}. & \end{cases} \tag{39.20}$$

If however the random component $Z_{k,\omega}(t)$ is significant, we need to infer its generalized covariance $K(\tau)$ and to filter subsequently $Z(t) = Z_{k,\omega}(t) + m_\omega(t)$ using the system

$$\begin{cases} \sum_{\beta=1}^{n} w_\beta\, \boxed{K(\mathbf{x}_\alpha - \mathbf{x}_\beta, t_\alpha - t_\beta)} & -\sum_{l=0}^{L} \mu_l\, f_l(\mathbf{x}_\alpha) - \nu_1 \sin(\omega\, t_\alpha) - \nu_2 \cos(\omega\, t_\alpha) \\ & = K(\mathbf{x}_\alpha - \mathbf{x}_0) \qquad \text{for } \alpha = 1,\dots,n \\ \sum_{\beta=1}^{n} w_\beta\, f_l(\mathbf{x}_\beta) = f_l(\mathbf{x}_0) & \qquad\qquad \text{for } l = 0,\dots,L \\ \sum_{\beta=1}^{n} w_\beta \sin(\omega\, t_\beta) = 0 & \\ \sum_{\beta=1}^{n} w_\beta \cos(\omega\, t_\beta) = 0, & \end{cases} \tag{39.21}$$

where the terms $K(\mathbf{x}_\alpha - \mathbf{x}_\beta, t_\alpha - t_\beta)$ are modeled as the sum of the generalized covariances $K(\mathbf{x}_\alpha - \mathbf{x}_\beta)$ and $K(t_\alpha - t_\beta)$.

Dual kriging

We now examine the question of defining an interpolation function based on the kriging system. The interpolator is the product of the vector of samples $\mathbf{z}$ with a vector of weights $\mathbf{w}_\mathbf{x}$ which depends on the location in the domain,

$$z^\star(\mathbf{x}) \;=\; \mathbf{z}^\top \mathbf{w}_\mathbf{x}. \tag{39.22}$$

The weight vector is solution of the kriging system

$$\begin{pmatrix} \mathbf{K} & \mathbf{F} \\ \mathbf{F}^\top & \mathbf{0} \end{pmatrix} \begin{pmatrix} \mathbf{w}_\mathbf{x} \\ -\boldsymbol{\mu}_\mathbf{x} \end{pmatrix} = \begin{pmatrix} \mathbf{k}_\mathbf{x} \\ \mathbf{f}_\mathbf{x} \end{pmatrix}, \tag{39.23}$$

in which all terms dependent on the estimation location have been subscripted with an $\mathbf{x}$. In this formulation we need to solve the system each time for each new interpolation location. As the left hand matrix does not depend on $\mathbf{x}$, let us define its inverse (assuming its existence) as

$$\begin{pmatrix} \mathbf{T} & \mathbf{U} \\ \mathbf{U}^\top & \mathbf{V} \end{pmatrix}. \tag{39.24}$$

The kriging system is

$$\begin{pmatrix} \mathbf{w_x} \\ -\boldsymbol{\mu}_\mathbf{x} \end{pmatrix} = \begin{pmatrix} \mathbf{T} & \mathbf{U} \\ \mathbf{U}^\top & \mathbf{V} \end{pmatrix} \begin{pmatrix} \mathbf{k_x} \\ \mathbf{f_x} \end{pmatrix}. \tag{39.25}$$

The interpolator can thus be written

$$z^\star(\mathbf{x}) = \mathbf{z}^\top \mathbf{T}\,\mathbf{k_x} + \mathbf{z}^\top \mathbf{U}\,\mathbf{f_x}. \tag{39.26}$$

Defining $\mathbf{b}^\top = \mathbf{z}^\top \mathbf{T}$ and $\mathbf{d}^\top = \mathbf{z}^\top \mathbf{U}$ the interpolator is a function of the right hand side of the kriging system

$$z^\star(\mathbf{x}) = \mathbf{b}^\top \mathbf{k_x} + \mathbf{d}^\top \mathbf{f_x}. \tag{39.27}$$

Contrarily to the weights $\mathbf{w_x}$, the weights $\mathbf{b}$ and $\mathbf{d}$ do not depend on the target point $\mathbf{x}$.

Combining the data vector $\mathbf{z}$ with a vector of zeroes, we can set up the system

$$\begin{pmatrix} \mathbf{T} & \mathbf{U} \\ \mathbf{U}^\top & \mathbf{V} \end{pmatrix} \begin{pmatrix} \mathbf{z} \\ \mathbf{0} \end{pmatrix} = \begin{pmatrix} \mathbf{b} \\ \mathbf{d} \end{pmatrix}, \tag{39.28}$$

which, once inverted, yields the *dual system* of kriging

$$\begin{pmatrix} \mathbf{K} & \mathbf{F} \\ \mathbf{F}^\top & \mathbf{0} \end{pmatrix} \begin{pmatrix} \mathbf{b} \\ \mathbf{d} \end{pmatrix} = \begin{pmatrix} \mathbf{z} \\ \mathbf{0} \end{pmatrix}. \tag{39.29}$$

There is no reference to any interpolation point $\mathbf{x}$ in this system: it needs to be solved only once for a given region.

It should be noted that when the variable to investigate is equal to one of the deterministic functions, e.g. $z(\mathbf{x})= f_l(\mathbf{x})$, the interpolator is $z^\star(\mathbf{x})= \mathbf{w}^\top \mathbf{f}_l$. As the weights are constrained to satisfy $\mathbf{w}^\top \mathbf{f}_l= f_l(\mathbf{x})$ we see that $z^\star(\mathbf{x})= f_l(\mathbf{x}) = z(\mathbf{x})$. Thus the interpolator is exact for $f_l(\mathbf{x})$. This clarifies the meaning of the constraints in kriging: the resulting weights are able to interpolate exactly each one of the deterministic functions.

Splines

The mathematical *spline* function took its name from the draftmen's mechanical spline, which is "a thin reedlike strip that was used to draw curves needed in the fabrication of cross-sections of ships' hulls" [347]). We shall restrain discussion to the case of smoothing thin-plate splines. A general framework for splines (with an annotated bibliography) is presented in CHAMPION et al. [40]. The equivalence between splines and kriging is analyzed in more depth in MATHERON [212] and Wahba [347]).

The model for the smoothing splines can be written as

$$Z(\mathbf{x}) = \underbrace{\phi(\mathbf{x})}_{\text{smooth}} + \underbrace{Y(x)}_{\text{white noise}} \tag{39.30}$$

where $\phi(\mathbf{x})$ is estimated by a *smooth* function g minimizing

$$\sum_{\alpha=1}^{n} \left(\phi(\mathbf{x}_\alpha) - g(\mathbf{x}_\alpha)\right)^2 + \lambda\, J_p(g) \tag{39.31}$$

with $J_p(g)$ a measure of roughness (in terms of pth degree derivatives) and $\lambda > 0$ a smoothing parameter.

The $g(\mathbf{x})$ function is written as the sum of two terms

$$g(\mathbf{x}) = \sum_{\alpha=1}^{n} b_\alpha\, \psi(\mathbf{x} - \mathbf{x}_\alpha) + \sum_{l=0}^{L} d_l\, f_l(\mathbf{x}), \tag{39.32}$$

where

$$\psi(\mathbf{x} - \mathbf{x}_\alpha) = K(\mathbf{x}-\mathbf{x}_\alpha) = |\mathbf{h}|^2 \log |\mathbf{h}| \tag{39.33}$$

is a $p-1$ conditional positive definite function (in geostatistics this generalized covariance is known as the *spline covariance model*).

The weights b_α and d_l are solution of

$$\begin{pmatrix} \mathbf{K} + \lambda\mathbf{I} & \mathbf{F} \\ \mathbf{F}^\top & 0 \end{pmatrix} \begin{pmatrix} \mathbf{b} \\ \mathbf{d} \end{pmatrix} = \begin{pmatrix} \mathbf{z} \\ 0 \end{pmatrix}. \tag{39.34}$$

This system is equivalent to the dual kriging system (39.29). In the random function model we can understand the term $\lambda\mathbf{I}$ as a nugget-effect (white noise) added to the variances at data locations, but not to the variance at the estimation location. So, in geostatistical terms, the system (39.34) represents a filtering of the nugget-effect component on the basis of a non-stationary linear model of regionalization.

In the spline approach the parameters are obtained by *generalized cross validation* (GCV), which is a predictive mean square error criterion (leave-one-out technique) to estimate the degree $p = k+1$ of the spline and the smoothing parameter λ.

Synthetic examples are discussed in WAHBA [347] on pages 46–47 and 48–50. DUBRULE [97] presents an example from oil exploration from the point of view of geostatistics. HUTCHINSON & GESSLER [147] have treated the same data set with splines and show that they can obtain equivalent results; in particular, they provide prediction errors from a Bayesian model which are analogous to the kriging standard deviations.

Appendix

Matrix Algebra

This review consists of a brief exposition of the principal results of matrix algebra, scattered with a few exercises and examples. For a more detailed account we recommend the book by STRANG [318].

Data table

In multivariate data analysis we handle tables of numbers (matrices) and the one to start with is generally the data table $\mathbf{Z}$

$$\begin{array}{cc} & \begin{array}{c} Variables \\ (columns) \end{array} \\ \begin{array}{c} Samples \\ (rows) \end{array} & \begin{pmatrix} z_{1,1} & \dots & z_{1,i} & \dots & z_{1,N} \\ \vdots & & \vdots & & \vdots \\ z_{\alpha,1} & & z_{\alpha,i} & & z_{\alpha,N} \\ \vdots & & \vdots & & \vdots \\ z_{n,1} & \dots & z_{n,i} & \dots & z_{n,N} \end{pmatrix} \end{array} = [z_{\alpha i}] = \mathbf{Z}. \qquad \text{(I.1)}$$

The element $z_{\alpha i}$ denotes a numerical value placed at the crossing of the row number α (index of the sample) and the column number i (index of the variable).

Matrix, vector, scalar

A matrix is a rectangular array of numbers,

$$\mathbf{A} = [a_{ij}], \qquad \text{(I.2)}$$

with indices

$$i = 1, \dots, n, \qquad j = 1, \dots, m. \qquad \text{(I.3)}$$

We speak of a matrix of order $n \times m$ to designate a matrix of n rows and m columns.

A vector (by convention: a column vector) of dimension n is a matrix of order $n \times 1$, i.e. a matrix having only one column.

A scalar is a number (one could say: a matrix of order 1×1).

Concerning notation, matrices are symbolized by bold capital letters and vectors by bold small caps.

Sum

Two matrices of the same order can be added

$$\mathbf{A}+\mathbf{B} \;=\; [a_{ij}]+[b_{ij}]=[a_{ij}+b_{ij}]. \tag{I.4}$$

The sum of the matrices is performed by adding together the elements of the two matrices having the same index i and j.

Multiplication by a scalar

A matrix can be multiplied equally from the right or from the left by a scalar

$$\mathbf{A}\,\lambda \;=\; \lambda\,\mathbf{A}=[\lambda\,a_{ij}], \tag{I.5}$$

which amounts to multiply all elements a_{ij} by the scalar λ.

Multiplication of two matrices

The product $\mathbf{AB}$ of two matrices $\mathbf{A}$ and $\mathbf{B}$ can only be performed if $\mathbf{B}$ has as many rows as $\mathbf{A}$ has columns. Let $\mathbf{A}$ be of order $n \times m$ and $\mathbf{B}$ of order $m \times l$. The multiplication of $\mathbf{A}$ by $\mathbf{B}$ produces a matrix $\mathbf{C}$ of order $n \times l$

$$\underset{(n\times m)}{\mathbf{A}} \cdot \underset{(m\times l)}{\mathbf{B}} \;=\; \left[\sum_{j=1}^{m} a_{ij}\,b_{jk}\right]=[c_{ik}]=\underset{(n\times l)}{\mathbf{C}}, \tag{I.6}$$

where $i=1,\ldots,n$ and $k=1,\ldots,l$.

The product $\mathbf{BA}$ of these matrices is only possible if n is equal to l and the result is then a matrix of order $m \times m$.

Transposition

The transpose $\mathbf{A}^{\top}$of a matrix $\mathbf{A}$ of order $n \times m$ is obtained by inverting the sequence of the indices, such that the rows of $\mathbf{A}$ become the columns of $\mathbf{A}^{\top}$

$$\underset{(n\times m)}{\mathbf{A}} \;=\; [a_{ij}], \qquad \underset{(m\times n)}{\mathbf{A}^{\top}} \;=\; [a_{ji}]. \tag{I.7}$$

The transpose of the product of two matrices is equal to the product of the transposed matrices in reverse order

$$(\mathbf{AB})^{\top} \;=\; \mathbf{B}^{\top}\mathbf{A}^{\top}. \tag{I.8}$$

EXERCISE I.1 *Let* $\mathbf{1}$ *be the vector of dimension* n *whose elements are all equal to* 1. *Carry out the products* $\mathbf{1}^{\top}\mathbf{1}$ *and* $\mathbf{1}\mathbf{1}^{\top}$.

EXERCISE I.2 *Calculate the matrices resulting from the products*

$$\frac{1}{n}\mathbf{Z}^\top\mathbf{1} \qquad \textit{and} \qquad \frac{1}{n}\mathbf{1}\mathbf{1}^\top\mathbf{Z}, \tag{I.9}$$

where $\mathbf{Z}$ *is the* $n \times N$ *matrix of the data.*

Square matrix

A square matrix has as many rows as columns.

Diagonal matrix

A diagonal matrix $\mathbf{D}$ is a square matrix whose only non zero elements are on the diagonal

$$\mathbf{D} \;=\; \begin{pmatrix} d_{11} & 0 & 0 \\ 0 & \ddots & 0 \\ 0 & 0 & d_{nn} \end{pmatrix}. \tag{I.10}$$

In particular we mention the identity matrix

$$\mathbf{I} \;=\; \begin{pmatrix} 1 & 0 & 0 \\ 0 & \ddots & 0 \\ 0 & 0 & 1 \end{pmatrix}, \tag{I.11}$$

which does not modify a matrix of the same order when multiplying it from the right or from the left

$$\mathbf{A}\,\mathbf{I} \;=\; \mathbf{I}\,\mathbf{A} = \mathbf{A}. \tag{I.12}$$

Orthogonal matrix

A square matrix $\mathbf{A}$ is orthogonal if it verifies

$$\mathbf{A}^\top\mathbf{A} \;=\; \mathbf{A}\,\mathbf{A}^\top = \mathbf{I}. \tag{I.13}$$

Symmetric matrix

A square matrix $\mathbf{A}$ is symmetric if it is equal to its transpose

$$\mathbf{A} \;=\; \mathbf{A}^\top. \tag{I.14}$$

EXAMPLE I.3 *An example of a symmetric matrix is the variance-covariance matrix* $\mathbf{V}$*, containing the experimental variances* s_{ii} *on the diagonal and the experimental covariances* s_{ij} *off the diagonal*

$$\mathbf{V} = [s_{ij}] \quad = \quad \frac{1}{n} \, (\mathbf{Z} - \mathbf{M})^{\mathsf{T}} \, (\mathbf{Z} - \mathbf{M}) \,, \tag{I.15}$$

where $\mathbf{M}$ *is the rectangular* $n \times m$ *matrix of the means (solution of the exercise I.2), whose column elements are equal to the mean of the variable corresponding to a given column.*

EXAMPLE I.4 *Another example of a symmetric matrix is the matrix* $\mathbf{R}$ *of correlations*

$$\mathbf{R} = [r_{ij}] \quad = \quad \mathbf{D}_{s^{-1}} \, \mathbf{V} \, \mathbf{D}_{s^{-1}}, \tag{I.16}$$

where $\mathbf{D}_{s^{-1}}$ *is a diagonal matrix containing the inverses of the experimental standard deviations* $s_i = \sqrt{s_{ii}}$ *of the variables*

$$\mathbf{D}_{s^{-1}} \quad = \quad \begin{pmatrix} \frac{1}{\sqrt{s_{11}}} & 0 & 0 \\ 0 & \ddots & 0 \\ 0 & 0 & \frac{1}{\sqrt{s_{NN}}} \end{pmatrix}. \tag{I.17}$$

Linear independence

A set of vectors $\{\mathbf{a}_1, \ldots, \mathbf{a}_m\}$ is said to be linearly independent, if there exists no trivial set of scalars $\{x_1, \ldots, x_m\}$ such as

$$\sum_{j=1}^{m} \mathbf{a}_j \, x_j \quad = \quad \mathbf{0}. \tag{I.18}$$

In other words, the linear independence of the columns of a matrix $\mathbf{A}$ is acquired, if only the nul vector $\mathbf{x} = \mathbf{0}$ satisfies the equation $\mathbf{A}\mathbf{x} = \mathbf{0}$.

Rank of a matrix

A rectangular matrix can be subdivided into the set of column vectors which make it up. Similarly, we can also consider a set of "row vectors" of this matrix, which are defined as the column vectors of its transpose.

The rank of the columns of a matrix is the maximum number of linearly independent column vectors of this matrix. The rank of the rows is defined in an analog manner. It can be shown that the rank of the rows is equal to the rank of the columns.

The rank of a rectangular $n \times m$ matrix $\mathbf{A}$ is thus lower or equal to the smaller of its two dimensions

$$\mathrm{rank}(\mathbf{A}) \quad \leq \quad \min(n, m). \tag{I.19}$$

The rank of the matrix $\mathbf{A}$ indicates the dimension of the vector spaces[1] $\mathcal{M}(\mathbf{A})$ and $\mathcal{N}(\mathbf{A})$ spanned by the columns and the rows of $\mathbf{A}$

$$\mathcal{M}(\mathbf{A}) = \{\mathbf{y} : \mathbf{y} = \mathbf{A}\mathbf{x}\}, \qquad \mathcal{N}(\mathbf{A}) = \{\mathbf{x} : \mathbf{x} = \mathbf{y}^\top \mathbf{A}\}, \tag{I.20}$$

where $\mathbf{x}$ is a vector of dimension m and $\mathbf{y}$ is a vector of dimension n.

Inverse matrix

A square $n \times n$ matrix $\mathbf{A}$ is singular, if $\text{rank}(\mathbf{A}) < n$, and non singular, if $\text{rank}(\mathbf{A}) = n$.

If $\mathbf{A}$ is non singular, an inverse matrix $\mathbf{A}^{-1}$ exists, such as

$$\mathbf{A}\,\mathbf{A}^{-1} = \mathbf{A}^{-1}\,\mathbf{A} = \mathbf{I} \tag{I.21}$$

and $\mathbf{A}$ is said to be invertible.

The inverse $\mathbf{Q}^{-1}$ of an orthogonal matrix $\mathbf{Q}$ is its transpose $\mathbf{Q}^\top$.

Determinant of a matrix

The determinant of a square $n \times n$ matrix $\mathbf{A}$ is

$$\det(\mathbf{A}) = \sum (-1)^{N(k_1,\ldots,k_n)} \prod_{i=1}^{n} a_{ik_i}, \tag{I.22}$$

where the sum is taken over all the permutations $(k_1, \ldots, k_n)$ of the integers $(1, \ldots, n)$ and where $N(k_1, \ldots, k_n)$ is the number of transpositions of two integers necessary to pass from the starting set $(1, \ldots, n)$ to a given permutation $(k_1, \ldots, k_n)$ of this set.

In the case of a 2×2 matrix there is the well-known formula

$$\mathbf{A} = \begin{pmatrix} a & b \\ c & d \end{pmatrix}, \qquad \det(\mathbf{A}) = a\,d - b\,c. \tag{I.23}$$

A non-zero determinant indicates that the corresponding matrix is invertible.

Trace

The trace of a square $n \times n$ matrix is the sum of its diagonal elements,

$$\text{tr}(\mathbf{A}) = \sum_{i=1}^{n} a_{ii}. \tag{I.24}$$

[1] do not mix up: the dimension of a vector and the dimension of a vector space.

Eigenvalues

Let $\mathbf{A}$ be square $n \times n$. The characteristic equation

$$\det(\lambda\mathbf{I} - \mathbf{A}) = 0 \tag{I.25}$$

has n in general complex solutions λ called the *eigenvalues* of $\mathbf{A}$.

The sum of the eigenvalues is equal to the trace of the matrix

$$\mathrm{tr}(\mathbf{A}) = \sum_{p=1}^{n} \lambda_p. \tag{I.26}$$

The product of the eigenvalues is equal to the determinant of the matrix

$$\det(\mathbf{A}) = \prod_{p=1}^{n} \lambda_p. \tag{I.27}$$

If $\mathbf{A}$ is symmetric, all its eigenvalues are real.

Eigenvectors

Let $\mathbf{A}$ be a square matrix and λ an eigenvalue of $\mathbf{A}$. Then vectors $\mathbf{x}$ and $\mathbf{y}$ exist, which are not equal to the zero vector $\mathbf{0}$, satisfying

$$(\lambda\mathbf{I} - \mathbf{A})\mathbf{x} = 0, \qquad \mathbf{y}^\top(\lambda\mathbf{I} - \mathbf{A}) = 0, \tag{I.28}$$

i.e.

$$\mathbf{A}\,\mathbf{x} = \lambda\mathbf{x}, \qquad \mathbf{y}^\top\mathbf{A} = \lambda\,\mathbf{y}^\top. \tag{I.29}$$

The vectors $\mathbf{x}$ are the eigenvectors of the columns of $\mathbf{A}$ and the $\mathbf{y}$ are the eigenvectors of the rows of $\mathbf{A}$.

When $\mathbf{A}$ is symmetric, it is not necessary to distinguish between the eigenvectors of the columns and of the rows.

EXERCISE I.5 *Show that the eigenvalues of a squared symmetric matrix $\mathbf{A}^2 = \mathbf{A}\mathbf{A}$ are equal to the square of the eigenvalues of $\mathbf{A}$. And that any eigenvector of $\mathbf{A}$ is an eigenvector of $\mathbf{A}^2$.*

Positive definite matrix

Definition: a symmetric $n \times n$ matrix $\mathbf{A}$ is positive definite, iff (if and only if) for any non zero vector $\mathbf{x}$ the quadratic form

$$\mathbf{x}^\top\mathbf{A}\mathbf{x} > 0. \tag{I.30}$$

Similarly $\mathbf{A}$ is said to be positive semi-definite (non negative definite), iff $\mathbf{x}^\top\mathbf{A}\mathbf{x} \geq 0$ for any vector $\mathbf{x}$. Furthermore, $\mathbf{A}$ is said to be indefinite, iff $\mathbf{x}^\top\mathbf{A}\mathbf{x} > 0$ for some $\mathbf{x}$ and $\mathbf{x}^\top\mathbf{A}\mathbf{x} < 0$ for other $\mathbf{x}$.

We notice that this definition is similar to the definition of a positive definite function, e.g. the covariance function $C(\mathbf{h})$.

We now list three very useful criteria for positive semi-definite matrices.

First Criterion: $\mathbf{A}$ is positive semi-definite, iff a matrix $\mathbf{W}$ exists such as $\mathbf{A} = \mathbf{W}^\top \mathbf{W}$.

In this context $\mathbf{W}$ is sometimes written $\sqrt{\mathbf{A}}$.

Second Criterion: $\mathbf{A}$ is positive semi-definite, iff all its n eigenvalues $\lambda_p \geq 0$.

Third Criterion: $\mathbf{A}$ is positive semi-definite, iff all its principal minors are non negative.

A principal minor is the determinant of a principal submatrix of $\mathbf{A}$. A principal submatrix is obtained by leaving out k columns ($k = 0, 1, \ldots, n-1$) and the corresponding rows which cross them at the diagonal elements. The combinatorial of the principal minors to be checked makes this criterion less interesting for applications with $n > 3$.

EXAMPLE I.6 *Using the first criterion we note that the variance-covariance matrix is positive semi-definite by construction:*

$$\mathbf{V} \;=\; \frac{1}{n}\,(\mathbf{Z}-\mathbf{M})^\top\,(\mathbf{Z}-\mathbf{M}) = \mathbf{W}^\top\mathbf{W} \tag{I.31}$$

with $\mathbf{W} = \dfrac{1}{\sqrt{n}}\,(\mathbf{Z}-\mathbf{M})$.

EXAMPLE I.7 *It is easy to check with the third criterion that the left hand* $(n+1) \times (n+1)$ *matrix of the ordinary kriging system, expressed with covariances, is not positive semi-definite (contrarily to the left hand matrix of simple kriging).*

The principal submatrix $\mathbf{S}$ *obtained by leaving out all columns and rows crossing diagonal elements, except for the last two,*

$$\mathbf{S} \;=\; \begin{pmatrix} C(\mathbf{x}_n{-}\mathbf{x}_n) & 1 \\ 1 & 0 \end{pmatrix}, \tag{I.32}$$

has a determinant equal to -1 *for* $C(\mathbf{x}_n{-}\mathbf{x}_n) = 1$.

The computation of the eigenvalues of the left hand matrix of OK yields one negative eigenvalue (resulting from the universality condition) and n *positive (or zero) eigenvalues, on the basis of covariances. Thus built up using a covariance function, this matrix is indefinite, while it is negative (semi-)definite with variogram values.*

Decomposition into eigenvalues and eigenvectors

With a symmetric matrix $\mathbf{A}$ the eigenvalues together with eigenvectors normed to unity form the system

$$\mathbf{A}\,\mathbf{Q} \;=\; \mathbf{Q}\,\mathbf{\Lambda} \qquad \text{with} \quad \mathbf{Q}^{\top}\mathbf{Q} = \mathbf{I}, \tag{I.33}$$

where $\mathbf{\Lambda}$ is the diagonal matrix of eigenvalues and $\mathbf{Q}$ is the orthogonal matrix of eigenvectors.

As $\mathbf{Q}^{\top} = \mathbf{Q}^{-1}$, this results in a decomposition of the symmetric matrix $\mathbf{A}$

$$\mathbf{A} \;=\; \mathbf{Q}\,\mathbf{\Lambda}\,\mathbf{Q}^{\top}. \tag{I.34}$$

Singular value decomposition

Multivariate analysis being the art to decompose tables of numbers, a decomposition which can be applied to any rectangular matrix (in the same spirit as the decomposition of a symmetric matrix into eigenvalues and eigenvectors) is to play a central role in data analysis.

The decomposition into singular values μ_p of a rectangular $n \times m$ matrix $\mathbf{A}$ of rank r can be written as:

$$\underset{(n \times m)}{\mathbf{A}} \;=\; \underset{(n \times n)}{\mathbf{Q}_1} \cdot \underset{(n \times m)}{\mathbf{\Sigma}} \cdot \underset{(m \times m)}{\mathbf{Q}_2^{\top}} \tag{I.35}$$

where $\mathbf{Q}_1$ and $\mathbf{Q}_2$ are orthogonal matrices and where $\mathbf{\Sigma}$ is a rectangular matrix with r positive values μ_p on the diagonal (the set of elements with equal indices) and zeroes elsewhere. For example, in the case $n > m$ and $r = m$, the matrix $\mathbf{\Sigma}$ will have the following structure

$$\mathbf{\Sigma} \;=\; \begin{pmatrix} \mu_1 & 0 & 0 \\ 0 & \ddots & 0 \\ 0 & 0 & \mu_r \\ 0 & 0 & 0 \\ \vdots & \vdots & \vdots \\ 0 & 0 & 0 \end{pmatrix}. \tag{I.36}$$

Such a decomposition always exists and can be obtained by computing the eigenvalues λ_p of $\mathbf{A}\mathbf{A}^{\top}$ and of $\mathbf{A}^{\top}\mathbf{A}$, which are identical, and positive or zero. The singular values are defined as the square roots of the non zero eigenvalues

$$\mu_p \;=\; \sqrt{\lambda_p}. \tag{I.37}$$

In this decomposition $\mathbf{Q}_1$ is the matrix of eigenvectors of $\mathbf{A}\mathbf{A}^{\top}$, while $\mathbf{Q}_2$ is the matrix of eigenvectors of $\mathbf{A}^{\top}\mathbf{A}$.

EXERCISE I.8 *What is the singular value decomposition of a symmetric matrix ?*

Moore-Penrose generalized inverse

An inverse matrix exists for any square non singular matrix. It is interesting to generalize the concept of an inverse to singular matrices as well as to rectangular matrices.

An $m \times n$ matrix $\mathbf{X}$ is a Moore-Penrose generalized inverse of a rectangular $n \times m$ matrix $\mathbf{A}$ if it verifies the following four conditions

$$\mathbf{A}\,\mathbf{X}\,\mathbf{A} = \mathbf{A}, \tag{I.38}$$

$$\mathbf{X}\,\mathbf{A}\,\mathbf{X} = \mathbf{X}, \tag{I.39}$$

$$(\mathbf{A}\,\mathbf{X})^{\top} = \mathbf{A}\,\mathbf{X}, \tag{I.40}$$

$$(\mathbf{X}\,\mathbf{A})^{\top} = \mathbf{X}\,\mathbf{A}. \tag{I.41}$$

Such a matrix $\mathbf{X}$ is denoted $\mathbf{A}^{+}$.

EXERCISE I.9 *Is the inverse $\mathbf{A}^{-1}$ of a square non singular matrix $\mathbf{A}$ a Moore-Penrose generalized inverse ?*

The Moore-Penrose inverse is obtained from a singular value decomposition, reversing the order of the two orthogonal matrices, transposing $\boldsymbol{\Sigma}$ and inverting each singular value:

$$\mathbf{A}^{+} = \mathbf{Q}_2\,\boldsymbol{\Sigma}^{+}\,\mathbf{Q}_1^{\top}. \tag{I.42}$$

The matrix $\boldsymbol{\Sigma}^{+}$ is of order $m \times n$ and has, taking as an example $n > m$ and $r = m$, the structure

$$\boldsymbol{\Sigma}^{+} = \begin{pmatrix} \mu_1^{-1} & 0 & 0 & 0 & \dots & 0 \\ 0 & \ddots & 0 & 0 & \dots & 0 \\ 0 & 0 & \mu_r^{-1} & 0 & \dots & 0 \end{pmatrix}$$

$$= \begin{pmatrix} \lambda_1^{-1/2} & 0 & 0 & 0 & \dots & 0 \\ 0 & \ddots & 0 & 0 & \dots & 0 \\ 0 & 0 & \lambda_r^{-1/2} & 0 & \dots & 0 \end{pmatrix}. \tag{I.43}$$

EXAMPLE I.10 (SIMPLE KRIGING WITH A DUPLICATED SAMPLE) *As a geostatistical application one might wonder if the Moore-Penrose inverse can be used to solve a kriging system whose left hand matrix is singular. We shall not treat the problem in a general manner and only solve a very simple exercise.*

A value is to be estimated at $\mathbf{x}_0$ using two values located at the same point $\mathbf{x}_1$ of the domain as represented on Figure I.1. The simple kriging system for this situation is

$$\begin{pmatrix} a & a \\ a & a \end{pmatrix} \begin{pmatrix} w_1 \\ w_2 \end{pmatrix} = \begin{pmatrix} b \\ b \end{pmatrix}, \tag{I.44}$$

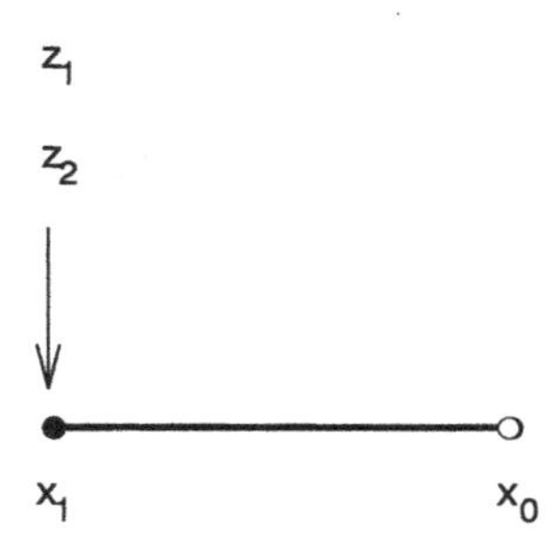

Figure I.1: Estimation point $\mathbf{x}_0$ and two samples at location $\mathbf{x}_1$.

where a is the variance of the data and where b is the covariance between the points $\mathbf{x}_1$ and $\mathbf{x}_0$.

The matrix $\mathbf{A}$ being singular we resort to the Moore-Penrose generalized inverse. Let

$$\mathbf{A}\mathbf{A}^\top = \mathbf{A}^\top\mathbf{A} = \begin{pmatrix} 2a^2 & 2a^2 \\ 2a^2 & 2a^2 \end{pmatrix} = \begin{pmatrix} c & c \\ c & c \end{pmatrix} = \mathbf{C}. \tag{I.45}$$

We have

$$\det(\mathbf{C}) = 0 \quad \Rightarrow \quad \lambda_2 = 0, \tag{I.46}$$

$$\operatorname{tr}(\mathbf{C}) = 2c \quad \Rightarrow \quad \lambda_1 = 2c, \tag{I.47}$$

and a matrix of eigenvectors (normed to one)

$$\mathbf{Q} = \begin{pmatrix} \frac{1}{\sqrt{2}} & \frac{1}{\sqrt{2}} \\ \frac{1}{\sqrt{2}} & -\frac{1}{\sqrt{2}} \end{pmatrix}. \tag{I.48}$$

The solution of the system, relying on the generalized inverse, is

$$\mathbf{w} = \mathbf{A}^+\mathbf{b} = \mathbf{Q}\,\boldsymbol{\Sigma}^+\,\mathbf{Q}^\top\mathbf{b} = \begin{pmatrix} \frac{b}{\sqrt{2c}} \\ \frac{b}{\sqrt{2c}} \end{pmatrix} = \begin{pmatrix} \frac{b}{2a} \\ \frac{b}{2a} \end{pmatrix}. \tag{I.49}$$

With centered data we have in the end the following estimation at $\mathbf{x}_0$

$$z^\star(\mathbf{x}_0) = \frac{b}{a}\left(\frac{z_1(\mathbf{x}_1) + z_2(\mathbf{x}_1)}{2}\right). \tag{I.50}$$

This solution does make sense as the estimator takes the average of the two sample values z_1 and z_2 measured at the location $\mathbf{x}_1$ and multiplies this average with the simple kriging weight obtained when only one information is available in the neighborhood of $\mathbf{x}_0$.

Linear Regression Theory

These are a few standard results of regression theory in the notation used in the rest of the book.

Best least squares estimator

The best estimation, in the least squares sense, of a random variable of interest Z_0 on the basis of a function of N random variables Z_i, $i = 1, \ldots, N$ is given by the conditional expectation of Z_0 knowing the variables Z_i

$$\mathrm{E}[\, Z_0 \,|\, Z_1, \ldots, Z_N \,] \quad = \quad m_0(\mathbf{z}), \tag{II.1}$$

where $\mathbf{z}$ is the vector of the explanative variables

$$\mathbf{z} \quad = \quad (Z_1, \ldots, Z_N)^\top. \tag{II.2}$$

In order to prove that no other function $f(\mathbf{z})$ is better than the conditional mean $m_0(\mathbf{z})$, we need to show that the mean square estimation error is larger for $f(\mathbf{z})$

$$\mathrm{E}[\,(Z_0 - f(\mathbf{z}))^2\,] \geq \mathrm{E}[\,(Z_0 - m_0(\mathbf{z}))^2\,]. \tag{II.3}$$

The mean square error using the arbitrary function f can be expanded

$$\begin{aligned} \mathrm{E}[\,(Z_0 - f(\mathbf{z}))^2\,] \quad &= \quad \mathrm{E}[\,(Z_0 - m_0(\mathbf{z}) + m_0(\mathbf{z}) - f(\mathbf{z}))^2\,] \\ &= \quad \mathrm{E}[\,(Z_0 - m_0(\mathbf{z}))^2\,] + \mathrm{E}[\,(m_0(\mathbf{z}) - f(\mathbf{z}))^2\,] \\ & \qquad + 2\,\mathrm{E}[\,(Z_0 - m_0(\mathbf{z})) \cdot (m_0(\mathbf{z}) - f(\mathbf{z}))\,]. \end{aligned} \tag{II.4}$$

The last term is in fact zero:

$$\begin{aligned} &\mathrm{E}[\,(Z_0 - m_0(\mathbf{z})) \cdot (m_0(\mathbf{z}) - f(\mathbf{z}))\,] \\ &\quad = \mathrm{E}[\,\mathrm{E}[\,(Z_0 - m_0(\mathbf{z})) \cdot (m_0(\mathbf{z}) - f(\mathbf{z}))\,|\,\mathbf{z}\,]\,] \\ &\quad = \mathrm{E}[\,\underbrace{\mathrm{E}[\,(Z_0 - m_0(\mathbf{z}))\,|\,\mathbf{z}\,]}_{0}\,(m_0(\mathbf{z}) - f(\mathbf{z}))\,] \\ &\quad = 0. \end{aligned} \tag{II.5}$$

Thus the mean square error related to an arbitrary function f is equal to the mean square error based on the conditional mean m_0 plus the mean square difference between m_0 and f.

Best linear estimator

The computation of the conditional expectation requires in general the knowledge of the joint distribution $F(Z_0, Z_1, \ldots, Z_N)$ which can be difficult to infer. An alternative, which we know from the previous section to be less effective, is to use a linear function $b + \mathbf{c}^\top \mathbf{z}$, where b is a constant and $\mathbf{c} = (c_1, \ldots, c_N)^\top$ is a vector of coefficients for the variables Z_i. The question is now to determine the linear function which is best in the least squares sense.

We first expand the mean squared estimation error around the mean m_0 of Z_0:

$$\begin{aligned} \mathrm{E}[\,(Z_0 - b - \mathbf{c}^\top \mathbf{z})^2\,] &= \mathrm{E}[\,(Z_0 - m_0 + m_0 - b - \mathbf{c}^\top \mathbf{z})^2\,] \\ &= \mathrm{E}[\,(Z_0 - m_0 - \mathbf{c}^\top \mathbf{z})^2\,] + \mathrm{E}[\,(m_0 - b)^2\,] \\ &\quad + 2\,\mathrm{E}[\,(Z_0 - m_0 - \mathbf{c}^\top \mathbf{z}) \cdot (m_0 - b)\,]. \end{aligned} \tag{II.6}$$

As the third term is zero,

$$\begin{aligned} \mathrm{E}[\,(Z_0 - m_0 - \mathbf{c}^\top \mathbf{z}) \cdot (m_0 - b)\,] &= \underbrace{\mathrm{E}[\,(Z_0 - m_0 - \mathbf{c}^\top \mathbf{z})\,]}_{0} \cdot (m_0 - b) \\ &= 0, \end{aligned} \tag{II.7}$$

the mean squared error is equal to

$$\mathrm{E}[\,(Z_0 - b - \mathbf{c}^\top \mathbf{z})^2\,] = \mathrm{E}[\,(Z_0 - m_0 - \mathbf{c}^\top \mathbf{z})^2\,] + (m_0 - b)^2 \tag{II.8}$$

and reduces to the first term for the optimal choice of $b = m_0$.

Previously we have derived $\mathbf{V}\,\mathbf{a} = \mathbf{v}_0$ as the equations yielding the solution $\mathbf{a}$ of multiple linear regression problem. We can now show that $\mathbf{a}$ is the vector of coefficients for the best linear estimator of Z_0. Suppose without loss of generality that the means are zero for all variables: $m_i = 0$ for $i = 0, 1, \ldots, N$. The mean squared estimation error is expanded again,

$$\begin{aligned} \mathrm{E}[\,(Z_0 - \mathbf{c}^\top \mathbf{z})^2\,] &= \mathrm{E}[\,(Z_0 - \mathbf{a}^\top \mathbf{z} + \mathbf{a}^\top \mathbf{z} - \mathbf{c}^\top \mathbf{z})^2\,] \\ &= \mathrm{E}[\,(Z_0 - \mathbf{a}^\top \mathbf{z})^2\,] + \mathrm{E}[\,(\mathbf{a}^\top \mathbf{z} - \mathbf{c}^\top \mathbf{z})^2\,] \\ &\quad + 2\,\mathrm{E}[\,(Z_0 - \mathbf{a}^\top \mathbf{z}) \cdot (\mathbf{a}^\top \mathbf{z} - \mathbf{c}^\top \mathbf{z})^\top\,], \end{aligned} \tag{II.9}$$

and the third term is zero,

$$\begin{aligned} &\mathrm{E}[\,(Z_0 - \mathbf{a}^\top \mathbf{z}) \cdot (\mathbf{a}^\top \mathbf{z} - \mathbf{c}^\top \mathbf{z})^\top\,] \\ &= \mathrm{E}[\,(Z_0 - \mathbf{a}^\top \mathbf{z}) \cdot (\mathbf{z}^\top \mathbf{a} - \mathbf{z}^\top \mathbf{c})\,] \\ &= \mathrm{E}[\,Z_0\,\mathbf{z}^\top \mathbf{a} - Z_0\,\mathbf{z}^\top \mathbf{c} - \mathbf{a}^\top \mathbf{z}\mathbf{z}^\top \mathbf{a} + \mathbf{a}^\top \mathbf{z}\mathbf{z}^\top \mathbf{c}\,] \\ &= \mathbf{v}_0^\top \mathbf{a} - \mathbf{v}_0^\top \mathbf{c} - \mathbf{a}^\top \mathbf{V}\,\mathbf{a} + \mathbf{a}^\top \mathbf{V}\,\mathbf{c} \\ &= \mathbf{v}_0^\top (\mathbf{a} - \mathbf{c}) - \mathbf{a}^\top \mathbf{V}\,(\mathbf{a} - \mathbf{c}) \\ &= 0 \qquad\qquad (\text{because } \mathbf{V}\,\mathbf{a} = \mathbf{v}_0). \end{aligned} \tag{II.10}$$

The mean squared estimation error

$$\mathrm{E}[\,(Z_0 - \mathbf{c}^\top \mathbf{z})^2\,] = \mathrm{E}[\,(Z_0 - \mathbf{a}^\top \mathbf{z})^2\,] + \mathrm{E}[\,(\mathbf{a}^\top \mathbf{z} - \mathbf{c}^\top \mathbf{z})^2\,] \tag{II.11}$$

is clearly lowest for $\mathbf{c} = \mathbf{a}$. Thus the multiple regression estimator provides the best linear estimation in the least squares sense

$$Z_0^\star = m_0 + \sum_{i=1}^{N} a_i \, (Z_i - m_i). \tag{II.12}$$

The estimator $Z_0^\star$ is identical with the conditional expectation $m_1(\mathbf{z})$ in the case of a multigaussian distribution function.

Projection

Let $\mathbf{Z}$ be the matrix of N centered auxiliary variables $\mathbf{z}_i$ and $\mathbf{z}_i$ be the vectors of n sample values of each variable, z_i^α. The sample values of the centered variable of interest $\mathbf{z}_0$ are not part of $\mathbf{Z}$. The variable of interest is estimated by multiple linear regression with the linear combination of $\mathbf{Z}\,\mathbf{a} = \mathbf{z}_0^\star$.

The variance-covariance matrix of the auxiliary variables is

$$\mathbf{V} = \frac{1}{n}\,\mathbf{Z}^\top \mathbf{Z}. \tag{II.13}$$

The vector of of covariances between the variable of interest and the auxiliary variables is

$$\mathbf{v}_0 = \frac{1}{n}\,\mathbf{Z}^\top \mathbf{z}_0. \tag{II.14}$$

Thus the multiple linear regression is

$$\mathbf{V}\,\mathbf{a} = \mathbf{v}_0 \quad \Longleftrightarrow \quad \mathbf{Z}^\top \mathbf{Z}\,\mathbf{a} = \mathbf{Z}^\top \mathbf{z}_0, \tag{II.15}$$

and assuming that $\mathbf{V}$ is invertible,

$$\mathbf{a} = (\mathbf{Z}^\top \mathbf{Z})^{-1}\,\mathbf{Z}^\top \mathbf{z}_0, \tag{II.16}$$

$$\mathbf{z}_0^\star = \mathbf{Z}\,(\mathbf{Z}^\top \mathbf{Z})^{-1}\,\mathbf{Z}^\top \mathbf{z}_0. \tag{II.17}$$

The matrix $\mathbf{P} = \mathbf{Z}\,(\mathbf{Z}^\top \mathbf{Z})^{-1}\,\mathbf{Z}^\top$ is symmetric and idempotent: $\mathbf{P} = \mathbf{P}^2$. It is a projection matrix, which projects orthogonally a vector of order n onto the space spanned by the columns of $\mathbf{Z}$

$$\mathbf{z}_0^\star = \mathbf{Z}\,\mathbf{a} = \mathbf{P}\,\mathbf{z}_0. \tag{II.18}$$

The vector of estimated values $\mathbf{z}_0^\star$ is the result of this projection. The vector of estimation errors,

$$\mathbf{e} = \mathbf{z}_0 - \mathbf{z}_0^\star, \tag{II.19}$$

connects $\mathbf{z}_0$ with its projection and is by construction orthogonal to $\mathbf{z}_0^\star$. We have

$$\mathbf{e} = \mathbf{z}_0 - \mathbf{Z}\,\mathbf{a} = (\mathbf{I} - \mathbf{P})\,\mathbf{z}_0, \tag{II.20}$$

and the square of the length of the error vector is given by its Euclidean norm,

$$\begin{aligned} \|\mathbf{e}\|^2 &= \mathbf{z}_0^\top \mathbf{z}_0 - 2\,\mathbf{z}_0^\top \mathbf{P}\,\mathbf{z}_0 + \mathbf{z}_0^\top \mathbf{P}\,\mathbf{P}\,\mathbf{z}_0 \\ &= \mathbf{z}_0^\top (\mathbf{I} - \mathbf{P})\,\mathbf{z}_0. \end{aligned} \tag{II.21}$$

Updating the variance-covariance matrix

The effect on the variance-covariance structure of adding or removing a sample from the data set is interesting to analyze when comparing the multiple regression with the data values.

An updating formula for the inverse in RAO (1973, p33) is useful in this context. For a non-singular $N \times N$ matrix $\mathbf{A}$, vectors $\mathbf{b}$, $\mathbf{c}$ of order N the inverse of $\mathbf{A} + \mathbf{b}\,\mathbf{c}^{\top}$ can be computed from the inverse of $\mathbf{A}$ by

$$(\mathbf{A} + \mathbf{b}\,\mathbf{c}^{\top})^{-1} \quad = \quad \mathbf{A}^{-1} - \frac{\mathbf{A}^{-1}\,\mathbf{b}\,\mathbf{c}^{\top}\,\mathbf{A}^{-1}}{1 + \mathbf{c}^{\top}\,\mathbf{A}^{-1}\mathbf{b}}. \tag{II.22}$$

Applying this formula to $\mathbf{A} = n\,\mathbf{V} = \mathbf{Z}^{\top}\mathbf{Z}$, the effect of removing an N-variate sample $\mathbf{z}_\alpha$ from a row of the centered data matrix $\mathbf{Z}$ is computed by

$$\begin{aligned}
(\mathbf{Z}_{(\alpha)}^{\top}\,\mathbf{Z}_{(\alpha)})^{-1} &= (\mathbf{Z}^{\top}\,\mathbf{Z} - \mathbf{z}_\alpha\,\mathbf{z}_\alpha^{\top})^{-1} \\
&= (\mathbf{Z}^{\top}\,\mathbf{Z})^{-1} + \frac{(\mathbf{Z}^{\top}\,\mathbf{Z})^{-1}\,\mathbf{z}_\alpha\,\mathbf{z}_\alpha^{\top}\,(\mathbf{Z}^{\top}\,\mathbf{Z})^{-1}}{1 - \mathbf{z}_\alpha^{\top}\,(\mathbf{Z}^{\top}\,\mathbf{Z})^{-1}\,\mathbf{z}_\alpha} \\
&= (\mathbf{Z}^{\top}\,\mathbf{Z})^{-1} + \frac{(\mathbf{Z}^{\top}\,\mathbf{Z})^{-1}\,\mathbf{z}_\alpha\,\mathbf{z}_\alpha^{\top}\,(\mathbf{Z}^{\top}\,\mathbf{Z})^{-1}}{1 - p_{\alpha\alpha}},
\end{aligned} \tag{II.23}$$

where $p_{\alpha\alpha}$ is the element number α of the diagonal of the projection matrix $\mathbf{P}$.

Cross validation

The error vector $\mathbf{e} = \mathbf{z}_0 - \mathbf{z}_0^{\star}$ is not a suitable basis for analyzing the goodness of fit of the multiple linear regression vector $\mathbf{z}_0^{\star}$. Indeed each estimated value $z_0^{\alpha\star}$ is obtained using the same weight vector $\mathbf{a}$. This vector is actually computed from covariances set up with the sample values z_0^α and z_i^α, $i = 1, \ldots, N$.

Cross-validation consists in computing an estimated value $z_{0[\alpha]}^{\star}$ leaving the values related to the sample number α out when determining the weights $\mathbf{a}_{[\alpha]}$

$$z_{0[\alpha]}^{\star} \quad = \quad \mathbf{z}_\alpha^{\top}\,\mathbf{a}_{[\alpha]}. \tag{II.24}$$

The notation $[\alpha]$ indicates that the sample values of index α have not been included into the calculations for establishing the weights $\mathbf{a}_{[\alpha]}$.

A recursive relation between $\mathbf{a}_{[\alpha]}$ and $\mathbf{a}_\alpha$ exists,

$$\begin{aligned}
\mathbf{a}_{[\alpha]} &= (\mathbf{Z}_{[\alpha]}^{\top}\,\mathbf{Z}_{[\alpha]})^{-1}\,\mathbf{Z}_{[\alpha]}^{\top}\mathbf{z}_0^{[\alpha]} \\
&= (\mathbf{Z}_{[\alpha]}^{\top}\,\mathbf{Z}_{[\alpha]})^{-1}\,(\mathbf{Z}^{\top}\,\mathbf{z}_0 - \mathbf{z}_\alpha\,z_0^\alpha) \\
&= \mathbf{a} - (\mathbf{Z}^{\top}\,\mathbf{Z})^{-1}\,\mathbf{z}_\alpha\,z_0^\alpha + \frac{(\mathbf{Z}^{\top}\,\mathbf{Z})^{-1}\,\mathbf{z}_\alpha\,\mathbf{z}_\alpha^{\top}\,\mathbf{a}}{1 - p_{\alpha\alpha}} - \frac{(\mathbf{Z}^{\top}\,\mathbf{Z})^{-1}\,\mathbf{z}_\alpha\,p_{\alpha\alpha}\,z_0^\alpha}{1 - p_{\alpha\alpha}} \\
&= \mathbf{a} - \frac{(\mathbf{Z}^{\top}\,\mathbf{Z})^{-1}\,\mathbf{z}_\alpha\,e_\alpha}{1 - p_{\alpha\alpha}},
\end{aligned} \tag{II.25}$$

where $e_\alpha = z_0^\alpha - z_0^{\alpha\star}$.

From this a recursive formula for the cross-validation error $e_{[\alpha]}$ is derived,

$$\begin{aligned} e_{[\alpha]} &= z_0^\alpha - \mathbf{z}_\alpha^\top \mathbf{a}_{[\alpha]} \\ &= z_0^\alpha - \mathbf{z}_\alpha^\top \left(\mathbf{a} - \frac{(\mathbf{Z}^\top \mathbf{Z})^{-1} \mathbf{z}_\alpha e_\alpha}{1 - p_{\alpha\alpha}} \right) \\ &= e_\alpha - \frac{p_{\alpha\alpha} e_\alpha}{1 - p_{\alpha\alpha}} \\ &= \frac{e_\alpha}{1 - p_{\alpha\alpha}}. \end{aligned} \tag{II.26}$$

An estimate of the mean squared cross validation error is now easily computed from the elements of the vector $\mathbf{e}$ and the diagonal elements $p_{\alpha\alpha}$ of the projection matrix,

$$\frac{1}{n} \sum_{\alpha=1}^{n} (e_{[\alpha]})^2 = \frac{1}{n} \sum_{\alpha=1}^{n} \left(\frac{e_\alpha}{1 - p_{\alpha\alpha}} \right)^2, \tag{II.27}$$

which can serve to evaluate the goodness of fit of the multiple linear regression.

Covariance and Variogram Models

This is a list of a few isotropic models commonly found in the literature. Unless specified otherwise, they are valid for vectors $\mathbf{h}$ in at least up to 3D.

Notation

- $\mathbf{h}$ is the spatial separation vector,
- $b \geq 0$ is the linear parameter,
- $a > 0$ and α are non-linear parameters,
- J_ν are Bessel functions of order ν,
- K_ν are Basset functions (modified Bessel of third kind).

Covariance Models

Nugget-effect

$$C_{nug}(\mathbf{h}) = \begin{cases} b & \text{when } |\mathbf{h}| = 0 \\ 0 & \text{when } |\mathbf{h}| > 0. \end{cases} \tag{III.1}$$

Spherical

Reference: [198], p57; [200], p86.

$$C_{sph}(\mathbf{h}) = \begin{cases} b\left(1 - \dfrac{3}{2}\dfrac{|\mathbf{h}|}{a} + \dfrac{1}{2}\dfrac{|\mathbf{h}|^3}{a^3}\right) & \text{for } 0 \leq |\mathbf{h}| \leq a \\ 0 & \text{for } |\mathbf{h}| > a. \end{cases} \tag{III.2}$$

Triangular (only for 1D)

Reference: [362], vol. 1, p131.

$$C_{tri}(h) = \begin{cases} b\left(1 - \dfrac{|h|}{a}\right) & \text{for } 0 \leq |h| \leq a \\ 0 & \text{for } |h| > a. \end{cases} \tag{III.3}$$

Cubic

Reference: [48], p.132; [359], p.290.

$$C_{cub}(\mathbf{h}) = \begin{cases} b\left(1 - \left(\dfrac{|\mathbf{h}|}{a}\right)^2 \left[7 - \dfrac{|\mathbf{h}|}{a}\left[\dfrac{35}{4} - \left(\dfrac{|\mathbf{h}|}{a}\right)^2 \left[\dfrac{7}{2} - \dfrac{3}{4}\left(\dfrac{|\mathbf{h}|}{a}\right)^2\right]\right]\right]\right) \\ \qquad\qquad\qquad\qquad\qquad\qquad \text{for } 0 \leq |\mathbf{h}| \leq a \\ 0 \qquad\qquad \text{for } |\mathbf{h}| > a. \end{cases} \tag{III.4}$$

Stable

Reference: [362], vol. 1, p364.

$$C_{exp\text{-}\alpha}(\mathbf{h}) = b \exp\left(-\frac{|\mathbf{h}|^\alpha}{a}\right) \qquad \text{with } 0 < \alpha \leq 2. \tag{III.5}$$

Exponential

$$C_{exp}(\mathbf{h}) = b \exp\left(-\frac{|\mathbf{h}|}{a}\right). \tag{III.6}$$

Gaussian

$$C_{gaus}(\mathbf{h}) = b \exp\left(-\frac{|\mathbf{h}|^2}{a}\right). \tag{III.7}$$

Hole-effect

Reference: [362], vol. 1, p366.

$$C_{hol}(\mathbf{h}) = b \exp\left(-\frac{|\mathbf{h}|}{a}\right) \cos\omega|\mathbf{h}| \qquad \text{with } a > \frac{1}{\sqrt{3}\,\omega}. \tag{III.8}$$

Whittle-Matern

Reference: [198], p43; [362], vol. 1, p363; [111], p64.

$$C_{wm}(\mathbf{h}) = b \left(\frac{|\mathbf{h}|}{a}\right)^{\nu} K_{\nu}\left(\frac{|\mathbf{h}|}{a}\right) \qquad \text{with } \nu \geq 0. \tag{III.9}$$

Bessel

Reference: [198], p42; [362], vol. 1, p366.

$$C_{bes}(\mathbf{h}) = b \left(\frac{|\mathbf{h}|}{a}\right)^{-(n-2)/2} J_{(n-2)/2}\left(\frac{|\mathbf{h}|}{a}\right) \tag{III.10}$$

with n equal to the number of spatial dimensions.

Cauchy

Reference: [362], vol. 1, p365;[51], p86.

$$C_{cau}(\mathbf{h}) = b \left[1 + \left(\frac{|\mathbf{h}|}{a}\right)^{2}\right]^{-\alpha} \qquad \text{with } \alpha > 0. \tag{III.11}$$

Variogram models

Power

Reference: [198], p128; [362], vol. 1, p406.

$$\gamma_{pow\text{-}\alpha}(\mathbf{h}) = b\,|\mathbf{h}|^{\alpha} \qquad \text{with } 0 < \alpha < 2. \tag{III.12}$$

De Wijsian-a^2

Reference: [196], vol. 1, p75.

$$\gamma_{wijs\text{-}a^2}(\mathbf{h}) = \frac{3}{2}\, b \log(|\mathbf{h}|^2 + a^2) \qquad \text{with } a \neq 0 \quad \text{and} \quad 0 \leq b < 1. \tag{III.13}$$

De Wijsian

Reference: [196], vol. 1, p75.

$$\gamma_{wijs}(\mathbf{h}) = 3\, b \log|\mathbf{h}| \qquad \text{for } |\mathbf{h}| \neq 0 \quad \text{and} \quad 0 \leq b < 1, \tag{III.14}$$

where b is called the *absolute dispersion*.

Additional Exercices

EXERCISE IV.1 *(by C. DALY) Simple kriging with an exponential covariance model in one dimension. Let $Z(\alpha)$ be a second-order stationary random function defined at points α with $\alpha = 0, 1, \ldots, n$ located at regular intervals on a line. We wish to estimate a value for the next point $\alpha{+}1$ on the line.*
i) Set up the simple kriging equations for an exponential covariance function

$$C(h) = b \, \exp\left(-\frac{|h|}{a}\right), \tag{IV.1}$$

where a is a range parameter.
ii) The solution of the system is

$$w_\alpha = 0 \;\; \text{for} \;\; \alpha = 0, \ldots, n{-}1, \qquad\qquad w_n = c. \tag{IV.2}$$

Compute the value of c.

In time series analysis the random function

$$Z(n) = p\, Z(n-1) + \varepsilon_n \tag{IV.3}$$

is called an autoregressive process of order one AR(1), where p is a weight and ε_n is a nugget-effect model

$$\operatorname{cov}(\varepsilon_n, \varepsilon_m) = \delta_{nm}\, \sigma^2 = \begin{cases} \sigma^2 & \text{if n=m} \\ 0 & \text{otherwise.} \end{cases} \tag{IV.4}$$

We restrict the exercise to stationary AR(1) processes with $0 < p < 1$.
iii) Show that $Z(n)$ can be expressed as a sum of values ε_α where $\alpha = -\infty, \ldots, n$.
iv) Compute the covariance $\operatorname{cov}(Z(n), Z(m))$ and the correlation coefficient for $n, m \in \mathbb{Z}$.
v) Compare the Box & Jenkins AR(1) estimator

$$Z^\star(n+1) = p\, Z(n) \tag{IV.5}$$

with the simple kriging solution obtained in ii).

EXERCISE IV.2 *$Z(\mathbf{x})$ is a second-order stationary random function with a zero mean, split into two uncorrelated zero mean components, which has the regionalization model*

$$Z(\mathbf{x}) = Y^S(\mathbf{x}) + Y^L(\mathbf{x}), \tag{IV.6}$$

where $Y^S(\mathbf{x})$ is a component with a short range covariance $C^S(\mathbf{h})$ and where $Y^L(\mathbf{x})$ is a component with a long range covariance $C^L(\mathbf{h})$,

$$C(\mathbf{h}) = C^S(\mathbf{h}) + C^L(\mathbf{h}). \tag{IV.7}$$

Data is located on a regular grid and simple kriging (i.e. without condition on the weights) is performed at the nodes of this grid to estimate short and long range components, using a neighborhood incorporating all data.

What relation can be established between the weights λ_α^S and λ_α^L of the two components at each point in the neighborhood ?

EXERCISE IV.3 *$Z(\mathbf{x})$ is a locally second-order stationary random function composed of uncorrelated zero mean components $Y^S(\mathbf{x})$ and $Y^L(\mathbf{x})$ as well as a drift which is approximately constant in any local neighborhood of the domain.*

Show that the sum of the krigings of the components and of the kriging of the mean is equal to the ordinary kriging of $Z(\mathbf{x})$,

$$Y_S^\star(\mathbf{x}_0) + Y_L^\star(\mathbf{x}_0) + m_l^\star(\mathbf{x}_0) = Z^\star(\mathbf{x}_0). \tag{IV.8}$$

Solutions to Exercises

EXERCISE 2.2 Applying the definition (2.18) of the variance:

$$\begin{aligned} \operatorname{var}(a\,Z) &= \mathrm{E}\Big[\,(a\,Z - \mathrm{E}[\,a\,Z\,])^2\Big] \;=\; \mathrm{E}[\,(a\,Z - a\,m)^2\,] \\ &= \mathrm{E}[\,a^2\,(Z-m)^2\,] \;=\; a^2 \operatorname{var}(Z), \end{aligned} \qquad (\mathrm{V.1})$$

and

$$\operatorname{var}(Z+b) \;=\; \mathrm{E}\Big[\,(Z+b-\mathrm{E}[\,Z\,]-b)^2\Big] \;=\; \operatorname{var}(Z). \qquad (\mathrm{V.2})$$

EXERCISE 2.3 Applying the definition (2.18) of the variance, multiplying out the squares and taking the expectations:

$$\begin{aligned} \operatorname{var}(Z_i+Z_j) &= \mathrm{E}\Big[\,(Z_i+Z_j-\mathrm{E}[\,Z_i+Z_j\,])^2\Big] \;=\; \mathrm{E}[\,(Z_i+Z_j-m_i-m_j)^2\,] \\ &= \mathrm{E}[\,((Z_i-m_i)+(Z_j-m_j))^2\,] \\ &= \mathrm{E}[\,(Z_i-m_i)^2\,]+\mathrm{E}[\,(Z_j-m_j)^2\,]-2\,\mathrm{E}[\,(Z_i-m_i)\cdot(Z_j-m_j)\,] \\ &= \operatorname{var}(Z_i)+\operatorname{var}(Z_j)-2\operatorname{cov}(Z_i,Z_j), \end{aligned} \qquad (\mathrm{V.3})$$

which is equal to the sum of the variances only, when the two variables are uncorrelated.

EXERCISE 17.4 We have $\frac{1}{n}\widetilde{\mathbf{Z}}^\top\mathbf{Y} = \frac{1}{n}\widetilde{\mathbf{Z}}^\top\widetilde{\mathbf{Z}}\,\widetilde{\mathbf{Q}}$, because $\mathbf{Y} = \widetilde{\mathbf{Z}}\,\widetilde{\mathbf{Q}}$.

$\mathbf{R}\,\widetilde{\mathbf{Q}} = \widetilde{\mathbf{Q}}\,\widetilde{\boldsymbol{\Lambda}}$ is the eigendecomposition of $\mathbf{R}$.

Therefore $\operatorname{cov}(\widetilde{z}_i, y_p) = \widetilde{\lambda}_p\,\widetilde{q}_{ip}$ and dividing by the standard deviation of y_p, which is $\sqrt{\widetilde{\lambda}_p}$, we obtain the correlation coefficient between the variable and the factor.

If the standardized variable is uncorrelated (orthogonal) with all others, an eigenvector $\mathbf{q}_p$ exists with all elements zero except for the element q_{ip} corresponding to that variable. As this element q_{ip} is 1 because of normation and the variable is identical with the factor, the eigenvalue has also to be 1.

EXERCISE 17.5

$$\begin{aligned} \mathbf{R} &= \operatorname{corr}(\widetilde{\mathbf{Z}},\widetilde{\mathbf{Z}}) = \widetilde{\mathbf{Q}}\,\sqrt{\widetilde{\boldsymbol{\Lambda}}}\,\sqrt{\widetilde{\boldsymbol{\Lambda}}}\,\widetilde{\mathbf{Q}}^\top \\ &= \operatorname{corr}(\widetilde{\mathbf{Z}},\mathbf{Y})\,[\operatorname{corr}(\,\widetilde{\mathbf{Z}},\mathbf{Y}\,)\,]^\top. \end{aligned} \qquad (\mathrm{V.4})$$

EXERCISE 18.1 The orthogonality constraints are not active like in PCA.

EXERCISE 18.2 Non active orthogonality constraints.

EXERCISE 18.3 Multiply the equation by the inverse of λ.

EXERCISE 20.2 $C_{12}(\mathbf{h}) = a_1\, C_{22}(\mathbf{h}+\mathbf{r}_1) + a_2\, C_{22}(\mathbf{h}+\mathbf{r}_2)$

EXERCISE 21.1

$$\begin{aligned} f(\omega) &= \frac{1}{2\pi}\int_{-\infty}^{+\infty} b\,\mathrm{e}^{-a|h|-\mathrm{i}\omega h}\,dh \\ &= \frac{b}{2\pi}\Big[\int_{-\infty}^{0}\mathrm{e}^{(a-\mathrm{i}\omega)h}\,dh + \int_{0}^{\infty}\mathrm{e}^{-(a+\mathrm{i}\omega)h}\,dh\Big] \\ &= \frac{b}{2\pi}\Big[\frac{1}{a-\mathrm{i}\omega}+\frac{1}{a+\mathrm{i}\omega}\Big] = \frac{b}{\pi}\frac{a}{a^2+\omega^2} \\ &\geq 0 \qquad \text{for any } \omega. \end{aligned} \tag{V.5}$$

EXERCISE 21.2 The inequality between the spectral densities,

$$\frac{a_i\, a_j}{(a_i+\omega^2)\,(a_j+\omega^2)} \;\geq\; \left(\frac{\dfrac{a_i+a_j}{2}}{\left(\dfrac{a_i+a_j}{2}\right)^2+\omega^2}\right)^2, \tag{V.6}$$

is rewritten as

$$\frac{a_i\, a_j}{\left(\dfrac{a_i+a_j}{2}\right)^2} \;\geq\; \frac{(a_i+\omega^2)\,(a_j+\omega^2)}{\left(\left(\dfrac{a_i+a_j}{2}\right)^2+\omega^2\right)^2}. \tag{V.7}$$

This inequality is false for $a_i \neq a_j$ when $\omega \to \infty$ because the left hand side is a constant < 1 while the right hand is dominated for large values of ω by a term ω^4 appearing both in the numerator and in the denominator. Thus the set of direct and cross covariance function is not an authorized covariance function matrix.

In this exercise the sills were implicitly set to one (which implies a linear correlation coefficient equal to one). YAGLOM [362], vol. I, p. 315, gives the example of a bivariate exponential covariance function model in which the range parameters a are linked to the sill parameters b in such a way that to a given degree of uncorrelatedness corresponds an interval of permissible ranges.

EXERCISE 22.1

$$\sum_{i=1}^{N}\sum_{j=1}^{N} w_i\, w_j\, b_{ij} \;\geq\; 0 \quad\text{and}\quad \sum_{\alpha=1}^{n}\sum_{\beta=1}^{n} w_\alpha\, w_\beta\, \rho(\mathbf{x}_\alpha-\mathbf{x}_\beta) \;\geq\; 0, \tag{V.8}$$

because $\mathbf{B}$ is positive semi-definite and $\rho(\mathbf{h})$ is a normalized covariance function.

Thus

$$\Big(\sum_{i=1}^{N}\sum_{j=1}^{N} w_i\, w_j\, b_{ij}\Big) \cdot \Big(\sum_{\alpha=1}^{n}\sum_{\beta=1}^{n} w_\alpha\, w_\beta\, \rho(\mathbf{x}_\alpha - \mathbf{x}_\beta)\Big) \tag{V.9}$$

$$= \sum_{i=1}^{N}\sum_{j=1}^{N}\sum_{\alpha=1}^{n}\sum_{\beta=1}^{n} w_\alpha^i\, w_\beta^j\, b_{ij}\, \rho(\mathbf{x}_\alpha - \mathbf{x}_\beta) \geq 0 \tag{V.10}$$

with $w_\alpha^i = w_i\, w_\alpha$.

EXERCISE 22.2 $\gamma(\mathbf{h})$ is a conditionally negative definite function and $-\gamma(\mathbf{h})$ is conditionally positive definite. $-\mathbf{B}\,\gamma(\mathbf{h})$ is a conditionally positive definite function matrix. The demonstration is in the same spirit as for a covariance function matrix.

EXERCISE 23.2

The diagonal elements of the coregionalization matrix are positive and the determinant is

$$\begin{vmatrix} 2 & -1 \\ -1 & 1 \end{vmatrix} = 2 - 1 = 1. \tag{V.11}$$

The principal minors are positive and the matrix is positive definite. The correlation coefficient is $\rho = -1/\sqrt{2} = -.71$.

EXERCISE 23.3 As there is no information about $Z_1(x)$ at the point x_2, the second row of the system and the second column of the left hand matrix vanish.

EXERCISE 23.4 The three vectors of weights for the three target points are

$$\begin{pmatrix} w_1 \\ w_2 \\ w_3 \end{pmatrix} = \begin{pmatrix} 0 \\ 0 \\ 0 \end{pmatrix}, \begin{pmatrix} 0 \\ 0 \\ -5/16 \end{pmatrix}, \begin{pmatrix} 0 \\ 0 \\ -1 \end{pmatrix}. \tag{V.12}$$

EXERCISE 24.1 With $\sigma_{ZZ} = \sigma_{SS} = 1$ we have the system

$$\begin{pmatrix} \mathbf{R} & \rho_{ZS}\,\mathbf{R} & \rho_{ZS}\,\mathbf{r}_0 \\ \rho_{ZS}\,\mathbf{R} & \mathbf{R} & \mathbf{r}_0 \\ \rho_{ZS}\,\mathbf{r}_0^\top & \mathbf{r}_0^\top & 1 \end{pmatrix} \begin{pmatrix} \mathbf{w}_Z \\ \mathbf{w}_S \\ w_0 \end{pmatrix} = \begin{pmatrix} \mathbf{r}_0 \\ \rho_{ZS}\,\mathbf{r}_0 \\ \rho_{ZS} \end{pmatrix}. \tag{V.13}$$

Can the vector $(\mathbf{w}_Z, 0, w_0)^\top$ be a solution? The system would reduce to

$$\begin{cases} \mathbf{w}_Z\,\mathbf{R} & + \quad w_0\,\rho_{ZS}\,\mathbf{r}_0 & = \quad \mathbf{r}_0 \\ \mathbf{w}_Z\,\rho_{ZS}\,\mathbf{R} & + \quad w_0\,\mathbf{r}_0 & = \quad \rho_{ZS}\,\mathbf{r}_0 \\ \mathbf{w}_Z\,\rho_{ZS}\,\mathbf{r}_0^\top & + \quad w_0 & = \quad \rho_{ZS}. \end{cases} \tag{V.14}$$

For $\rho_{ZS} = 0$ we have a trivial reduction to collocated cokriging with $w_0 = 0$ and $\mathbf{w}_z$ being the simple kriging weights. For $\rho_{ZS} = \pm 1$ we also are back to it with $w_0 = \pm 1$ and $\mathbf{w}_Z = 0$.

For $\rho_{ZS} \neq 0$, $\rho_{ZS} \neq \pm 1$ let us multiply the first equation block with ρ_{ZS} and substract it from the second. We get $w_0\, \rho_{ZS}^2\, \mathbf{r}_0 = w_0\, \mathbf{r}_0$ so that w_0 would be zero. This is not the collocated cokriging solution.

EXERCISE 26.1 When all coregionalization matrices $\mathbf{B}_u$ are proportional to one matrix $\mathbf{B}$ we have:

$$\mathbf{C}(\mathbf{h}) = \sum_{u=0}^{S} a_u\, \mathbf{B}\, \rho_u(\mathbf{h}) = \mathbf{B} \sum_{u=0}^{S} a_u\, \rho_u(\mathbf{h}), \tag{V.15}$$

where the a_u are the coefficients of proportionality.

EXERCISE 26.2 As a matrix $\mathbf{B}_u$ is positive semi-definite by definition, the positivity of its second order minors implies an inequality between direct and cross sills,

$$|b_{ij}^u| \leq \sqrt{b_{ii}^u\, b_{jj}^u}, \tag{V.16}$$

from which the assertions are easily deduced.

EXERCISE 29.1 For $w_\alpha^{\mathrm{Re}} = w_\alpha^{\mathrm{Im}}$ we have

$$2 \sum_{\alpha=1}^{n} \sum_{\beta=1}^{n} w_\alpha^{\mathrm{Re}}\, w_\beta^{\mathrm{Re}}\, C^{\mathrm{Re}}(\mathbf{x}_\alpha - \mathbf{x}_\beta) \geq 0 \tag{V.17}$$

with any set of weights w_α^{Re}. Thus $C^{\mathrm{Re}}(\mathbf{h})$ is a positive definite function.

EXERCISE 29.2

$$\mathrm{var}(\,(1+\mathrm{i})\, Z(\mathbf{0}) + (1-\mathrm{i})\, Z(\mathbf{h})\,) = 4C^{\mathrm{Re}}(\mathbf{0}) + 4C^{\mathrm{Im}}(\mathbf{h}) \geq 0, \tag{V.18}$$

$$\mathrm{var}(\,(1-\mathrm{i})\, Z(\mathbf{0}) + (1+\mathrm{i})\, Z(\mathbf{h})\,) = 4C^{\mathrm{Re}}(\mathbf{0}) - 4C^{\mathrm{Im}}(\mathbf{h}) \geq 0, \tag{V.19}$$

and we have $C^{\mathrm{Re}}(\mathbf{0}) \geq |C^{\mathrm{Im}}(\mathbf{h})|$.

EXERCISE 29.3 The estimation variance is:

$$\begin{aligned}
\mathrm{var}(Z(\mathbf{x}_0) - Z_{\mathrm{CK}}^\star(\mathbf{x}_0)) &= \mathrm{E}\Big[\, (Z(\mathbf{x}_0) - Z_{\mathrm{CK}}^\star(\mathbf{x}_0)) \cdot \overline{(Z(\mathbf{x}_0) - Z_{\mathrm{CK}}^\star(\mathbf{x}_0))}\,\Big] \\
&= \mathrm{E}\Big[\Big((U(\mathbf{x}_0) - U_{\mathrm{CK}}^\star(\mathbf{x}_0)) + \mathrm{i}\,(V(\mathbf{x}_0) - V_{\mathrm{CK}}^\star(\mathbf{x}_0))\Big) \\
&\qquad \times \Big((U(\mathbf{x}_0) - U_{\mathrm{CK}}^\star(\mathbf{x}_0)) - \mathrm{i}\,(V(\mathbf{x}_0) - V_{\mathrm{CK}}^\star(\mathbf{x}_0))\Big)\Big] \\
&= \mathrm{var}(U(\mathbf{x}_0) - U_{\mathrm{CK}}^\star(\mathbf{x}_0)) \\
&\quad + \mathrm{var}(V(\mathbf{x}_0) - V_{\mathrm{CK}}^\star(\mathbf{x}_0)).
\end{aligned} \tag{V.20}$$

EXERCISE 29.4 The estimation variance is:

$$\mathrm{var}(U(\mathbf{x}_0) - U_{\mathrm{CK}}^\star(\mathbf{x}_0)) = C_{UU}(\mathbf{x}_0 - \mathbf{x}_0) + \sum_{\alpha=1}^{n} \sum_{\beta=1}^{n} \mu_\alpha^1\, \mu_\beta^1\, C_{UU}(\mathbf{x}_\alpha - \mathbf{x}_\beta)$$

$$
\begin{aligned}
&+\sum_{\alpha=1}^{n}\sum_{\beta=1}^{n} \nu_\alpha^1\, \nu_\beta^1\, C_{VV}(\mathbf{x}_\alpha - \mathbf{x}_\beta)\\
&+2\sum_{\alpha=1}^{n}\sum_{\beta=1}^{n} \mu_\alpha^1\, \nu_\beta^1\, C_{UV}(\mathbf{x}_\alpha - \mathbf{x}_\beta)\\
&-2\sum_{\alpha=1}^{n} \mu_\alpha^1\, C_{UU}(\mathbf{x}_\alpha - \mathbf{x}_0)\\
&-2\sum_{\alpha=1}^{n} \nu_\alpha^1\, C_{UV}(\mathbf{x}_\alpha - \mathbf{x}_0).
\end{aligned}
\tag{V.21}
$$

EXERCISE 29.5 With an even cross covariance function $C_{UV}(\mathbf{h}) = C_{VU}(\mathbf{h})$ the kriging variance of complex kriging is equal to

$$
\begin{aligned}
\sigma^2_{\mathrm{CC}} &= \sigma^2 - \mathbf{C}^{\mathrm{Re}^\top}\, \mathbf{w}^{\mathrm{Re}}\\
&= \sigma^2 - (\mathbf{c}_{UU} + \mathbf{c}_{VV})^\top\, \mathbf{w}^{\mathrm{Re}}\\
&= \sigma^2 - \mathbf{w}^{\mathrm{Re}^\top}\, (\mathbf{C}_{UU} + \mathbf{C}_{VV})\, \mathbf{w}^{\mathrm{Re}},
\end{aligned}
\tag{V.22}
$$

because the weights $\mathbf{w}^{\mathrm{Re}}$ satisfy the equation system

$$
\mathbf{C}^{\mathrm{Re}^\top}\, \mathbf{w}^{\mathrm{Re}} = \mathbf{c}^{\mathrm{Re}} \iff (\mathbf{C}_{UU} + \mathbf{C}_{VV})\, \mathbf{w}^{\mathrm{Re}} = \mathbf{c}_{UU} + \mathbf{c}_{VV}.
\tag{V.23}
$$

The kriging variance of the separate kriging is

$$
\sigma^2_{\mathrm{CC}} = \sigma^2 - \mathbf{c}_{UU}^\top\, \mathbf{w}_{\mathrm{K}}^1 - \mathbf{c}_{VV}^\top\, \mathbf{w}_{\mathrm{K}}^2,
\tag{V.24}
$$

where the weights are solution of the two simple kriging systems

$$
\mathbf{C}_{UU}\, \mathbf{w}_{\mathrm{K}}^1 = \mathbf{c}_{UU} \quad \text{and} \quad \mathbf{C}_{VV}\, \mathbf{w}_{\mathrm{K}}^2 = \mathbf{w}_{\mathrm{K}}^2.
\tag{V.25}
$$

The difference between the two kriging variances is

$$
\begin{aligned}
\sigma^2_{\mathrm{CC}} - \sigma^2_{\mathrm{KS}} &= -\mathbf{w}^{\mathrm{Re}^\top}\, (\mathbf{c}_{UU} + \mathbf{c}_{VV}) + \mathbf{c}_{UU}^\top\, \mathbf{w}_{\mathrm{K}}^1 + \mathbf{c}_{VV}^\top\, \mathbf{w}_{\mathrm{K}}^2\\
&= -\mathbf{w}^{\mathrm{Re}^\top}\, (\mathbf{C}_{UU}\, \mathbf{w}_{\mathrm{K}}^1 + \mathbf{C}_{VV}\, \mathbf{w}_{\mathrm{K}}^2) + \mathbf{c}_{UU}^\top\, \mathbf{w}_{\mathrm{K}}^1 + \mathbf{c}_{VV}^\top\, \mathbf{w}_{\mathrm{K}}^2\\
&= (\mathbf{w}_{\mathrm{K}}^1 - \mathbf{w}^{\mathrm{Re}})^\top\, \mathbf{C}_{UU}\, \mathbf{w}_{\mathrm{K}}^1 + (\mathbf{w}_{\mathrm{K}}^2 - \mathbf{w}^{\mathrm{Re}})^\top\, \mathbf{C}_{VV}\, \mathbf{w}_{\mathrm{K}}^2\\
&= Q^2 + (\mathbf{w}_{\mathrm{K}}^1 - \mathbf{w}^{\mathrm{Re}})^\top\, \mathbf{C}_{UU}\, \mathbf{w}^{\mathrm{Re}}\\
&\quad + (\mathbf{w}_{\mathrm{K}}^2 - \mathbf{w}^{\mathrm{Re}})^\top\, \mathbf{C}_{VV}\, \mathbf{w}^{\mathrm{Re}},
\end{aligned}
\tag{V.26}
$$

where Q^2 is the sum of two quadratic forms

$$
\begin{aligned}
Q^2 &= (\mathbf{w}_{\mathrm{K}}^1 - \mathbf{w}^{\mathrm{Re}})^\top\, \mathbf{C}_{UU}\, (\mathbf{w}_{\mathrm{K}}^1 - \mathbf{w}^{\mathrm{Re}})\\
&\quad + (\mathbf{w}_{\mathrm{K}}^2 - \mathbf{w}^{\mathrm{Re}})^\top\, \mathbf{C}_{VV}\, (\mathbf{w}_{\mathrm{K}}^2 - \mathbf{w}^{\mathrm{Re}}),
\end{aligned}
\tag{V.27}
$$

which are nonnegative because $\mathbf{C}_{UU}$ and $\mathbf{C}_{VV}$ are positive definite.

The difference between the two kriging variances reduces to Q^2

$$\begin{aligned}\sigma^2_{\mathrm{CC}} - \sigma^2_{\mathrm{KS}} &= Q^2 + (\mathbf{c}_{UU} + \mathbf{c}_{VV})\,\mathbf{w}^{\mathrm{Re}} - \mathbf{w}^{\mathrm{Re}^\top}\,\mathbf{C}_{UU}\,\mathbf{w}^{\mathrm{Re}} - \mathbf{w}^{\mathrm{Re}^\top}\,\mathbf{C}_{VV}\,\mathbf{w}^{\mathrm{Re}} \\ &= Q^2. \qquad \text{(V.28)}\end{aligned}$$

The variance of complex kriging is larger than the variance of the separate kriging of the real and imaginary parts when the cross covariance function is even. Equality is achieved with intrinsic correlation.

EXERCISE 33.2 To compute the cross covariance between $\varphi(U)$ and $\psi(Y)$ we first compute the second moment:

$$\begin{aligned}\mathrm{E}[\,\varphi(U)\,\psi(Y)\,] &= \mathrm{E}[\,\mathrm{E}[\,\varphi(U)\,\psi(Y)\mid Y\,]] \qquad \text{(V.29)} \\ &= \mathrm{E}[\,\psi(Y)\,\mathrm{E}[\,\varphi(U)\mid Y\,]] \qquad \text{(V.30)}\end{aligned}$$

and using (33.39) we get

$$\begin{aligned}&= \sum_{k=0}^{\infty} \frac{\varphi_k\,\rho^k}{k!}\,\mathrm{E}[\,\psi(Y)\,H_k(Y)\,] \qquad \text{(V.31)} \\ &= \sum_{k=0}^{\infty} \frac{\varphi_k\,\psi_k}{k!}\,\rho^k \qquad \text{(V.32)}\end{aligned}$$

The cross covariance is

$$\begin{aligned}\mathrm{cov}(\varphi(U),\psi(Y)) &= \left(\sum_{k=0}^{\infty} \frac{\varphi_k\,\psi_k}{k!}\,\rho^k\right) - \varphi_0\,\psi_0 \qquad \text{(V.33)} \\ &= \sum_{k=1}^{\infty} \frac{\varphi_k\,\psi_k}{k!}\,\rho^k \qquad \text{(V.34)}\end{aligned}$$

EXERCISE 34.1 Being a square integrable function the conditional expectation can be developed into factors χ_l

$$\mathrm{E}[\,\chi_k(U)\mid V\,] = \sum_{l=0}^{\infty} c_{kl}\,\chi_l(V) \qquad \text{(V.35)}$$

with coefficients c_{kl} whose value is

$$\begin{aligned}c_{kl} &= \int \mathrm{E}[\,\chi_k(U)\mid V=v\,]\,\chi_l(v)\,F(dv) \qquad \text{(V.36)} \\ &= \mathrm{E}[\,\mathrm{E}[\,\chi_k(U)\mid V\,]\,\chi_l(V)\,] \qquad \text{(V.37)} \\ &= \mathrm{E}[\,\chi_k(U)\,\chi_l(V)\,] = \delta_{kl}\,T_k \quad \text{by (34.3).} \qquad \text{(V.38)}\end{aligned}$$

Inserting the coefficients into the sum we have

$$\sum_{l=0}^{\infty} \delta_{kl}\,T_l\,\chi_l(V) = T_k\,\chi_k(V). \qquad \text{(V.39)}$$

EXERCISE 35.3 RECOVERABLE RESERVES OF LEAD

1. The anamorphosis for lead is

$$z_1 = 13 + (-6.928)(-y) + \frac{2}{2}(y^2 - 1) = y^2 + 6.928y + 12 \quad (V.40)$$
$$= (y + 3.464)^2 \quad (V.41)$$

The anamorphosis function (a parabola) is valid for $y > -3.464$ as it needs to be an increasing function.

2. The variance of the blocks is

$$\sigma_{1v}^2 = (\varphi_1)^2 r_1^2 + \frac{(\varphi_2)^2}{2} r_1^4 \quad (V.42)$$

so that $r_1 \cong .755$.

3. The equation for z_{1c} is

$$z_{1c} = \varphi_0 + \varphi_1 r_1(-y_c) + \varphi_2 \frac{r_1^2}{2}(y_c^2 - 1) \quad (V.43)$$

so that $y_c \cong .47$.

4. The tonnage of lead is $T(z_{1c}) \cong .32$.

5. The quantity of metal is

$$Q_1(z_{1c}) = \varphi_0 T(z_{1c}) - \varphi_1 r_1 g(0.47) + \frac{\varphi_2}{2} r_1^2 \cdot 0.47 \cdot g(0.47) \quad (V.44)$$
$$\cong 6.1. \quad (V.45)$$

The mean grade of the selected ore is

$$m_1(z_{1c}) = \frac{Q_1(z_{1c})}{T(z_{1c})} \cong 19.1\% \quad (V.46)$$

to be compared to the average of 13% for the whole deposit.

ANAMORPHOSIS OF SILVER

1. The anamorphosis for silver is

$$z_2 = 4 - 1.71(-u) + \frac{0.384}{2}(u^2 - 1) \quad (V.47)$$
$$= 0.192(u^2 + 8.906u + 19.83) = 0.192(u + 4.453)^2 \quad (V.48)$$

The anamorphosis is only valid for values $u > -4.453$.

2. Let $A = (r_2)^2$. The variance of the blocks is

$$\sigma_{2v}^2 = (\psi_1)^2 A + \frac{(\psi_2)^2}{2} A^4 \tag{V.49}$$

So we need to solve

$$0.0737A^2 + 2.924A - 1.26 = 0 \tag{V.50}$$

Of the two solutions we can only accept $A = 0.4266$ so that $r_2 \cong 0.653$.

RECOVERABLE RESERVES OF SILVER WHEN SELECTING ON LEAD

1. We need to solve

$$3.5 = \varphi_1 \psi_1 \, r_1 r_2 \, \rho + \frac{1}{2} \varphi_2 \psi_2 \, (r_1 r_2)^2 \, \rho^2 \tag{V.51}$$

so that $\rho \cong 0.594$ and the product $r_2 \rho \cong 0.388$.

2. The recovered quantity of silver is

$$Q_2(z_{1c}) = \psi_0 \int_{y_c}^{\infty} g(y)\, dy - \sum_{k=1}^{2} \frac{\psi_k}{k!} (r_2 \rho)^k \, H_{k-1}(y_c) \, g(y_c) \tag{V.52}$$

$$= \psi_0 \, 0.32 - \psi_1 \, 0.388 g(0.47) + \frac{\psi_2}{2} (0.388)^2 \cdot 0.47 g(0.47) \tag{V.53}$$

$$\cong 1.526 \tag{V.54}$$

The mean grade of the selected ore is

$$m_2(z_{1c}) = \frac{Q_2(z_{1c})}{T(z_{1c})} \cong 4.77 \tag{V.55}$$

to be compared to the average of 4 for the whole deposit.

LOCAL ESTIMATION OF LEAD

1. The value of the change-of-support coefficient for panels is $r_1' \cong 0.497$.

2. The development for $z_1(V)$ is

$$12.76 = 13 + (-6.928) \cdot 0.497(-y_V) + \frac{2}{2} 0.247(y_V^2 - 1) \tag{V.56}$$

so that $y_V = 0$.

3. The proportion of blocks is

$$P^\star(z_{1c}) = \mathrm{E}[\, \mathbf{1}_{Y_1(\underline{v})>y_c} | Y_V = y_V \,] \tag{V.57}$$

$$= \mathrm{E}[\, \mathbf{1}_{R\, y_v + \sqrt{1-R^2}\, W > y_c} \,] \quad \text{with} \quad W \sim N(0,1) \tag{V.58}$$

$$= 1 - G\left(\frac{y_c - R\, y_V}{\sqrt{1-R^2}} \right). \tag{V.59}$$

p	1	2	3	4	5	6	10	15
$Q^\star_V(z_{1c})$	6.074	6.019	6.019	4.824	4.824	4.765	4.738	4.735

Table V.1: The value of $Q^\star_V(z_{1c})$ truncating at different orders p.

4. For a panel value of $z_1(V) = 12.76\%$ we have a proportion of blocks

$$P^\star(z_{1c}) = 1 - G\left(\frac{y_c}{\sqrt{1-R^2}}\right) \qquad (\text{as } y_V = 0) \tag{V.60}$$

$$= 0.27. \tag{V.61}$$

5. We have

$$\mathrm{E}[\, H_p(Y_{\underline{v}}) | Y_V\,] = R^p\, H_p(Y_V), \tag{V.62}$$

so that

$$Q^\star_V(z_{1c}) = \mathrm{E}[\, Q_{\underline{v}}(z_{1c}) | Y_V\,] = \sum_{p=0}^{\infty} \frac{Q_p}{p!}\, R^p\, H_p(Y_V). \tag{V.63}$$

6. With normalized Hermite polynomials this is

$$Q^\star_V(z_{1c}) = \sum_{p=0}^{\infty} \frac{Q_p}{\sqrt{p!}}\, R^p\, \eta_p(Y_V). \tag{V.64}$$

7. The value for $p = 6$ is $Q^\star_V(z_{1c}) = 4.765$.

Table V.1 gives results for different values of p obtained with a computer, showing the slow convergence.

The mean grade of lead above cut-off within that panel is

$$\frac{4.765}{0.27} = 17.65\% \tag{V.65}$$

EXERCISE 37.2 a)

$$\sigma^2_{\mathrm{OK}} = c_{00} - \mathbf{k}_0^\top \mathbf{K}^{-1} \mathbf{k}_0 = c_{00} - \mathbf{k}_0^\top \begin{pmatrix} \mathbf{w}_{\mathrm{OK}} \\ -\mu_{\mathrm{OK}} \end{pmatrix}$$

$$= C(\mathbf{x}_0 - \mathbf{x}_0) - \sum_{\alpha=1}^{n} w_\alpha^{\mathrm{OK}}\, C(\mathbf{x}_\alpha - \mathbf{x}_0) + \mu_{\mathrm{OK}}. \tag{V.66}$$

b)

$$\begin{aligned} u &= \frac{-1}{\mathbf{1}^\top \mathbf{R}\,\mathbf{1}}; \qquad \mathbf{v} = \frac{\mathbf{R}\,\mathbf{1}}{\mathbf{1}^\top \mathbf{R}\,\mathbf{1}}; \qquad \mathbf{A} = \mathbf{R} - \frac{\mathbf{R}\,\mathbf{1}\,(\mathbf{R}\,\mathbf{1})^\top}{\mathbf{1}^\top \mathbf{R}\,\mathbf{1}}; \\ m^\star &= (\mathbf{z}^\top, 0)\,\mathbf{K}^{-1} \begin{pmatrix} \mathbf{o} \\ 1 \end{pmatrix} = (\mathbf{z}^\top, 0) \begin{pmatrix} \mathbf{v} \\ u \end{pmatrix} = \mathbf{z}^\top \mathbf{v} = \frac{\mathbf{z}^\top \mathbf{R}\,\mathbf{1}}{\mathbf{1}^\top \mathbf{R}\,\mathbf{1}}. \end{aligned} \tag{V.67}$$

c)

$$\begin{aligned} \mathcal{K}_n(\mathbf{z}, \mathbf{s}) &= \mathbf{z}^\top \mathbf{A}\,\mathbf{s} = \mathbf{1}^\top \mathbf{R}\,\mathbf{1} \left(\frac{\mathbf{z}^\top \mathbf{R}\,\mathbf{s}}{\mathbf{1}^\top \mathbf{R}\,\mathbf{1}} - \frac{\mathbf{z}^\top \mathbf{R}\,\mathbf{1}}{\mathbf{1}^\top \mathbf{R}\,\mathbf{1}} \cdot \frac{\mathbf{1}^\top \mathbf{R}\,\mathbf{s}}{\mathbf{1}^\top \mathbf{R}\,\mathbf{1}} \right) \\ &= \mathbf{1}^\top \mathbf{R}\,\mathbf{1}\,(\mathrm{E}_n[\mathbf{z}, \mathbf{s}] - \mathrm{E}_n[\mathbf{z}]\,\mathrm{E}_n[\mathbf{s}]) \\ &= \mathbf{1}^\top \mathbf{R}\,\mathbf{1}\,\mathrm{cov}_n(\mathbf{z}, \mathbf{s}). \end{aligned} \tag{V.68}$$

d)

$$\text{Let} \quad \mathbf{z}' = \begin{pmatrix} \mathbf{z} \\ 0 \end{pmatrix}, \qquad \mathbf{s}' = \begin{pmatrix} \mathbf{s} \\ 0 \end{pmatrix}, \qquad \mathbf{F}^{-1} = \begin{pmatrix} \mathbf{K} & \mathbf{s}' \\ \mathbf{s}'^\top & 0 \end{pmatrix} = \begin{pmatrix} \mathbf{A}_\mathrm{F} & \mathbf{v}_\mathrm{F} \\ \mathbf{v}_\mathrm{F}^\top & u_\mathrm{F} \end{pmatrix},$$

$$\text{with} \quad \mathbf{v}_\mathrm{F} = \frac{\mathbf{K}^{-1}\,\mathbf{s}'}{\mathbf{s}'^\top\,\mathbf{K}^{-1}\,\mathbf{s}'} \qquad \text{and} \qquad \mathbf{A}_\mathrm{F} = \mathbf{K}^{-1} - \frac{\mathbf{K}^{-1}\,\mathbf{s}'\,(\mathbf{K}^{-1}\,\mathbf{s}')^\top}{\mathbf{s}'^\top\,\mathbf{K}^{-1}\,\mathbf{s}'}. \tag{V.69}$$

Then

$$\begin{aligned} b^\star &= (\mathbf{z}^\top, 0, 0)\,\mathbf{F}^{-1} \begin{pmatrix} \mathbf{o} \\ 0 \\ 1 \end{pmatrix} = (\mathbf{z}^\top, 0, 0) \begin{pmatrix} \mathbf{v}_\mathrm{F} \\ u_\mathrm{F} \end{pmatrix} = \mathbf{z}'^\top\,\mathbf{v}_\mathrm{F} \\ &= \frac{\mathbf{z}'^\top\,\mathbf{K}^{-1}\,\mathbf{s}'}{\mathbf{s}'^\top\,\mathbf{K}^{-1}\,\mathbf{s}'} = \frac{\mathcal{K}_n(\mathbf{z}, \mathbf{s})}{\mathcal{K}_n(\mathbf{s}, \mathbf{s})} = \frac{\mathrm{cov}_n(\mathbf{z}, \mathbf{s})}{\mathrm{cov}_n(\mathbf{s}, \mathbf{s})}, \end{aligned} \tag{V.70}$$

and

$$\begin{aligned} a^\star &= (\mathbf{z}^\top, 0, 0)\,\mathbf{F}^{-1} \begin{pmatrix} \mathbf{o} \\ 1 \\ 0 \end{pmatrix} = (\mathbf{z}^\top, 0)\,\mathbf{A}_\mathrm{F} \begin{pmatrix} \mathbf{o} \\ 1 \end{pmatrix} \\ &= \mathbf{z}' \left(\mathbf{K}^{-1} - \frac{\mathbf{K}^{-1}\,\mathbf{s}'\,(\mathbf{K}^{-1}\,\mathbf{s}')^\top}{\mathbf{s}'^\top\,\mathbf{K}^{-1}\,\mathbf{s}'} \right) \begin{pmatrix} \mathbf{o} \\ 1 \end{pmatrix} \\ &= \mathbf{z}'\,\mathbf{K}^{-1} \begin{pmatrix} \mathbf{o} \\ 1 \end{pmatrix} - \frac{\mathbf{z}'\,\mathbf{K}^{-1}\,\mathbf{s}'}{\mathbf{s}'\,\mathbf{K}^{-1}\,\mathbf{s}'} \cdot \mathbf{s}'\,\mathbf{K}^{-1} \begin{pmatrix} \mathbf{o} \\ 1 \end{pmatrix} \\ &= \mathrm{E}_n[\mathbf{z}] - b^\star\,\mathrm{E}_n[\mathbf{s}]. \end{aligned} \tag{V.71}$$

EXERCISE 37.3 a) $\mathbf{v}_0 = -\mathbf{K}^{-1}\,\mathbf{k}_0\,u_{00}$ and

$$\begin{aligned} c_{00}\,u_{00} - \mathbf{k}_0^\top\,\mathbf{K}^{-1}\,\mathbf{k}_0\,u_{00} &= 1 \\ u_{00}\left(c_{00} - \mathbf{k}_0^\top\,\mathbf{K}^{-1}\,\mathbf{k}_0 \right) &= 1 \\ u_{00} &= \frac{1}{\sigma_\mathrm{OK}^2}. \end{aligned} \tag{V.72}$$

b)

$$\mathcal{K}_{n+1}(\mathbf{z}_0, \mathbf{s}_0) = (z_0, \mathbf{z}^\top, 0) \begin{pmatrix} u_{00} & \mathbf{v}_0^\top \\ \mathbf{v}_0 & \mathbf{A}_0 \end{pmatrix} (s_0, \mathbf{s}^\top, 0)$$

$$\text{with} \qquad \mathbf{A}_0 = \mathbf{K}^{-1} + u_{00}\, \mathbf{K}^{-1}\, \mathbf{k}_0\, (\mathbf{K}^{-1}\, \mathbf{k}_0)^\top. \qquad \text{(V.73)}$$

$$\begin{aligned} \mathcal{K}_{n+1}(\mathbf{z}_0, \mathbf{s}_0) &= z_0\, u_{00}\, s_0 + (\mathbf{z}^\top, 0)\, \mathbf{v}_0\, s_0 + z_0\, \mathbf{v}_0^\top \begin{pmatrix} \mathbf{s} \\ 0 \end{pmatrix} + (\mathbf{z}^\top, 0)\, \mathbf{A}_0 \begin{pmatrix} \mathbf{s} \\ 0 \end{pmatrix} \\ &= u_{00}\, z_0\, s_0 - u_{00}\, (\mathbf{z}^\top, 0)\, \mathbf{K}^{-1}\, \mathbf{k}_0\, s_0 - u_{00}\, z_0\, \mathbf{K}^{-1}\, \mathbf{k}_0 \begin{pmatrix} \mathbf{s} \\ 0 \end{pmatrix} \\ &\quad + \mathcal{K}_n(\mathbf{z}, \mathbf{s}) + u_{00}\, (\mathbf{z}^\top, 0)\, \mathbf{K}^{-1}\, \mathbf{k}_0\, (\mathbf{K}^{-1}\, \mathbf{k}_0)^\top \begin{pmatrix} \mathbf{s} \\ 0 \end{pmatrix} \\ &= u_{00}\, (z_0 - (\mathbf{z}^\top, 0)\, \mathbf{K}^{-1}\, \mathbf{k}_0) \cdot (s_0 - (\mathbf{s}^\top, 0)\, \mathbf{K}^{-1}\, \mathbf{k}_0) \\ &\quad + \mathcal{K}_n(\mathbf{z}, \mathbf{s}). \end{aligned} \qquad \text{(V.74)}$$

c)

$$\mathcal{K}_{n+1}(\mathbf{z}_0, \mathbf{s}_0) = (z_0 - (\mathbf{z}^\top, 0)\, \mathbf{K}^{-1}\, \mathbf{k}_0) \cdot \frac{\Delta(s_0|\mathbf{s})}{\sigma^2_{\mathrm{OK}}} + \mathcal{K}_n(\mathbf{z}, \mathbf{s}), \qquad \text{(V.75)}$$

and

$$\begin{aligned} (u_{00}, \mathbf{v}_0^\top) \cdot \begin{pmatrix} \mathbf{s}_0 \\ 0 \end{pmatrix} &= u_{00}\, s_0 + \mathbf{v}_0^\top \begin{pmatrix} \mathbf{s} \\ 0 \end{pmatrix} = u_{00}\, s_0 - u_{00}\, \mathbf{K}^{-1}\, \mathbf{k}_0 \begin{pmatrix} \mathbf{s} \\ 0 \end{pmatrix} \\ &= u_{00}\, \Delta(s_0|\mathbf{s}) = \frac{\Delta(s_0|\mathbf{s})}{\sigma^2_{\mathrm{OK}}}. \end{aligned} \qquad \text{(V.76)}$$

EXERCISE 37.4 a) For $\alpha = 1$ we have

$$(u_{11}, \mathbf{v}_1^\top) \begin{pmatrix} \mathbf{s} \\ 0 \end{pmatrix} = \frac{\Delta(s_1|\mathbf{s}_{[1]})}{\sigma^2_{[1]}}, \qquad \text{(V.77)}$$

and for $\alpha = n$

$$(v_{1,n}, \ldots, v_{n-1,n}, u_{nn}, v_{n+1,n}) \begin{pmatrix} \mathbf{s} \\ 0 \end{pmatrix} = \frac{\Delta(s_n|\mathbf{s}_{[n]})}{\sigma^2_{[n]}}, \qquad \text{(V.78)}$$

as well as for $\alpha \neq 1$ and $\alpha \neq n$:

$$(v_{1,\alpha}, \ldots, v_{\alpha-1,\alpha}, u_{\alpha\alpha}, v_{\alpha+1,\alpha}, \ldots, v_{n+1,\alpha}) \begin{pmatrix} \mathbf{s} \\ 0 \end{pmatrix} = \frac{\Delta(s_\alpha|\mathbf{s}_{[\alpha]})}{\sigma^2_{[\alpha]}}. \qquad \text{(V.79)}$$

b)

$$
\begin{aligned}
b^{\star} &= \frac{\mathcal{K}_n(\mathbf{z},\mathbf{s})}{\mathcal{K}_n(\mathbf{s},\mathbf{s})} = (\mathbf{z}^\top, 0)\, \frac{\mathbf{K}^{-1}\begin{pmatrix}\mathbf{s}\\ 0\end{pmatrix}}{\mathcal{K}_n(\mathbf{s},\mathbf{s})} \\
&= \sum_{\alpha=1}^{n} z_\alpha \, \frac{\Delta(s_\alpha|\mathbf{s}_{[\alpha]})}{\sigma^2_{[\alpha]}\, \mathcal{K}_n(\mathbf{s},\mathbf{s})}.
\end{aligned}
\tag{V.80}
$$

EXERCISE I.1

$$
\mathbf{1}^\top\mathbf{1} = n, \tag{V.81}
$$

$$
\mathbf{1}\mathbf{1}^\top = \underbrace{\begin{pmatrix} 1 & \dots & 1 \\ \vdots & \ddots & \vdots \\ 1 & \dots & 1 \end{pmatrix}}_{(n\times n)}. \tag{V.82}
$$

EXERCISE I.2 $\frac{1}{n}\mathbf{Z}^\top\mathbf{1} = \mathbf{m}$ is the vector of the means m_i of the variables.

$\frac{1}{n}\mathbf{1}\mathbf{1}^\top\mathbf{Z} = \mathbf{M}$ is an $n\times N$ matrix containing in each row the transpose of the vector of means; each column $\mathbf{m}_i$ of $\mathbf{M}$ has N elements equal to the mean m_i of the variable number i.

EXERCISE I.8 The classical eigenvalue decomposition.

EXERCISE IV.1

i)

$$
\sum_{\beta=1}^{n} w_\beta\, \mathrm{e}^{-\frac{|\alpha-\beta|}{a}} = \mathrm{e}^{-\frac{|\alpha-(n+1)|}{a}} \qquad \alpha = 1,\dots,n. \tag{V.83}
$$

ii)

$$
w_n\, \mathrm{e}^{-\frac{|\alpha-n|}{a}} = \mathrm{e}^{-\frac{|\alpha-(n+1)|}{a}} \qquad \alpha = 1,\dots,n. \tag{V.84}
$$

$$
\begin{aligned}
\Rightarrow \qquad w_n &= \frac{\mathrm{e}^{-\frac{|\alpha-(n+1)|}{a}}}{\mathrm{e}^{-\frac{|\alpha-n|}{a}}} \qquad \alpha = 1,\dots,n \\
&= \frac{\mathrm{e}^{-\frac{(n+1)}{a}}}{\mathrm{e}^{-\frac{n}{a}}} = \mathrm{e}^{-\frac{1}{a}} = c.
\end{aligned}
\tag{V.85}
$$

iii)

$$\begin{aligned}
Z(n) &= p\,Z(n-1) + \varepsilon_n = p\,[\,p\,Z(n-2) + \varepsilon_{n-1}\,] + \varepsilon_n \\
&= p^2\,Z(n-2) + p\,\varepsilon_{n-1} + \varepsilon_n \\
&= p^m\,Z(n-m) + \sum_{\alpha=0}^{m-1} p^{\alpha}\varepsilon_{n-\alpha}.
\end{aligned} \tag{V.86}$$

As $\lim_{m\to\infty} p^m\,Z(n-m) = 0$, because $|p| < 1$,

$$Z(n) = \sum_{\alpha=0}^{\infty} p^{\alpha}\,\varepsilon_{n-\alpha} = \sum_{\alpha=-\infty}^{n} p^{n-\alpha}\varepsilon_n. \tag{V.87}$$

iv) It is easy to show that $\mathrm{E}[\,Z(n)\,] = 0$. Then

$$\begin{aligned}
\mathrm{cov}(Z(n), Z(m)) &= \mathrm{E}[\,Z(n)\,Z(m)\,] = \mathrm{E}\Big[\,\sum_{\alpha=0}^{\infty}\sum_{\beta=0}^{\infty} p^{\alpha}\,p^{\beta}\,\varepsilon_{n-\alpha}\,\varepsilon_{n-\beta}\,\Big] \\
&= \sum_{\alpha=0}^{\infty}\sum_{\beta=0}^{\infty} p^{\alpha}\,p^{\beta}\mathrm{E}[\,\varepsilon_{n-\alpha}\,\varepsilon_{n-\beta}\,] \\
&= \sum_{\alpha=0}^{\infty} p^{\alpha}\Big(\sum_{\beta=0}^{\infty} p^{\beta}\,\sigma^2\,\delta_{n-\alpha,m-\beta}\Big).
\end{aligned} \tag{V.88}$$

$\mathrm{E}[\,\varepsilon_{n-\alpha}\,\varepsilon_{n-\beta}\,] = \sigma^2$, if $n-\alpha = m-\beta$ (i.e. if $\beta = m-n+\alpha$).

If $n \le m$:

$$\begin{aligned}
\mathrm{cov}(Z(n), Z(m)) &= \sum_{\alpha=0}^{\infty} p^{\alpha}\,p^{(m-n)+\alpha}\,\sigma^2 = \sigma^2\,p^{m-n}\sum_{\alpha=0}^{\infty} p^{2\alpha} \\
&= \frac{\sigma^2\,p^{m-n}}{1-p^2},
\end{aligned} \tag{V.89}$$

and if $n > m$, we have $\alpha \ge n-m$, so:

$$\mathrm{cov}(Z(n), Z(m)) = \sum_{\alpha=n-m}^{\infty} p^{\alpha}\,p^{(m-n)+\alpha}\,\sigma^2 \tag{V.90}$$

$$\begin{aligned}
&= \sum_{i=0}^{\infty} p^{(n-m)+i}\,p^{(m-n)+(n-m)+i}\,\sigma^2 \\
&= \sigma^2\,p^{n-m}\sum_{i=0}^{\infty} p^{2i} \\
&= \frac{\sigma^2 p^{n-m}}{1-p^2}.
\end{aligned} \tag{V.91}$$

Thus $\mathrm{cov}(Z(n), Z(m)) = \dfrac{\sigma^2\,p^{|n-m|}}{1-p^2}$ and $r_{nm} = p^{|n-m|}$.

v) $p = w_n$, thus the range

$$a = -\frac{1}{\log p}, \tag{V.92}$$

and the sill

$$b = \mathrm{var}(Z(n)) = \frac{\sigma^2}{1-p^2}. \tag{V.93}$$

EXERCISE IV.2 As we have exact interpolation, the kriging weights for $Z^\star(\mathbf{x}_0)$ are linked by

$$w_\alpha = \delta_{\mathbf{x}_\alpha,\mathbf{x}_0}. \tag{V.94}$$

As the estimators are linear we have

$$w_\alpha^P = \delta_{\mathbf{x}_\alpha,\mathbf{x}_0} - w_\alpha^G. \tag{V.95}$$

EXERCISE IV.3 Ordinary kriging can be written as

$$\underbrace{\begin{pmatrix} C_{1,1} & \dots & C_{1,N} & 1 \\ \vdots & \ddots & \vdots & \vdots \\ C_{N,1} & \dots & C_{N,N} & 1 \\ 1 & \dots & 1 & 0 \end{pmatrix}}_{\mathbf{A}} \left[\underbrace{\begin{pmatrix} w_1^P \\ \vdots \\ w_N^P \\ \mu_P \end{pmatrix}}_{\mathbf{x}_P} + \underbrace{\begin{pmatrix} w_1^G \\ \vdots \\ w_N^G \\ \mu_G \end{pmatrix}}_{\mathbf{x}_G} + \underbrace{\begin{pmatrix} w_1^{m_0} \\ \vdots \\ w_N^{m_0} \\ \mu_{m_0} \end{pmatrix}}_{\mathbf{x}_{m_0}} \right]$$
$$= \underbrace{\begin{pmatrix} C_{0,1}^P \\ \vdots \\ C_{0,N}^P \\ 0 \end{pmatrix}}_{\mathbf{b}_P} + \underbrace{\begin{pmatrix} C_{0,1}^G \\ \vdots \\ C_{0,N}^G \\ 0 \end{pmatrix}}_{\mathbf{b}_G} + \underbrace{\begin{pmatrix} 0 \\ \vdots \\ 0 \\ 1 \end{pmatrix}}_{\mathbf{b}_{m_0}} \tag{V.96}$$

with $\mathbf{A}\,\mathbf{x}_P = \mathbf{b}_P, \quad \mathbf{A}\,\mathbf{x}_G = \mathbf{b}_G, \quad \mathbf{A}\,\mathbf{x}_{m_0} = \mathbf{b}_{m_0}$.

We then have

$$z^\star(\mathbf{x}_0) = \mathbf{z}^\top \mathbf{w}_P + \mathbf{z}^\top \mathbf{w}_G + \mathbf{z}^\top \mathbf{w}_{m_0} = \mathbf{z}^\top(\mathbf{w}_P + \mathbf{w}_G + \mathbf{w}_{m_0}) = \mathbf{z}^\top \mathbf{w}, \tag{V.97}$$

where $\mathbf{z}$ is the vector of the n data and $\mathbf{w}_P$, $\mathbf{w}_G$, $\mathbf{w}_{m_0}$, $\mathbf{w}$ are the vectors of the n corresponding weights.

References

This classification of selected references is aimed at the reader who wants to get a start into the subject for a specific topic or application. The list is by no means exhaustive nor always up-to-date. References are grouped under the three headings: concepts, applications and books. Indications about sources of computer software are given at the end.

Concepts

Change of support: Matheron [207, 215, 216]; Demange et al. [87]; Lantuéjoul [175]; Cressie [65]; Gelfand [105].

Classification: Sousa [307]; Oliver & Webster [236]; Raspa et al. [256]; Ambroise et al. [7].

Cokriging: Matheron [200, 211]; Maréchal [192]; François-Bongarçon [101]; Myers [231, 232]; Doyen [95]; Stein et al. [312]; Wackernagel [339]; Goovaerts [122].

Cokriging of compositional data: Pawlowsky [244]; Pawlowsky et al. [245].

Cokriging of variables and their derivatives: Chauvet et al. [46]; Thiébaux & Pedder [324]; Renard & Ruffo [261]; Lajaunie [169]; Lajaunie et al. [171]; Kitanidis [159].

Collocated cokriging: Xu et al. [360]; Journel [155]; Chilès & Delfiner [51]; Rivoirard [271].

Coregionalization analysis, factor cokriging: Matheron [214]; Sandjivy [288]; Wackernagel [336, 337]; Wackernagel et al. [343]; Goulard [125, 126]; Royer [280]; Daly et al. [74]; Grunski & Agterberg [129]; Goovaerts [120]; Grzebyk & Wackernagel [131]; Goovaerts [121]; Vargas-Guzman et al. [327]; Bailey & Krzanowski [20]; Desbarats & Dimitrakopoulos [88].

Correlation functions on the sphere: Gaspari & Cohn [104]; Gneiting [113].

Cross covariance function, cross variogram: Yaglom [361, 362]; Matheron [198].

Cross validation: Cook & Weisberg [58]; Dubrule [96]; Castelier [37, 38]; Christensen et al. [57].

Deconvolution by kriging: JEULIN & RENARD [152].

Dynamic graphics: HASLETT et al. [139]; HASLETT & POWER [140]; SPARKS et al. [308].

Empirical Orthogonal Functions: STORCH & NAVARRA [334]; STORCH & ZWIERS [335].

External drift: DELHOMME [84]; DELFINER et al. [82]; GALLI et al. [102]; CHILÈS & GUILLEN [49]; MARÉCHAL [193]; GALLI & MEUNIER [103]; RENARD & NAI-HSIEN [260]; CASTELIER [37, 38]; HUDSON & WACKERNAGEL [146]; MONESTIEZ et al. [226].

Fractals and geostatistics: BRUNO & RASPA [33].

Generalized covariance, variogram, IRF-k: MATHERON [203]; DELFINER [80]; DELFINER & MATHERON [83]; CHILÈS [48]; KITANIDIS [158]; DOWD [94]; CHAUVET [44]; PARDO-IGUZQUIZA [242]; GNEITING [111]; KÜNSCH et al. [165]; GNEITING et al. [114].

Isofactorial models: MATHERON [206, 209, 215, 216, 217, 221, 14, 15, 16]; KLEINGELD [160]; HU [142]; LAJAUNIE & LANTUÉJOUL [172]; CHILÈS & DELFINER [51].

Kalman filtering and geostatistics: THIÉBAUX [322]; HUANG HC & CRESSIE [145]; MARDIA et al. [189]; WIKLE & CRESSIE [356]; SÉNÉGAS et al. [298]; WOLF et al. [358]; BERTINO [26]; BERTINO et al. [27].

Kriging of spatial components: MATHERON [200, 214]; SANDJIVY [287]; GALLI et al. [102]; CHILÈS & GUILLEN [49].

Kriging weights: MATHERON [200]; RIVOIRARD [264]; BARNES & JOHNSON [22]; MATHERON [219]; CHAUVET [43].

Lognormal kriging: MATHERON [195]; MATHERON [196]; SICHEL [302]; ORFEUIL [238]; MATHERON [204]; RIVOIRARD [268]; ROTH [275].

Multiple time series: ROUHANI & WACKERNAGEL [278]; JAQUET & CARNIEL [151].

Multivariate fitting of variograms/cross covariances: GOULARD [125, 126]; BOURGAULT & MARCOTTE [29]; GOULARD & VOLTZ [127]; GRZEBYK [130]; VER HOEF & CRESSIE [330]; HAAS [135]; YAO & JOURNEL [363]; YAO & JOURNEL [364].

Noise filtering: SWITZER & GREEN [320]; BERMAN [25]; MA & ROYER [185]; DALY [71]; DALY et al. [74, 73].

Nonlinear geostatistics, disjunctive kriging: MATHERON [209, 213, 218, 221]; ORFEUIL [239]; LANTUÉJOUL [175, 176]; RIVOIRARD [266, 267, 269, 270]; LIAO [182]; PETITGAS [247]; LAJAUNIE [168]; CRESSIE [65]; DIGGLE et al. [91].

Orthogonal polynomials: SZEGÖ [321]; ASKEY [18].

Robust variogram estimation: GENTON [106].

Sampling: LAJAUNIE et al. [173]; BUESO et al. [35].

Sensitivity of kriging: WARNES [349]; ARMSTRONG & WACKERNAGEL [17].

Simulation: MATHERON [203]; JOURNEL [154]; DELFINER [80]; DELFINER & CHILÈS [81]; ARMSTRONG & DOWD [13]; CHILÈS & DELFINER [51]; GNEITING [112]; LANTUÉJOUL [178].

Space-time drift, trigonometric kriging: SÉGURET & HUCHON [297]; SÉGURET [295]; SÉGURET [296].

Space-time modeling: STEIN [313]; COX & ISHAM [59, 60]; HASLETT [138]; CHRISTAKOS [54]; GOODALL & MARDIA [118]; HAAS [134]; KYRIAKIDIS & JOURNEL [166]; GNEITING & SCHLATHER [115].

Spatial correlation mapping: SAMPSON & GUTTORP [286]; MONESTIEZ & SWITZER [229]; MONESTIEZ et al. [228]; GUTTORP & SAMPSON [133]; BROWN et al. [32]; LE et al. [180]; PERRIN [246].

Splines and kriging: MATHERON [212]; DUBRULE [97]; WAHBA [347]; CRESSIE [64]; HUTCHINSON & GESSLER [147]; MARDIA et al. [191].

Universal kriging: MATHERON [199]; SABOURIN [284]; CHILÈS [48]; CHAUVET & GALLI [45]; ARMSTRONG [10]; CHAUVET [44].

Variables linked by partial differential equations: MATHERON [201]; DONG [93]; MATHERON et al. [223]; CASTELIER [39]; ROTH & CHILÈS [276].

Variogram cloud: CHAUVET [42]; HASLETT et al. [139].

Variogram and generalized covariance function estimation: DELFINER [80]; KITANIDIS [158]; ZIMMERMANN [367]; PILZ et al. [249]; PARDO-IGUZQUIZA [242]; KÜNSCH et al. [165]; GENTON [107].

Applications

Demography: BALABDOUI et al. [21].

Design of computer experiments: SACKS et al. [285]; WALTER & PRONZATO [348].

Ecology: MONESTIEZ et al. [227]; WOLF et al. [358]; BERTINO et al. [27].

Electromagnetic compatibility: LEFÈBVRE et al. [181]; RANNOU et al. [252].

Epidemiology: OLIVER et al. [235, 237]; WEBSTER et al. [353].

Fisheries: PETITGAS [247]; RIVOIRARD et al. [272]; PETITGAS [248].

Forestry: MARBEAU [187]; FOUQUET & MANDALLAZ [77].

Geochemical exploration: SANDJIVY [287]; SANDJIVY [288]; WACKERNAGEL & BUTENUTH [340]; ROYER [280]; GRUNSKI & AGTERBERG [129]; LINDNER & WACKERNAGEL [183]; HASLETT et al. [139]; WACKERNAGEL & SANGUINETTI [344]; HASLETT & POWER [140]; BAILEY & KRZANOWSKI [20].

Geodesy: MEIER & KELLER [224]; SCHAFFRIN [290].

Geophysical exploration: GALLI et al. [102]; CHILÈS & GUILLEN [49]; DOYEN [95]; SCHULZ-OHLBERG [293]; RENARD & NAI-HSIEN [260]; SÉGURET & HUCHON [297]; SÉGURET [296].

Hydrogeology: DELHOMME [85]; CREUTIN & OBLED [67]; BRAS & RODRÍGUEZ-ITURBE [31]; DE MARSILY [78]; STEIN [313]; AHMED & DE MARSILY [3]; DAGAN [69]; DONG [93]; ROUHANI & WACKERNAGEL [278]; CHILÈS [50]; MATHERON et al. [223]; CASTELIER [39]; KITANIDIS [159].

Image analysis: SWITZER & GREEN [320]; BERMAN [25]; MA & ROYER [185]; DALY et al. [74, 73, 72]; CHICA-OLMO & ABARCA-HERNANDEZ [47].

Industrial hygienics: PRÉAT [250]; SCHNEIDER et al. [291]; WACKERNAGEL et al. [342]; VINCENT et al. [332]; LAJAUNIE et al. [173]; WACKERNAGEL et al. [345].

Material science: DALY et al. [74, 73, 72].

Meteorology and climatology: CHAUVET et al. [46]; THIÉBAUX & PEDDER [324]; COX & ISHAM [59, 60]; HASLETT [138]; THIÉBAUX et al. [323]; HUDSON & WACKERNAGEL [146]; HUANG & CRESSIE [145]; CASSIRAGA & GOMEZ-HERNANDEZ [36]; AMANI & LEBEL [6]; RASPA et al. [257]; BIAU et al. [28].

Mining: JOURNEL & HUIJBREGTS [156]; PARKER [243]; SOUSA [307]; WELLMER [354].

Petroleum and gas exploration: DELHOMME et al. [86]; DELFINER et al. [82]; GALLI et al. [102]; MARÉCHAL [193]; GALLI & MEUNIER [103]; JAQUET [150]; RENARD & RUFFO [261]; YARUS & CHAMBERS [366]; YAO et al. [365].

Pollution: ORFEUIL [239]; LAJAUNIE [167]; BROWN et al. [32].

Soil science: WEBSTER [350]; WACKERNAGEL et al. [346]; GOULARD [126]; OLIVER & WEBSTER [236]; WEBSTER & OLIVER [352]; GOULARD & VOLTZ [127]; STEIN, STARISKY & BOUMA [311]; GOOVAERTS [119, 120]; GOOVAERTS et al. [123]; PAPRITZ & FLÜHLER [240]; GOOVAERTS & WEBSTER [124]; WEBSTER et al. [351].

Volcanology: JAQUET & CARNIEL [151].

Books

Basic geostatistical texts: MATHERON [196, 198, 200, 220].

Introductory texts: DAVID [75]; JOURNEL & HUIJBREGTS [156]; MATHERON & ARMSTRONG [222]; AKIN & SIEMES [5]; ISAAKS & SHRIVASTAVA [148]; CRESSIE [64]; CHRISTAKOS [54]; BRUNO & RASPA [34]; RIVOIRARD [270]; YARUS & CHAMBERS [366]; KITANIDIS [159]; GOOVAERTS [121]; STOYAN et al. [317]; WELLMER [354]; ARMSTRONG [12]; CHAUVET [44]; CHILÈS & DELFINER [51]; STEIN [314]; HOHN [141]; RIVOIRARD et al. [272].

Proceedings: GUARASCIO et al. [132]; VERLY et al. [331]; ARMSTRONG [11]; SOARES [305]; ARMSTRONG & DOWD [13]; DIMITRAKOPOULOS [92]; ROUHANI et al. [277]; BAAFI & SCHOFIELD [19]; SOARES et al. [306]; GOMEZ-HERNANDEZ et al. [116]; KLEINGELD & KRIGE [161]; MONESTIEZ et al. [225].

Books of related interest: MATERN [194]; YAGLOM [361, 362]; BOX & JENKINS [30]; LUMLEY [184]; BENNETT [23]; ADLER [1]; RIPLEY [262]; BRAS & RODRÍGUEZ-ITURBE [31]; MARSILY [78]; RIPLEY [263]; RYTOV et al. [281, 282, 283]; THIÉBAUX & PEDDER [324]; ANSELIN [9]; WEBSTER & OLIVER [352]; HAINING [137]; DALEY [70]; CHRISTENSEN [56]; STOYAN & STOYAN [316]; STORCH & NAVARRA [334]; DIGGLE et al. [90]; STORCH & ZWIERS [335].

Introductions to probability and statistics: FELLER [99]; MORRISON [230]; CHATFIELD [41]; CHRISTENSEN [55]; SAPORTA [289]; STOYAN [315].

Books on multivariate analysis: RAO [254]; MORRISON [230]; MARDIA, KENT & BIBBY [190]; COOK & WEISBERG [58]; ANDERSON [8]; SEBER [294]; GREENACRE [128]; VOLLE [333]; GITTINS [110]; JOLLIFFE [153]; GIFI [108]; SAPORTA [289]; WHITTAKER [355].

Software

Many public domain and commercial software products exist nowadays that include geostatistics in one form or another. They can easily be found by checking corresponding pages about *spatial statistics*, *geostatistics* or *geographical information systems* on the worldwide computer web. See, for example: `http://www.ai-geostats.org`

We have been mainly using the following software which are all available for Unix/Linux workstations and MS Windows:

- *Isatis*. A general purpose 3D geostatistical package. Originally developed by Centre de Géostatistique and now commercialized by Géovariances. See: `http://www.geovariances.fr`
- *S-Plus*. A general purpose statistical language with spatial statistics packages. See: `http://lib.stat.cmu.edu`
- *R*. A general purpose statistical language, strongly ressembling *S-Plus*, yet in the public domain. Several contributed packages of geostatistical modules can be downloaded. See: `http://http://www.r-project.org`

Bibliography

[1] ADLER, R. *The Geometry of Random Fields.* Wiley, New York, 1981.

[2] AGTERBERG, F. P. Appreciation of contributions by Georges Matheron and John Tuckey to a mineral-resources research project. *Natural Resources Research 10* (2001), 287–295.

[3] AHMED, S., AND DE MARSILY, G. Comparison of geostatistical methods for estimating transmissivity using data on transmissivity and specific capacity. *Water Resources Research 23* (1987), 1717–1737.

[4] AITCHISON, J., AND BROWN, J. A. C. *The Lognormal Distribution: with Special Reference to its Uses in Economics.* Cambridge University Press, Cambridge, 1957.

[5] AKIN, H., AND SIEMES, H. *Praktische Geostatistik.* Springer-Verlag, Heidelberg, 1988.

[6] AMANI, A., AND LEBEL, T. Lagrangian kriging for the Sahelian rainfall at small time steps. *Journal of Hydrology 192* (1997), 125–157.

[7] AMBROISE, C., DANG, M., AND G, G. Clustering of spatial data by the EM algorithm. In Soares et al. [306], pp. 493–504.

[8] ANDERSON, T. W. *An Introduction to Multivariate Statistical Analysis.* Wiley, New York, 1984.

[9] ANSELIN, L. *Spatial Econometrics: Methods and Models.* Kluwer, Amsterdam, 1988.

[10] ARMSTRONG, M. Problems with universal kriging. *Mathematical Geology 16* (1984), 101–108.

[11] ARMSTRONG, M., Ed. *Geostatistics.* Kluwer, Amsterdam, 1989.

[12] ARMSTRONG, M. *Linear Geostatistics.* Springer-Verlag, Berlin, 1998.

[13] ARMSTRONG, M., AND DOWD, P. A., Eds. *Geostatistical Simulation.* Kluwer, Amsterdam, 1994.

[14] ARMSTRONG, M., AND MATHERON, G. Disjunctive kriging revisited: part I. *Mathematical Geology 18* (1986), 711–728.

[15] ARMSTRONG, M., AND MATHERON, G. Disjunctive kriging revisited: part II. *Mathematical Geology 18* (1986), 729–742.

[16] ARMSTRONG, M., AND MATHERON, G. Isofactorial models for granulodensimetric data. *Mathematical Geology 18* (1986), 743–757.

[17] ARMSTRONG, M., AND WACKERNAGEL, H. The influence of the covariance function on the kriged estimator. *Sciences de la Terre, Série Informatique 27* (1988), 245–262.

[18] ASKEY, R. *Orthogonal Polynomials and Special Functions*. Society for Industrial and Applied Mathematics, Philadelphia, 1975.

[19] BAAFI, E. Y., AND SCHOFIELD, N. A., Eds. *Geostatistics Wollongong '96*. Kluwer, Amsterdam, 1997.

[20] BAILEY, T. C., AND KRZANOWSKI, W. J. Extensions to spatial factor methods with illustrations in geochemistry. *Mathematical Geology 32* (2000), 657–682.

[21] BALABDOUI, F., BOCQUET-APPEL, J. P., LAJAUNIE, C., AND IRUDAYA RAJAN, S. Space-time evolution of the fertility transition in india 1961-1991. *International Journal of Population Geography 7* (2001), 129–148.

[22] BARNES, R. J., AND JOHNSON, T. B. Positive kriging. In Verly et al. [331], pp. 231–244.

[23] BENNETT, R. J. *Spatial Time Series*. Pion, London, 1979.

[24] BENZÉCRI, J. P. *L'Analyse des Données: l'Analyse des Correspondance*, vol. 2. Dunod, Paris, 1973.

[25] BERMAN, M. The statistical properties of three noise removal procedures for multichannel remotely sensed data. Tech. Rep. NSW/85/31/MB9, CSIRO, Lindfield, 1985.

[26] BERTINO, L. *Assimilation de Données pour la Prédiction de Paramètres Hydrodynamiques et Ecologiques: Cas de la Lagune de l'Oder*. Doctoral thesis, Ecole des Mines de Paris, Fontainebleau, 2001.

[27] BERTINO, L., EVENSEN, G., AND WACKERNAGEL, H. Combining geostatistics and Kalman filtering for data assimilation in an estuarine system. *Inverse problems 18* (2002), 1–23.

[28] BIAU, G., ZORITA, E., VON STORCH, H., AND WACKERNAGEL, H. Estimation of precipitation by kriging in the EOF space of the sea level pressure field. *Journal of Climate 12* (1999), 1070–1085.

[29] BOURGAULT, G., AND MARCOTTE, D. Multivariable variogram and its application to the linear model of coregionalization. *Mathematical Geology 23* (1991), 899–928.

[30] BOX, G. E. P., AND JENKINS, G. M. *Time Series Analysis: Forecasting and Control*. Holden-Day, San Francisco, 1970.

[31] BRAS, R. L., AND RODRÍGUEZ-ITURBE, I. *Random Functions and Hydrology*. Addison-Wesley, Reading, 1985.

[32] BROWN, P. J., LE, N. D., AND ZIDEK, J. V. Multivariate spatial interpolation and exposure to air pollutants. *The Canadian Journal of Statistics 22* (1994), 489–509.

[33] BRUNO, R., AND RASPA, G. Geostatistical characterization of fractal models of surfaces. In Armstrong [11], pp. 77–89.

[34] BRUNO, R., AND RASPA, G. *La Pratica della Geostatistica Lineare*. Guerini, Milano, 1994.

[35] BUESO, M. C., ANGULO, J. M., CRUZ-SANJULIÁN, J., AND GARCÍA-ARÓSTEGUI, F. Optimal spatial sampling design in a multivariate framework. *Mathematical Geology 31* (1999), 507–525.

[36] CASSIRAGA, E. F., AND GOMEZ-HERNANDEZ, J. Improved rainfall estimation by integration of radar data: a geostatistical approach. In Soares et al. [306], pp. 363–374.

[37] CASTELIER, E. La dérive externe vue comme une régression linéaire. Tech. Rep. N-34/92/G, Centre de Géostatistique, Ecole des Mines de Paris, Fontainebleau, 1992.

[38] CASTELIER, E. Dérive externe et régression linéaire. In *Cahiers de Géostatistique* (Paris, 1993), vol. 3, Ecole des Mines de Paris, pp. 47–59.

[39] CASTELIER, E. *Estimation d'un Champ de Perméabilité à partir de Mesures de Charge Hydraulique*. PhD thesis, Ecole des Mines de Paris, Fontainebleau, 1995.

[40] CHAMPION, R., LENARD, C. T., AND MILLS, T. M. An introduction to abstract splines. *Mathematical Scientist 21* (1996), 8–26.

[41] CHATFIELD, C. *Statistics for Technology*. Chapman & Hall, London, 1983.

[42] CHAUVET, P. The variogram cloud. In *17th APCOM* (New York, 1982), T. B. Johnson and R. J. Barnes, Eds., Society of Mining Engineers, pp. 757–764.

[43] CHAUVET, P. Réflexions sur les pondérateurs négatifs du krigeage. *Sciences de la Terre, Série Informatique 28* (1988), 65–113.

[44] CHAUVET, P. *Aide Mémoire de Géostatistique Linéaire*. Les Presses, Ecole des Mines de Paris, 1999.

[45] CHAUVET, P., AND GALLI, A. Universal kriging. Tech. Rep. C-96, Centre de Géostatistique, Ecole des Mines de Paris, Fontainebleau, 1982.

[46] CHAUVET, P., PAILLEUX, J., AND CHILÈS, J. P. Analyse objective des champs météorologiques par cokrigeage. *La Météorologie 6*, 4 (1976), 37–54.

[47] CHICA-OLMO, M., AND ABARCA-HERNANDEZ, F. Radiometric coregionalization of Landsat TM and SPOT HRV images. *International Journal of Remote Sensing 19* (1998), 997–1005.

[48] CHILÈS, J. *Géostatistique des Phénomènes Non Stationnaires (dans le plan)*. PhD thesis, Université de Nancy I, Nancy, 1977.

[49] CHILÈS, J., AND GUILLEN, A. Variogrammes et krigeages pour la gravimétrie et le magnétisme. *Sciences de la Terre, Série Informatique 20* (1984), 455–468.

[50] CHILÈS, J. P. Application du krigeage avec dérive externe à l'implantation d'un réseau de mesures piézométriques. *Sciences de la Terre, Série Informatique 30* (1991), 131–147.

[51] CHILÈS, J. P., AND DELFINER, P. *Geostatistics: Modeling Spatial Uncertainty*. Wiley, New York, 1999.

[52] CHILÈS, J. P., AND LIAO, H. T. Estimating the recoverable reserves of gold deposits: comparison between disjunctive kriging and indicator kriging. In Soares [305], pp. 1053–1064.

[53] CHOQUET, G. *Lectures on Analysis*, vol. 3. Benjamin, New York, 1967.

[54] CHRISTAKOS, G. *Random Field Models in Earth Sciences*. Academic Press, San Diego, 1992.

[55] CHRISTENSEN, R. *Plane Answers to Complex Questions: the Theory of Linear Models*. Springer-Verlag, New York, 1987.

[56] CHRISTENSEN, R. *Linear Models for Multivariate, Time Series and Spatial Data*. Springer-Verlag, New York, 1991.

[57] CHRISTENSEN, R., JOHNSON, W., AND PEARSON, L. M. Covariance function diagnostics for spatial linear models. *Mathematical Geology 25* (1993), 145–160.

[58] COOK, R. D., AND WEISBERG, S. *Residuals and Influence in Regression*. Chapman and Hall, New York, 1982.

[59] COX, D. R., AND ISHAM, V. A simple spatial-temporal model of rainfall. *Proceedings of the Royal Society of London A 415* (1988), 317–328.

[60] COX, D. R., AND ISHAM, V. Stochastic models of precipitation. In *Statistics for the Environment 2: Water Related Issues* (New York, 1994), V. Barnett and K. Turkman, Eds., Wiley, pp. 3–19.

[61] CRAMER, H. On the theory of stationary random processes. *Annals of Mathematics 41* (1940), 215–230.

[62] CRAMER, H. *Mathematical Methods of Statistics*. Princeton University Press, Princeton, 1945.

[63] CRESSIE, N. The origins of kriging. *Mathematical Geology 22* (1990), 239–252.

[64] CRESSIE, N. *Statistics for Spatial Data*, revised ed. Wiley, New York, 1993.

[65] CRESSIE, N. Change of support and the modifiable areal unit problem. *Geographical Systems 3* (1996), 159–180.

[66] CRESSIE, N., AND WIKLE, C. K. The variance-based cross-variogram: you can add apples and oranges. *Mathematical Geology 30* (1998), 789–800.

[67] CREUTIN, J. D., AND OBLED, C. Objective analysis and mapping techniques for rainfall fields: an objective analysis. *Water Resources Research 18* (1982), 413–431.

[68] DAGAN, G. Stochastic modeling of groundwater flow by unconditional and conditional probabilities: the inverse problem. *Water Resources Research 21* (1985), 65–72.

[69] DAGAN, G. *Flow and Transport in Porous Formations.* Springer-Verlag, Berlin, 1989.

[70] DALEY, R. *Atmospheric Data Analysis.* Cambridge University Press, Cambridge, 1991.

[71] DALY, C. *Application de la Géostatistique à quelques Problèmes de Filtrage.* PhD thesis, Ecole des Mines de Paris, Fontainebleau, 1991.

[72] DALY, C., JEULIN, D., AND BENOIT, D. Nonlinear statistical filtering and applications to segregation in steels from microprobe images. *Scanning Microscopy Supplement 6* (1992), 137–145.

[73] DALY, C., JEULIN, D., BENOIT, D., AND AUCLAIR, G. Application of multivariate geostatistics to macroprobe mappings in steels. *ISIJ International 30* (1990), 529–534.

[74] DALY, C., LAJAUNIE, C., AND JEULIN, D. Application of multivariate kriging to the processing of noisy images. In Armstrong [11], pp. 749–760.

[75] DAVID, M. *Geostatistical Ore Reserve Estimation.* Elsevier, Amsterdam, 1977.

[76] DAVIS, J. *Statistics and Data Analysis in Geology.* Wiley, New York, 1973.

[77] DE FOUQUET, C., AND MANDALLAZ, D. Using geostatistics for forest inventory with air cover: an example. In Soares et al. [306], pp. 875–886.

[78] DE MARSILY, G. *Quantitative Hydrogeology.* Academic Press, London, 1986.

[79] DE WIJS, H. J. Statistics of ore distribution, part I: frequency distribution of assay values. *Geologie en Mijnbouw 13* (1951), 365–375.

[80] DELFINER, P. Linear estimation of non stationary spatial phenomena. In Guarascio et al. [132], pp. 49–68.

[81] DELFINER, P., AND CHILÈS, J. P. Conditional simulation: a new Monte-Carlo approach to probabilistic evaluation of hydrocarbon in place. Tech. Rep. N-526, Centre de Géostatistique, Ecole des Mines de Paris, Fontainebleau, 1977.

[82] DELFINER, P., DELHOMME, J. P., AND PELISSIER-COMBESCURE, J. Application of geostatistical analysis to the evaluation of petroleum reservoirs with well logs. In *24th Annual Logging Symposium of the SPWLA* (Calgary, June 27–30 1983).

[83] DELFINER, P., AND MATHERON, G. Les fonctions aléatoires intrinsèques d'ordre k. Tech. Rep. C-84, Centre de Géostatistique, Ecole des Mines de Paris, Fontainebleau, 1980.

[84] DELHOMME, J. Réflexions sur la prise en compte simultanée des données de forages et des données sismiques. Tech. Rep. LHM/RC/79/41, Centre d'Informatique Géologique, Ecole des Mines de Paris, Fontainebleau, 1979.

[85] DELHOMME, J. P. Kriging in the hydrosciences. *Advances in Water Resources 1* (1978), 251–266.

[86] DELHOMME, J. P., BOUCHER, M., MEUNIER, G., AND JENSEN, F. Apport de la géostatistique à la description des stockages de gaz en aquifère. *Revue de l'Institut Français du Pétrole 36* (1981), 309–327.

[87] DEMANGE, C., LAJAUNIE, C., LANTUÉJOUL, C., AND RIVOIRARD, J. Global recoverable reserves: testing various change of support models on uranium data. In Matheron and Armstrong [222], pp. 187–208.

[88] DESBARATS, A. J., AND DIMITRAKOPOULOS, R. Geostatistical simulation of regionalized pore-size distributions using min/max autocorrelation factors. *Mathematical Geology 32* (2000), 919–942.

[89] DHI. *MIKE 12 Version 3.02 General Reference Manual.* Danish Hydraulic Institute, Copenhagen, 1992.

[90] DIGGLE, P. J., LIANG, K. Y., AND ZEGER, S. L. *Analysis of Longitudinal Data.* Clarendon Press, Oxford, 1994.

[91] DIGGLE, P. J., TAWN, J. A., AND MOYEED, R. A. Model-based geostatistics (with discussion). *Applied Statistics 47* (1998), 299–350.

[92] DIMITRAKOPOULOS, R., Ed. *Geostatistics for the Next Century.* Kluwer, Amsterdam, 1994.

[93] DONG, A. *Estimation Géostatistique des Phénomènes régis par des Equations aux Dérivées Partielles.* PhD thesis, Ecole des Mines de Paris, Fontainebleau, 1990.

[94] DOWD, P. A. Generalized cross-covariances. In Armstrong [11], pp. 151–162.

[95] DOYEN, P. M. Porosity from seismic data: a geostatistical approach. *Geophysics 53* (1988), 1263–1275.

[96] DUBRULE, O. Cross-validation of kriging in a unique neighborhood. *Mathematical Geology 15* (1983), 687–699.

[97] DUBRULE, O. Comparing splines and kriging. *Computers & Geosciences 10* (1984), 327–338.

[98] ESPOSITO, E. Statistical investigation about the radar backscatter intensity. Tech. Rep. S-434, Centre de Géostatistique, Ecole des Mines de Paris, Fontainebleau, 2002.

[99] FELLER, W. *An Introduction to Probability Theory and its Applications*, 3rd ed., vol. I. Wiley, New York, 1968.

[100] FEYNMAN, R. P., LEIGHTON, R. B., AND SANDS, M. *The Feynman Lectures on Physics*, vol. 1. Addison Wesley, Reading, 1963.

[101] FRANÇOIS-BONGARÇON, D. Les corégionalisations, le cokrigeage. Tech. Rep. C-86, Centre de Géostatistique, Ecole des Mines de Paris, Fontainebleau, 1981.

[102] GALLI, A., GERDIL-NEUILLET, F., AND DADOU, C. Factorial kriging analysis: a substitute to spectral analysis of magnetic data. In Verly et al. [331], pp. 543–557.

[103] GALLI, A., AND MEUNIER, G. Study of a gas reservoir using the external drift method. In Matheron and Armstrong [222], pp. 105–119.

[104] GASPARI, G., AND COHN, S. E. Construction of correlation functions in two and three dimensions. *Quaterly Journal of the Royal Meteorological Society 126* (1999), 723–762.

[105] GELFAND, A. E., ZHU, L., AND CARLIN, B. P. On the change of support problem for spatio-temporal data. *Biostatistics 2* (2001), 31–45.

[106] GENTON, M. G. Highly robust variogram estimation. *Mathematical Geology 30* (1998), 213–221.

[107] GENTON, M. G. The correlation structure of Matheron's classical variogram estimator under elliptically contoured distributions. *Mathematical Geology 32* (2000), 127–137.

[108] GIFI, A. *Nonlinear Multivariate Analysis*. Wiley, New York, 1990.

[109] GIHMAN, I. I., AND SKOROHOD, A. V. *The Theory of Stochastic Processes I*. Springer-Verlag, Berlin, 1974.

[110] GITTINS, R. *Canonical Analysis: a Review with Applications in Ecology*. Springer-Verlag, Berlin, 1985.

[111] GNEITING, T. *Symmetric Positive Definite Functions with Applications in Spatial Statistics*. PhD thesis, University of Bayreuth, Bayreuth, 1997.

[112] GNEITING, T. The correlation bias for two-dimensional simulation by turning bands. *Mathematical Geology 31* (1999), 195–211.

[113] GNEITING, T. Correlation functions for atmospheric data analysis. *Quaterly Journal of the Royal Meteorological Society 125* (1999), 2449–2464.

[114] GNEITING, T., SASVÁRI, Z., AND SCHLATHER, M. Analogies and correspondences between variograms and covariance functions. *Advances in Applied Probability 33* (2001), 617–630.

[115] GNEITING, T., AND SCHLATHER, M. Space-time covariance models. In *Encyclopedia of Environmetrics* (2001), A. H. El-Shaarawi and W. W. Piegorsch, Eds., vol. 4, Wiley, pp. 2041–2045.

[116] GOMEZ-HERNANDEZ, J., SOARES, A., AND FROIDEVAUX, R., Eds. *GeoENV II – Geostatistics for Environmental Applications*. Kluwer, Amsterdam, 1999.

[117] GONZÁLEZ DEL RÍO, J., FALCO, S., SIERRA, J. P., RODILLA, M., SÁNCHEZ-ARCILLA, A., ROMERO, I., RODRIGO, J., MARTÍNEZ, R., BENEDITO, V., APARISI, F., MÖSSO, C., AND MOVELLÁN, E. Nutrient behaviour in Ebro river estuary. *Hydrobiologia* (2000), 24p. Submitted.

[118] GOODALL, C., AND MARDIA, K. V. Challenges in multivariate spatio-temporal modeling. In *Proceedings of XVIIth International Biometrics Conference* (Hamilton, Ontario, 1994), vol. 1, pp. 1–17.

[119] GOOVAERTS, P. Factorial kriging analysis: a useful tool for exploring the structure of multivariate spatial information. *Journal of Soil Science 43* (1992), 597–619.

[120] GOOVAERTS, P. Study of spatial relationships between two sets of variables using multivariate geostatistics. *Geoderma 62* (1994), 93–107.

[121] GOOVAERTS, P. *Geostatistics for Natural Resources Evaluation.* Oxford University Press, Oxford, 1997.

[122] GOOVAERTS, P. Ordinary cokriging revisited. *Mathematical Geology 30* (1998), 21–42.

[123] GOOVAERTS, P., SONNET, P., AND NAVARRE, A. Factorial kriging analysis of springwater contents in the Dyle river basin, Belgium. *Water Resources Research 29* (1993), 2115–2125.

[124] GOOVAERTS, P., AND WEBSTER, R. Scale-dependent correlation between topsoil copper and cobalt concentrations in Scotland. *European Journal of Soil Science 45* (1994), 79–95.

[125] GOULARD, M. *Champs Spatiaux et Statistique Multidimensionnelle.* PhD thesis, Université des Sciences et Techniques du Languedoc, Montpellier, 1988.

[126] GOULARD, M. Inference in a coregionalization model. In Armstrong [11], pp. 397–408.

[127] GOULARD, M., AND VOLTZ, M. Linear coregionalization model: tools for estimation and choice of multivariate variograms. *Mathematical Geology 24* (1992), 269–286.

[128] GREENACRE, M. J. *Theory and Applications of Correspondence Analysis.* Academic Press, London, 1984.

[129] GRUNSKI, E. C., AND AGTERBERG, F. P. Spatial relationsships of multivariate data. *Mathematical Geology 24* (1992), 731–758.

[130] GRZEBYK, M. *Ajustement d'une Corégionalisation Stationnaire.* PhD thesis, Ecole des Mines de Paris, Fontainebleau, 1993.

[131] GRZEBYK, M., AND WACKERNAGEL, H. Challenges in multivariate spatio-temporal modeling. In *Proceedings of XVIIth International Biometrics Conference* (Hamilton, Ontario, 1994), vol. 1, pp. 19–33.

[132] GUARASCIO, M., DAVID, M., AND HUIJBREGTS, C., Eds. *Advanced Geostatistics in the Mining Industry*, vol. C24 of *NATO ASI Series.* Reidel, Dordrecht, 1976.

[133] GUTTORP, P., AND SAMPSON, P. D. Methods for estimate heterogeneous spatial covariance functions with environmental applications. In *Environmental Statistics* (Amsterdam, 1994), G. P. Patil and C. R. Rao, Eds., vol. 12 of *Handbook of Statistics*, North-Holland, pp. 661–689.

[134] HAAS, T. Local prediction of a spatio-temporal process with an application to wet sulfate deposition. *Journal of the American Statistical Association 90* (1995), 1189–1199.

[135] HAAS, T. Multivariate spatial prediction in the presence of non-linear trend and covariance non-stationarity. *Environmetrics 7* (1996), 145–165.

[136] HAGEN, D. The application of principal components analysis to seismic data sets. *Geoexploration 20* (1982), 93–111.

[137] HAINING, R. *Spatial Data Analysis in the Social and Environmental Sciences.* Cambridge University Press, Cambridge, 1990.

[138] HASLETT, J. Space time modeling in meteorology: a review. In *Proceedings of 47th Session* (Paris, 1989), International Statistical Institute, pp. 229–246.

[139] HASLETT, J., BRADLEY, R., CRAIG, P. S., WILLS, G., AND UNWIN, A. R. Dynamic graphics for exploring spatial data, with application to locating global and local anomalies. *The American Statistician 45* (1991), 234–242.

[140] HASLETT, J., AND POWER, G. M. Interactive computer graphics for a more open exploration of stream sediment geochemical data. *Computers & Geosciences 21* (1995), 77–87.

[141] HOHN, M. E. *Geostatistics and Petroleum Geology.* Kluwer, Dordrecht, 1999.

[142] HU, L. Y. *Mise en oeuvre du modèle gamma pour l'estimation des distributions spatiales.* Doctoral thesis, Ecole des Mines de Paris, Fontainebleau, 1988.

[143] HU, L. Y. Comparing gamma isofactorial disjunctive kriging and indicator kriging for estimating local spatial distributions. In Armstrong [11], pp. 335–346.

[144] HU, L. Y., AND LANTUÉJOUL, C. Recherche d'une fonction d'anamorphose pour la mise en oeuvre du krigeage disjonctif isofactoriel gamma. *Sciences de la Terre, Série Informatique 28* (1988), 145–173.

[145] HUANG, H. C., AND CRESSIE, N. Spatio-temporal prediction of snow water equivalent using the Kalman filter. *Computational Statistics 22* (1996), 159–175.

[146] HUDSON, G., AND WACKERNAGEL, H. Mapping temperature using kriging with external drift: theory and an example from Scotland. *International Journal of Climatology 14* (1994), 77–91.

[147] HUTCHINSON, M. F., AND GESSLER, P. E. Splines — more than just a smooth interpolator. *Geoderma 62* (1994), 45–67.

[148] ISAAKS, E. H., AND SRIVASTAVA, R. M. *Applied Geostatistics.* Oxford University Press, Oxford, 1989.

[149] ISO. *Acoustics: Guidelines for the measurement and assessment of exposure to noise in a working environment,* ISO/DIS 9612.2. International Organization for Standardization, Geneva, 1995.

[150] JAQUET, O. Factorial kriging analysis applied to geological data from petroleum exploration. *Mathematical Geology 21* (1989), 683–691.

[151] JAQUET, O., AND CARNIEL, R. Stochastic modelling at Stromboli: a volcano with remarkable memory. *Journal of Volcanology and Geothermal Research 105* (2001), 249–262.

[152] JEULIN, D., AND RENARD, D. Practical limits of the deconvolution of images by kriging. *Micros. Microanal. Microstruct. 3* (1992), 333–361.

[153] JOLLIFFE, I. T. *Principal Component Analysis*. Springer-Verlag, New York, 1986.

[154] JOURNEL, A. G. Geostatistics for conditional simulation of orebodies. *Economic Geology 69* (1974), 673–687.

[155] JOURNEL, A. G. Markov models for cross-covariances. *Mathematical Geology 31* (1999), 931–954.

[156] JOURNEL, A. G., AND HUIJBREGTS, C. J. *Mining Geostatistics*. Academic Press, London, 1978.

[157] KENDALL, M., AND STUART, A. *The Advanced Theory of Statistics*, vol. 1. Griffin, London, 1977.

[158] KITANIDIS, P. K. Statistical estimation of polynomial generalized covariance functions and hydrologic applications. *Water Resources Research 19* (1983), 909–921.

[159] KITANIDIS, P. K. *Introduction to Geostatistics: Applications to Hydrogeology*. Cambridge University Press, Cambridge, 1997.

[160] KLEINGELD, W. J. *La Géostatistique pour des Variables Discrètes*. Doctoral thesis, Ecole des Mines de Paris, Fontainebleau, 1987.

[161] KLEINGELD, W. J., AND KRIGE, D. G., Eds. *Geostats 2000 – Cape Town*. Geostatistical Association of South Africa, Cape Town, 2000.

[162] KRIGE, D. G. A statistical approach to some mine valuation and allied problems on the Witwatersrand. Master's thesis, University of Witwatersrand, 1951.

[163] KRIGE, D. G. A statistical analysis of some of the borehole values in the Orange free state gold field. *Journal of the Chemical and Metalurgical Society of South Africa 53*, 47–64 (1952).

[164] KRIGE, D. G., GUARASCIO, M., AND CAMISANI-CALZOLARI, F. A. Early South African geostatistical techniques in today's perspective. In Armstrong [11], pp. 1–19.

[165] KÜNSCH, H. R., PAPRITZ, A., AND BASSI, F. Generalized cross-covariances and their estimation. *Mathematical Geology 29* (1997), 779–799.

[166] KYRIAKIDIS, P. C., AND JOURNEL, A. G. Geostatistical space-time models: a review. *Mathematical Geology 31* (1999), 651–684.

[167] LAJAUNIE, C. A geostatistical approach to air pollution modeling. In Verly et al. [331], pp. 877–891.

[168] LAJAUNIE, C. L'estimation géostatistique non linéaire. Tech. Rep. C-152, Centre de Géostatistique, Ecole des Mines de Paris, Fontainebleau, 1993.

[169] LAJAUNIE, C. Kriging and mass balance. In Baafi and Schofield [19], pp. 80–91.

[170] LAJAUNIE, C., AND BÉJAOUI, R. Sur le krigeage des fonctions complexes. Tech. Rep. N-23/91/G, Centre de Géostatistique, Ecole des Mines de Paris, Fontainebleau, 1991.

[171] LAJAUNIE, C., COURRIOUX, G., AND MANUEL, L. Foliation fields and 3D cartography in geology. *Mathematical Geology 29* (1997), 571–584.

[172] LAJAUNIE, C., AND LANTUÉJOUL, C. Setting up the general methodology for discrete isofactorial models. In Armstrong [11], pp. 323–334.

[173] LAJAUNIE, C., WACKERNAGEL, H., THIÉRY, L., AND GRZEBYK, M. Sampling multiphase noise exposure time series. In Gomez-Hernandez et al. [116], pp. 101–112.

[174] LANDAU, L. D., AND LIFSCHITZ, E. M. *Quantum Mechanics.* Pergamon Press, Oxford, 1977.

[175] LANTUÉJOUL, C. Some stereological and statistical consequences derived from Cartier's formula. *Journal of Microscopy 151* (1988), 265–276.

[176] LANTUÉJOUL, C. Cours de sélectivité. Tech. Rep. C-140, Centre de Géostatistique, Ecole des Mines de Paris, Fontainebleau, 1990.

[177] LANTUÉJOUL, C. Ergodicity and integral range. *Journal of Microscopy 161* (1991), 387–403.

[178] LANTUÉJOUL, C. *Geostatistical Simulation: Models and Algorithms.* Springer-Verlag, Berlin, 2001.

[179] LARSEN, R. I. A new mathematical model of air pollutant concentration averaging time and frequency. *Journal of the Air Pollution Control Association 19* (1969), 24–30.

[180] LE, N. D., SUN, W., AND ZIDEK, J. V. Bayesian multivariate spatial interpolation with data missing by design. *Journal of the Royal Statistical Society B 59* (1997), 501–510.

[181] LEFÈBVRE, J., ROUSSEL, H., WALTER, E., LECOINTE, D., AND TABBARA, W. Prediction from wrong models: the kriging approach. *Antennas & Propagation Magazine IEEE 38* (1996), 35–45.

[182] LIAO, H. T. *Estimation des Réserves Récupérables de Gisements d'Or: Comparaison entre Krigeage Disjonctif et Krigeage des Indicatrices.* No. 202 in Documents du BRGM. Editions du BRGM, Orléans, 1991.

[183] LINDNER, S., AND WACKERNAGEL, H. Statistische Definition eines Lateritpanzer-Index für SPOT/Landsat-Bilder durch Redundanzanalyse mit bodengeochemischen Daten. In *Beiträge zur Mathematischen Geologie und Geoinformatik* (Köln, 1993), G. Peschel, Ed., vol. Bd 5, Sven-von-Loga Verlag, pp. 69–73.

[184] LUMLEY, J. L. *Stochastic Tools in Turbulence.* Academic Press, London, 1970.

[185] MA, Y. Z., AND ROYER, J. J. Local geostatistical filtering: application to remote sensing. *Sciences de la Terre, Série Informatique 27* (1988), 17–36.

[186] MALCHAIRE, J., AND PIETTE, A. A comprehensive strategy for the assessment of noise exposure and risk of hearing impairment. *Annals of Occupational Hygiene 41* (1997), 467–484.

[187] MARBEAU, J. P. *Géostatistique Forestière.* PhD thesis, Université de Nancy, Nancy, 1976.

[188] MARCOTTE, D., AND DAVID, M. The bi-gaussian approach: a simple method for recovery estimation. *Mathematical Geology 17* (1985), 625–644.

[189] MARDIA, K. V., GOODALL, C., REDFERN, E., AND ALONSO, F. J. The kriged Kalman filter (with discussion). *Test 7* (1998), 217–285.

[190] MARDIA, K. V., KENT, J. T., AND BIBBY, J. M. *Multivariate Analysis.* Academic Press, London, 1979.

[191] MARDIA, K. V., KENT, J. T., GOODALL, C. R., AND LITTLE, J. A. Kriging and splines with derivative information. *Biometrika 83* (1996), 207–221.

[192] MARÉCHAL, A. Cokrigeage et régression en corrélation intrinsèque. Tech. Rep. N-205, Centre de Géostatistique, Ecole des Mines de Paris, Fontainebleau, 1970.

[193] MARÉCHAL, A. Kriging seismic data in presence of faults. In Verly et al. [331], pp. 271–294.

[194] MATERN, B. *Spatial Variation.* Springer-Verlag, Berlin, 1960.

[195] MATHERON, G. Application des méthodes statistiques à l'évaluation des gisements. *Annales des Mines 144 (12)* (1955), 50–75.

[196] MATHERON, G. *Traité de Géostatistique Appliquée*, vol. 1. Technip, Paris, 1962.

[197] MATHERON, G. Principles of geostatistics. *Economic Geology 58* (1963), 1246–1266.

[198] MATHERON, G. *Les Variables Régionalisées et leur Estimation.* Masson, Paris, 1965.

[199] MATHERON, G. *Le Krigeage Universel: Recherche d'Opérateurs Optimaux en Présence d'une Dérive.* No. 1 in Les Cahiers du Centre de Morphologie Mathématique. Ecole des Mines de Paris, Fontainebleau, 1969.

[200] MATHERON, G. *The Theory of Regionalized Variables and its Applications.* No. 5 in Les Cahiers du Centre de Morphologie Mathématique. Ecole des Mines de Paris, Fontainebleau, 1970.

[201] MATHERON, G. La théorie des fonctions aléatoires intrinsèques généralisées. Tech. Rep. N-252, Centre de Géostatistique, Ecole des Mines de Paris, Fontainebleau, 1971.

[202] MATHERON, G. Leçons sur les fonctions aléatoires d'ordre 2. Tech. rep., Ecole des Mines de Paris, Paris, 1972.

[203] MATHERON, G. The intrinsic random functions and their applications. *Advances in Applied Probability 5* (1973), 439–468.

[204] MATHERON, G. Effet proportionnel et lognormalité ou: le retour du serpent de mer. Tech. Rep. N-374, Centre de Géostatistique, Ecole des Mines de Paris, Fontainebleau, 1974.

[205] MATHERON, G. Les fonctions de transfert des petits panneaux. Tech. Rep. N-395, Centre de Géostatistique, Ecole des Mines de Paris, Fontainebleau, 1974.

[206] MATHERON, G. Compléments sur les modèles isofactoriels. Tech. Rep. N-432, Centre de Géostatistique, Ecole des Mines de Paris, Fontainebleau, 1975.

[207] MATHERON, G. Forecasting block grade distributions: the transfert functions. In Guarascio et al. [132], pp. 237–251.

[208] MATHERON, G. Les concepts de base et l'évolution de la géostatistique minière. In Guarascio et al. [132], pp. 3–10.

[209] MATHERON, G. A simple substitute for conditional expectation: the disjunctive kriging. In Guarascio et al. [132], pp. 221–236.

[210] MATHERON, G. Comment translater les catastrophes: la structure des F.A.I. générales. Tech. Rep. N-617, Centre de Géostatistique, Ecole des Mines de Paris, Fontainebleau, 1979.

[211] MATHERON, G. Recherche de simplification dans un problème de cokrigeage. Tech. Rep. N-628, Centre de Géostatistique, Ecole des Mines de Paris, Fontainebleau, 1979.

[212] MATHERON, G. Splines and kriging: their formal equivalence. In *Down-to-Earth Statistics: Solutions Looking for Geological Problems* (New York, 1981), D. F. Merriam, Ed., no. 8 in Geology Contribution, Syracuse University, pp. 77–95.

[213] MATHERON, G. La sélectivité des distributions. Tech. Rep. N-686, Centre de Géostatistique, Ecole des Mines de Paris, Fontainebleau, 1982.

[214] MATHERON, G. Pour une analyse krigeante des données régionalisées. Tech. Rep. N-732, Centre de Géostatistique, Ecole des Mines de Paris, Fontainebleau, 1982.

[215] MATHERON, G. Modèle isofactoriel et changement de support. *Sciences de la Terre, Série Informatique 18* (1983), 71–123.

[216] MATHERON, G. Isofactorial models and change of support. In Verly et al. [331], pp. 449–467.

[217] MATHERON, G. Pour une méthodologie générale des modèles isofactoriels discrets. *Sciences de la Terre, Série Informatique 21* (1984), 1–64.

[218] MATHERON, G. The selectivity of distributions and "the second principle of geostatistics". In Verly et al. [331], pp. 421–433.

[219] MATHERON, G. Sur la positivité des poids de krigeage. Tech. Rep. N-30/86/G, Centre de Géostatistique, Ecole des Mines de Paris, Fontainebleau, 1986.

[220] MATHERON, G. *Estimating and Choosing*. Springer-Verlag, Berlin, 1989.

[221] MATHERON, G. Two classes of isofactorial models. In Armstrong [11], pp. 309–322.

[222] MATHERON, G., AND ARMSTRONG, M., Eds. *Geostatistical Case Studies.* Reidel, Dordrecht, 1987.

[223] MATHERON, G., ROTH, C., AND DE FOUQUET, C. Modélisation et cokrigeage de la charge et de la transmissivité avec conditions aux limites à distance finie. In *Cahiers de Géostatistique* (Fontainebleau, 1993), vol. 3, Ecole des Mines de Paris, pp. 61–76.

[224] MEIER, S., AND KELLER, W. *Geostatistik: eine Einführung in die Theorie der Zufallsprozesse.* Springer-Verlag, Vienna, 1990.

[225] MONESTIEZ, P., ALLARD, D., AND FROIDEVAUX, R., Eds. *GeoENV III – Geostatistics for Environmental Applications.* Kluwer, Amsterdam, 2001.

[226] MONESTIEZ, P., ALLARD, D., NAVARRO-SANCHEZ, I., AND COURAULT, D. Kriging with categorical external drift: use of thematic maps in spatial prediction and application to local climate interpolation for agriculture. In Gomez-Hernandez et al. [116], pp. 163–174.

[227] MONESTIEZ, P., GOULARD, M., CHARMET, G., AND BALFOURIER, F. Analysing spatial genetic structures of wild populations of perennial ryegrass (Lolium perenne). In Baafi and Schofield [19], pp. 1197–1208.

[228] MONESTIEZ, P., SAMPSON, P., AND GUTTORP, P. Modeling of heterogeneous spatial correlation structure by spatial deformation. In *Cahiers de Géostatistique* (Fontainebleau, 1993), vol. 3, Ecole des Mines de Paris, pp. 35–46.

[229] MONESTIEZ, P., AND SWITZER, P. Semiparametric estimation of nonstationary spatial covariance models by multidimensional scaling. Tech. Rep. 165, Stanford University, Stanford, 1991.

[230] MORRISON, D. F. *Multivariate Statistical Methods.* McGraw-Hill International, Auckland, 1978.

[231] MYERS, D. E. Matrix formulation of cokriging. *Mathematical Geology 14* (1982), 249–258.

[232] MYERS, D. E. Estimation of linear combinations and cokriging. *Mathematical Geology 15* (1983), 633–637.

[233] MYERS, D. E. Pseudo-cross variograms, positive-definiteness, and cokriging. *Mathematical Geology 23* (1991), 805–816.

[234] NAOURI, J. C. Analyse factorielle des correspondances continues. *Publications de l'Institut de Statistique de la Université de Paris XIX*, 1 (1970), 1–100.

[235] OLIVER, M. A., LAJAUNIE, C., WEBSTER, R., MUIR, K. R., AND MANN, J. R. Estimating the risk of childhood cancer. In Soares et al. [306], pp. 899–910.

[236] OLIVER, M. A., AND WEBSTER, R. A geostatistical basis for spatial weighting in multivariate classification. *Mathematical Geology 21* (1989), 15–35.

[237] OLIVER, M. A., WEBSTER, R., LAJAUNIE, C., MUIR, K. R., PARKES, S. E., CAMERON, A. H., STEVENS, M. C. G., AND MANN, J. R. Binomial cokriging for estimating and mapping the risk of childhood cancer. *IMA Journal of Mathematics Applied in Medecine and Biology 15* (1998), 279–297.

[238] ORFEUIL, J. P. Interprétation statistique du modèle de larsen. Tech. Rep. N-413, Centre de Géostatistique, Ecole des Mines de Paris, Fontainebleau, 1973.

[239] ORFEUIL, J. P. Etude, mise en oeuvre et test d'un modèle de prédiction à court terme de pollution athmosphérique. Tech. Rep. N-498, Centre de Géostatistique, Ecole des Mines de Paris, Fontainebleau, 1977.

[240] PAPRITZ, A., AND FLÜHLER, H. Temporal change of spatially autocorrelated soil properties: optimal estimation by cokriging. *Geoderma 62* (1994), 43.

[241] PAPRITZ, A., KÜNSCH, H. R., AND WEBSTER, R. On the pseudo cross-variogram. *Mathematical Geology 25* (1993), 1015–1026.

[242] PARDO-IGUZQUIZA, E. GCINFE: a computer program for inference of polynomial generalized covariance functions. *Computers & Geosciences 23* (1997), 163–174.

[243] PARKER, H. The volume-variance relationship: a useful tool for mine planning. *Engineering and Mining Journal 180*, 10 (1979), 106–123.

[244] PAWLOWSKY, V. Cokriging of regionalized compositions. *Mathematical Geology 21* (1989), 513–521.

[245] PAWLOWSKY, V., OLEA, R. A., AND DAVIS, J. C. Estimation of regionalized compositions: a comparison of three methods. *Mathematical Geology 27* (1995), 105–127.

[246] PERRIN, O. *Modèle de Covariance d'un Processus Non-Stationnaire par Déformation de l'Espace et Statistique*. PhD thesis, University of Paris I Panthéon-Sorbonne, Paris, 1997.

[247] PETITGAS, P. Use of a disjunctive kriging to model areas of high pelagic fish density in acoustic fisheries surveys. *Aquatic Living Resources 6* (1993), 201–209.

[248] PETITGAS, P. Geostatistics in fisheries survey design and stock assessment: models, variances and applications. *Fish and Fisheries 2* (2001), 231–249.

[249] PILZ, J., SPOECK, G., AND SCHIMECK, M. G. Taking account of uncertainty in spatial covariance estimation. In Baafi and Schofield [19], pp. 302–313.

[250] PRÉAT, B. Application of geostatistical methods for estimation of the dispersion variance of occupational exposures. *American Industrial Hygiene Association Journal 48* (1987), 877–884.

[251] PRIESTLEY, M. *Spectral Analysis and Time Series*. Academic Press, London, 1981.

[252] RANNOU, V., BROUAYE, F., HÉLIER, M., AND TABBARA, W. Kriging the quantile: application to a simple transmission line model. *Inverse Problems 18* (2002), 37–48.

[253] RAO, C. R. The use and interpretation of principal component analysis in applied research. *Sankhya Series A* (1964), 329–358.

[254] RAO, C. R. *Linear Statistical Inference and its Applications.* Wiley, New York, 1973.

[255] RASMUSSEN, E. K., SEHESTED-HANSEN, I., ERICHSEN, A. C., MUHLENSTEIN, D., AND DØRGE, J. 3D model system for hydrodynamics, eutrophication and nutrient transport. In *Environmental Coastal Regions III* (Southampton, 2000), G. E. Rodrigues, C. A. Brebbia, and E. Perez-Martell, Eds., WIT Press.

[256] RASPA, G., BRUNO, R., DOSI, P., PHILIPPI, N., AND PATRIZI, G. Multivariate geostatistics for soil classification. In Soares [305], pp. 793–804.

[257] RASPA, G., TUCCI, M., AND BRUNO, R. Reconstruction of rainfall fields by combining ground raingauges data with radar maps using external drift method. In Baafi and Schofield [19], pp. 1306–1315.

[258] REMACRE, A. Z. Conditionnement uniforme. *Sciences de la Terre, Série Informatique 18* (1984), 125–139.

[259] REMACRE, A. Z. Conditioning by the panel grade for recovery estimation of non-homogeneous orebodies. In Matheron and Armstrong [222], pp. 135–147.

[260] RENARD, D., AND NAI-HSIEN, M. Utilisation de dérives externes multiples. *Sciences de la Terre, Série Informatique 28* (1988), 281–301.

[261] RENARD, D., AND RUFFO, P. Depth, dip and gradient. In Soares [305], pp. 167–178.

[262] RIPLEY, B. D. *Spatial Statistics.* Wiley, New York, 1981.

[263] RIPLEY, B. D. *Stochastic Simulation.* Wiley, New York, 1987.

[264] RIVOIRARD, J. *Le Comportement des Poids de Krigeage.* PhD thesis, Ecole des Mines de Paris, Fontainebleau, 1984.

[265] RIVOIRARD, J. Convergence des développements en polynomes d'Hermite. *Sciences de la Terre, Série Informatique 24* (1985), 129–159.

[266] RIVOIRARD, J. Modèles à résidus d'indicatrices autokrigeables. *Sciences de la Terre, Série Informatique 28* (1988), 303–326.

[267] RIVOIRARD, J. Models with orthogonal indicator residuals. In Armstrong [11], pp. 91–107.

[268] RIVOIRARD, J. A review of lognormal estimators for in situ reserves. *Mathematical Geology 22* (1990), 213–221.

[269] RIVOIRARD, J. Relations between the indicators related to a regionalized variable. In Soares [305], pp. 273–284.

[270] RIVOIRARD, J. *Introduction to Disjunctive Kriging and Non-Linear Geostatistics.* Oxford University Press, Oxford, 1994.

[271] RIVOIRARD, J. Which models for collocated cokriging? *Mathematical Geology 33* (2001), 117–131.

[272] RIVOIRARD, J., SIMMONDS, J., FOOTE, K. G., FERNANDES, P., AND BEZ, N. *Geostatistics for Estimating Fish Abundance.* Blackwell Science, London, 2000.

[273] ROQUIN, C., DANDJINOU, T., FREYSSINET, P., AND PION, J. C. The correlation between geochemical data and SPOT satellite imagery of lateritic terrain in Southern Mali. *Journal of Geochemical Exploration 32* (1989), 149–168.

[274] ROQUIN, C., FREYSSINET, P., ZEEGERS, H., AND TARDY, Y. Element distribution patterns in laterites of southern Mali: Consequence for geochemical prospection and mineral exploration. *Applied Geochemistry 5* (1990), 303–315.

[275] ROTH, C. Is lognormal kriging suitable for local estimation? *Mathematical Geology 30* (1998), 999–1009.

[276] ROTH, C., AND CHILÈS, J. P. Modélisation géostatistique des écoulements souterrains: comment prendre en compte les lois physiques [with a detailed summary in english]. *Hydrogéologie* (1997), 23–32.

[277] ROUHANI, S., SRIVASTAVA, R. M., DESBARATS, A. J., CROMER, M. V., AND JOHNSON, A. I., Eds. *Geostatistics for Environmental and Geotechnical Applications*, vol. STP 1283. American Society for Testing and Materials, West Conshohocken, 1996.

[278] ROUHANI, S., AND WACKERNAGEL, H. Multivariate geostatistical approach to space-time data analysis. *Water Resources Research 26* (1990), 585–591.

[279] ROYER, J. J. Proximity analysis: a method for multivariate geodata processing. *Sciences de la Terre, Série Informatique 20* (1984), 223–243.

[280] ROYER, J. J. *Analyse Multivariable et Filtrage des Données Régionalisées.* PhD thesis, Institut National Polytechnique de Lorraine, Nancy, 1988.

[281] RYTOV, S. M., KRAVTSOV, Y. A., AND TATARSKII, V. I. *Principles of Statistical Radiophysics 1: Elements of Random Process Theory.* Springer-Verlag, Berlin, 1987.

[282] RYTOV, S. M., KRAVTSOV, Y. A., AND TATARSKII, V. I. *Principles of Statistical Radiophysics 2: Correlation Theory of Random Processes.* Springer-Verlag, Berlin, 1988.

[283] RYTOV, S. M., KRAVTSOV, Y. A., AND TATARSKII, V. I. *Principles of Statistical Radiophysics 3: Elements of Random Fields.* Springer-Verlag, Berlin, 1989.

[284] SABOURIN, R. Application of two methods for the interpretation of the underlying variogram. In Guarascio et al. [132], pp. 101–109.

[285] SACKS, J., WELCH, W. J., MITCHELL, T. J., AND WYNN, H. P. Design and analysis of computer experiments. *Statistical Science 4* (1989), 409–435.

[286] SAMPSON, P., AND GUTTORP, P. Nonparametric estimation of nonstationary spatial structure. *Journal of the American Statistical Association 87* (1992), 108–119.

[287] SANDJIVY, L. Analyse krigeante de données géochimiques. *Sciences de la Terre, Série Informatique 18* (1983), 141–172.

[288] SANDJIVY, L. The factorial kriging analysis of regionalized data. In Verly et al. [331], pp. 559–571.

[289] SAPORTA, G. *Probabilités, Analyse des Données et Statistique*. Technip, Paris, 1990.

[290] SCHAFFRIN, B. Kriging with soft unbiasedness. In Baafi and Schofield [19], pp. 69–79.

[291] SCHNEIDER, T., HOLM PETERSEN, O., AASBERG NIELSEN, A., AND WINDFELD, K. A geostatistical approach to indoor surface sampling strategies. *Journal of Aerosol Science 21* (1990), 555–567.

[292] SCHOENBERG, I. J. Metric spaces and completely monotone functions. *Annals of Mathematics 39* (1938), 811–841.

[293] SCHULZ-OHLBERG, J. Die Anwendung geostatistischer Verfahren zur Interpretation von gravimetrischen und magnetischen Felddaten. Tech. Rep. 1989–6, Deutsches Hydrographisches Institut, Hamburg, 1989.

[294] SEBER, G. A. F. *Multivariate Observations*. Wiley, New York, 1984.

[295] SÉGURET, S. *Géostatistique des Phénomènes à Tendance Périodique (dans l'Espace-Temps)*. PhD thesis, Ecole des Mines de Paris, Fontainebleau, 1991.

[296] SÉGURET, S. Analyse krigeante spatio-temporelle appliquée à des données aéromagnétiques. In *Cahiers de Géostatistique* (Fontainebleau, 1993), vol. 3, Ecole des Mines de Paris, pp. 115–138.

[297] SÉGURET, S., AND HUCHON, P. Trigonometric kriging: a new method for removing the diurnal variation from geomagnetic data. *Journal of Geophysical Research 32*, B13 (1990), 21.383–21.397.

[298] SÉNÉGAS, J., WACKERNAGEL, H., ROSENTHAL, W., AND WOLF, T. Error covariance modeling in sequential data assimilation. *Stochastic Environmental Research and Risk Assessment 15* (2001), 65–86.

[299] SERRA, J. Les structures gigognes: morphologie mathématique et interprétation métallogénique. *Mineralium Deposita 3* (1968), 135–154.

[300] SHILOV, G. E., AND GUREVICH, B. L. *Integral, Measure and Derivative: a Unified Approach*. Dover, New York, 1977.

[301] SICHEL, H. S. New methods in the statistical evaluation of mine sampling data. *Transactions of the Institution of Mining and Metallurgy* (1952).

[302] SICHEL, H. S. The estimation of means and associated confidence limits for small samples from lognormal populations. *Journal of the South African Institute of Mining and Metallurgy March* (1966), 106–122.

[303] SIERRA, J. P., GONZÁLEZ DEL RÍO, J., SÁNCHEZ-ARCILLA, A., FLOS, J., MOVELLÁN, E., RODILLA, M., MÖSSO, C., FALCO, S., ROMERO, I., AND CRUZADO, A. Spatial distribution of nutrients in the Ebro estuary and plume. *Continental Shelf Research* (2001), 30p. in press.

[304] SIERRA, J. P., GONZÁLEZ DEL RÍO, J., SÁNCHEZ-ARCILLA, A., MOVELLÁN, E., RODILLA, M., MÖSSO, C., MARTÍNEZ, R., FALCO, S., ROMERO, I., AND MAROTTA, L. Dynamics of the Ebro river estuary and plume in the Mediterranean Sea. In *Proceedings of 10th International Biennial Conference on Physics of Estuaries and Coastal Seas* (Norfolk, Virginia, 2000).

[305] SOARES, A., Ed. *Geostatistics Tróia '92*. Kluwer, Amsterdam, 1993.

[306] SOARES, A., GOMEZ-HERNANDEZ, J., AND FROIDEVAUX, R., Eds. *GeoENV I: Geostatistics for Environmental Applications*. Kluwer, Amsterdam, 1997.

[307] SOUSA, A. J. Geostatistical data analysis: an application to ore typology. In Armstrong [11], pp. 851–860.

[308] SPARKS, R., ADOLPHSON, A., AND PHATAK, A. Multivariate process monitoring using the dynamic biplot. *International Statistical Review 65* (1997), 325–349.

[309] SPECTOR, A., AND BHATTACHARYYA, B. K. Energy density spectrum and autocorrelation functions due to simple magnetic models. *Geophysical Prospecting 14* (1966), 242–272.

[310] SPECTOR, A., AND GRANT, F. S. Statistical models for interpreting aeromagnetic data. *Geophysics 35* (1970), 293–302.

[311] STEIN, A., STARISKY, I. G., AND J, B. Simulation of moisture deficits and areal interpolation by universal cokriging. *Water Resources Research 27* (1991), 1963–1973.

[312] STEIN, A., VAN EINJSBERGEN, A. C., AND BARENDREGT, L. G. Cokriging nonstationary data. *Mathematical Geology 23* (1991), 703–719.

[313] STEIN, M. L. A simple model for spatial-temporal processes. *Water Resources Research 22* (1986), 2107–2110.

[314] STEIN, M. L. *Interpolation of Spatial Data: Some Theory for Kriging*. Springer-Verlag, New York, 1999.

[315] STOYAN, D. *Stochastik für Ingenieure und Naturwissenschaftler*. Akademie-Verlag, Berlin, 1993.

[316] STOYAN, D., AND STOYAN, H. *Fractals, Random Shapes and Point Fields*. Wiley, New York, 1994.

[317] STOYAN, D., STOYAN, H., AND JANSEN, U. *Umweltstatistik: statistische Verarbeitung und Analyse von Umweltdaten*. Teubner, Stuttgart, 1997.

[318] STRANG, G. *Linear Algebra and its Applications*. Academic Press, London, 1984.

[319] STRANG, G. *Introduction to Applied Mathematics*. Wellesley-Cambridge Press, Wellesley, 1986.

[320] SWITZER, P., AND GREEN, A. A. Min/max autocorrelation factors for multivariate spatial imagery. Tech. Rep. 6, Department of Statistics, Stanford University, Stanford, 1984.

[321] SZEGÖ, G. *Orthogonal Polynomials*, 4th ed. American Mathematical Society, Providence, 1975.

[322] THIÉBAUX, H. J. The power of duality in spatial-temporal estimation. *Journal of Climate 10* (1997), 567–573.

[323] THIÉBAUX, H. J., MORONE, L. L., AND WOBUS, R. L. Global forecast error correlation. part 1: isobaric wind and geopotential. *Monthly Weather Review 118* (1990), 2117–2137.

[324] THIÉBAUX, H. J., AND PEDDER, M. A. *Spatial Objective Analysis: with Applications in Atmospheric Science*. Academic Press, London, 1987.

[325] TYLER, D. E. On the optimality of the simultaneous redundancy transformations. *Psychometrika 47* (1982), 77–86.

[326] VAN DOORN, E. *Stochastic Monotonicity and Queueing Applications of Birth-Death Processes*. Springer-Verlag, New York, 1981.

[327] VARGAS-GUZMÁN, J. A., WARRICK, A. W., AND MYERS, D. E. Scale effect on principal component analysis for vector random functions. *Mathematical Geology 31* (1999), 701–722.

[328] VARGAS-GUZMÁN, J. A., AND YEH, T. C. J. Sequential kriging and cokriging: two powerful geostatistical approaches. *Stochastic Environmental Research and Risk Assessment 13* (1999), 416–435.

[329] VENABLES, W. N., AND RIPLEY, B. D. *Modern Applied Statistics with S-Plus*. Springer-Verlag, New York, 1994.

[330] VER HOEF, J. M., AND CRESSIE, N. Multivariable spatial prediction. *Mathematical Geology 25* (1993), 219–240.

[331] VERLY, G., DAVID, M., AND JOURNEL, A. G., Eds. *Geostatistics for Natural Resources Characterization*, vol. C-122 of *NATO ASI Series C-122*. Reidel, Dordrecht, 1984.

[332] VINCENT, R., GRZEBYK, M., WACKERNAGEL, H., AND LAJAUNIE, C. Application de la géostatistique à l'hygiène industrielle: Evaluation d'un cas d'exposition professionnelle au trichloroéthylène. *Cahiers de Notes Documentaires 174*, ND 2094-174-99 (1999), 5–13.

[333] VOLLE, M. *Analyse des Données*. Economica, Paris, 1985.

[334] VON STORCH, H., AND NAVARRA, A., Eds. *Analysis of Climate Variablility*. Springer-Verlag, Berlin, 1995.

[335] VON STORCH, H., AND ZWIERS, F. *Statistical Analysis in Climate Research*. Cambridge University Press, Cambridge, 1999.

[336] WACKERNAGEL, H. *L'inférence d'un Modèle Linéaire en Géostatistique Multivariable*. PhD thesis, Ecole des Mines de Paris, Fontainebleau, 1985.

[337] WACKERNAGEL, H. Geostatistical techniques for interpreting multivariate spatial information. In *Quantitative Analysis of Mineral and Energy Resources* (Dordrecht, 1988), C. F. Chung, Ed., vol. C-223 of *NATO ASI Series*, Reidel, pp. 393–409.

[338] WACKERNAGEL, H. Description of a computer program for analyzing multivariate spatially distributed data. *Computers & Geosciences 15* (1989), 593–598.

[339] WACKERNAGEL, H. Cokriging versus kriging in regionalized multivariate data analysis. *Geoderma 62* (1994), 83–92.

[340] WACKERNAGEL, H., AND BUTENUTH, C. Caractérisation d'anomalies géochimiques par la géostatistique multivariable. *Journal of Geochemical Exploration 32* (1989), 437–444.

[341] WACKERNAGEL, H., LAJAUNIE, C., THIÉRY, L., AND GRZEBYK, M. Evaluation de l'exposition sonore en milieu professionnel: Application de méthodes géostatistiques à l'estimation du Leq et conséquences sur les stratégies de mesurage. Tech. Rep. MAV–NT-373/LT, Institut National de Recherche et de Sécurité, Vandoeuvre-Les-Nancy, 1998.

[342] WACKERNAGEL, H., LAJAUNIE, C., THIÉRY, L., VINCENT, R., AND GRZEBYK, M. Applying geostatistics to exposure monitoring data in industrial hygiene. In Soares et al. [306], pp. 463–476.

[343] WACKERNAGEL, H., PETITGAS, P., AND TOUFFAIT, Y. Overview of methods for coregionalization analysis. In Armstrong [11], pp. 409–420.

[344] WACKERNAGEL, H., AND SANGUINETTI, H. Gold prospecting with factorial cokriging in the Limousin, France. In *Computers in Geology: 25 years of progress* (Oxford, 1992), J. C. Davis and U. C. Herzfeld, Eds., vol. 5 of *Studies in Mathematical Geology*, Oxford University Press, pp. 33–43.

[345] WACKERNAGEL, H., THIÉRY, L., AND GRZEBYK, M. The Larsen model from a de Wijsian perspective. In Gomez-Hernandez et al. [116], pp. 125–135.

[346] WACKERNAGEL, H., WEBSTER, R., AND OLIVER, M. A. A geostatistical method for segmenting multivariate sequences of soil data. In *Classification and Related Methods of Data Analysis* (Amsterdam, 1988), H. H. Bock, Ed., Elsevier (North-Holland), pp. 641–650.

[347] WAHBA, G. *Spline Models for Observational Data.* Society for Industrial and Applied Mathematics, Philadelphia, 1990.

[348] WALTER, E., AND PRONZATO, L. *Identification of Parametric Models from Experimental Data.* Springer-Verlag, Berlin, 1997.

[349] WARNES, J. J. A sensitivity analysis of universal kriging. *Mathematical Geology 18* (1986), 653–676.

[350] WEBSTER, R. Optimally partitioning soil transects. *Journal of Soil Science 29* (1978), 388–402.

[351] WEBSTER, R., ATTEIA, O., AND DUBOIS, J. P. Coregionalization of trace metals in the soil in the Swiss jura. *European Journal of Soil Science 45* (1994), 205–218.

[352] WEBSTER, R., AND OLIVER, M. A. *Statistical Methods in Soil and Land Resource Survey*. Oxford University Press, Oxford, 1990.

[353] WEBSTER, R., OLIVER, M. A., MUIR, K. R., AND MANN, J. R. Kriging the local risk of a rare disease from a register of diagnoses. *Geographical Analysis 26* (1990), 168–185.

[354] WELLMER, F. W. *Statistical Evaluations in Exploration for Mineral Deposits*. Springer-Verlag, Berlin, 1998.

[355] WHITTAKER, J. *Graphical Models in Applied Multivariate Analysis*. Wiley, New York, 1990.

[356] WIKLE, C. K., AND CRESSIE, N. A dimension-reduced approach to space-time Kalman filtering. *Biometrika 86* (1999), 815–829.

[357] WILD, P., HORDAN, R., LEPLAY, A., AND VINCENT, R. Confidence intervals for probabilities of exceeding threshold limits with censored log-normal data. *Environmetrics 7* (1996), 247–259.

[358] WOLF, T., SÉNÉGAS, J., BERTINO, L., AND WACKERNAGEL, H. Application of data assimilation to three-dimensional hydrodynamics: the case of the Odra lagoon. In Monestiez et al. [225], pp. 157–168.

[359] WU, Z. Compactly supported positive definite radial functions. *Advances in Computational Mathematics 4* (1995), 283–292.

[360] XU, W., TRAN, T. T., SRIVASTAVA, R. M., AND JOURNEL, A. G. Integrating seismic data in reservoir modeling: the collocated cokriging alternative. In *Proceedings of 67th Annual Technical Conference of the Society of Petroleum Engineers* (Washington, 1992), no. 24742 in SPE, pp. 833–842.

[361] YAGLOM, A. M. *An Introduction to the Theory of Stationary Random Functions*. Dover, New York, 1962.

[362] YAGLOM, A. M. *Correlation Theory of Stationary and Related Random Functions*. Springer-Verlag, Berlin, 1986.

[363] YAO, T., AND JOURNEL, A. G. Automatic modelling of (cross)covariance tables using Fast Fourier Transform. *Mathematical Geology 30* (1998), 589–615.

[364] YAO, T., AND JOURNEL, A. G. Note and corrections to: Automatic modelling of (cross)covariance tables using Fast Fourier Transform. *Mathematical Geology 32* (2000), 147–148.

[365] YAO, T., MUKERJI, T., JOURNEL, A. G., AND MAVKO, G. Scale matching with factorial kriging for improved porosity estimation from seismic data. *Mathematical Geology 31* (1999), 23–46.

[366] YARUS, J. M., AND CHAMBERS, R. L., Eds. *Stochastic Modelling and Geostatistics: Principles, Methods and Case Studies*. American Association of Petroleum Geologists, Tulsa, 1994.

[367] ZIMMERMANN, D. L. Computationally efficient restricted maximum likelihood estimation of generalized covariance functions. *Mathematical Geology 21* (1989), 655–672.

Index

Printing: Strauss GmbH, Mörlenbach
Binding: Schäffer, Grünstadt